BASIC ELECTRICITY

A TEXT-LAB MANUAL

SIXTH EDITION

PAUL B. ZBAR ▪ GORDON ROCKMAKER

BASIC ELECTRICITY

A TEXT-LAB MANUAL

SIXTH EDITION

GLENCOE

McGraw-Hill

New York, New York   Columbus, Ohio   Woodland Hills, California   Peoria, Illinois

Other Glencoe Books of Interest

Basic Electronics: A Text-Lab Manual, Sixth Edition by Paul B. Zbar, Albert P. Malvino, and Michael A. Miller
Electricity-Electronics Fundamentals: A Text-Lab Manual, Fourth Edition by Paul B. Zbar and Joseph G. Sloop
Industrial Electronics: A Text-Lab Manual, Fourth Edition by Paul B. Zbar and Richard Koelker
Basic Television: Theory and Servicing: A Text-Lab Manual, Third Edition by Paul B. Zbar and Peter W. Orne
Industrial Electricity: Principles and Practices, Third Edition by James Adams and Gordon Rockmaker

Library of Congress Cataloging-in-Publication Data
Zbar, Paul B.
    Basic electricity: a text-lab manual / Paul B. Zbar, Gordon
Rockmaker. —6th ed.
        p.   cm.
    ISBN 0-07-072861-5
    1. Electronics—Laboratory manuals. I. Rockmaker, Gordon.
II Title.
TK7818.Z18 1991
621.31'078—dc20   91–11784
                                                          CIP

Imprint 1997

Send all inquiries to:
Glencoe/McGraw-Hill
936 Eastwind Drive
Westerville, OH 43081

ISBN 0-07-072861-5

Printed in the United States of America.

5 6 7 8 9 10 11 12 13 14 15   066   03 02 01 00 99 98 97

SERIES PREFACE

Electronics is at the core of a wide variety of specialized technologies that have been developing over several decades. Challenged by rapidly expanding technology and the need for increasing numbers of technicians, the Consumer Electronics Group Product Service Committee of the Electronic Industries Association (EIA) and various publishers have been active in creating and developing educational materials to meet these challenges.

In recent years, a great many consumer electronic products have been introduced and the traditional audio and television receivers have become more complex. As a result, the pressing need for training programs to permit students of various backgrounds and abilities to enter this growing industry has induced EIA to sponsor the preparation of an expanding range of materials. Three branches of study have been developed in two specific formats. The tables list the books in each category; the paragraphs following them explain these materials and suggest how best to use them to achieve the desired results.

The foreword to the first edition of the EIA-cosponsored basic series states: "The aim of this basic instructional series is to supply schools with a well-integrated, standardized training program, fashioned to produce a technician tailored to industry's needs." This is still the objective of the varied training program that has been developed through joint industry-educator-publisher cooperation.

Peter McCloskey, President
Electronic Industries Association

THE BASIC ELECTRICITY-ELECTRONICS SERIES

Title	Author	Publisher
Electricity-Electronics Fundamentals	Zbar/Sloop	Glencoe (Macmillan/McGraw-Hill)
Basic Electricity	Zbar/Rockmaker	Glencoe (Macmillan/McGraw-Hill)
Basic Electronics	Zbar/Malvino/Miller	Glencoe (Macmillan/McGraw-Hill)

The laboratory text-manuals in the Basic Electricity-Electronics Series provide in-depth, detailed, completely up-to-date technical material by combining a comprehensive discussion of the objectives, theory, and underlying principles with a closely coordinated program of experiments. Electricity-Electronics Fundamentals provides material of an introductory course especially suitable for preparing service technicians; it can also be used for other broad-based courses. Basic Electricity and Basic Electronics are planned for 270-hour courses, one to follow the other, providing a more thorough background for all levels of technician training. A related instructor's guide is available for each course.

THE TELEVISION-AUDIO SERVICING SERIES

Title	Author	Publisher
Television Servicing with Basic Electronics	Sloop	Electronic Industries Association
Advanced Color Television Servicing	Sloop	Electronic Industries Association
Audio Servicing—Theory and Practice	Wells	Glencoe (Macmillan/McGraw-Hill)
Audio Servicing—Text-Lab Manual	Wells	Glencoe (Macmillan/McGraw-Hill)
Basic Television: Theory and Servicing	Zbar and Orne	Glencoe (Macmillan/McGraw-Hill)
Cable Television Technology	Deschler	Glencoe (Macmillan/McGraw-Hill)

The Television-Audio Servicing Series includes materials in two categories: those designed to prepare apprentice technicians to perform in-home servicing and other apprenticeship functions, and those designed to prepare technicians to perform more sophisticated and complicated servicing such as bench-type servicing in the shop.

Television Servicing with Basic Electronics (text, student workbook, instructor's guide) covers the basics, the math, and the test equipment required in Television Servicing. The book uses a diagnostic troubleshooting method.

Advanced bench-type diagnosis servicing techniques are covered in *Advanced Color Television Servicing* (text, student workbook, instructor's guide). Written primarily for color television servicing courses in schools and in industry, this set follows the logical diagnostic troubleshooting approach consistent with the manufacturers' approach to bench servicing.

Audio Servicing (theory and practice, text-lab manual, and instructor's guide) covers each component of a modern home stereo with an easy-to-follow block diagram and a diagnosis approach consistent with the latest industry techniques.

Basic Television: Theory and Servicing provides a series of experiments, with preparatory theory, designed to provide the in-depth, detailed training necessary to produce skilled television service technicians for both home and bench servicing of all types of television. A related instructor's guide is also available.

Cable Television Technology (text, instructor's guide) covers all aspects of cable television operation, from the traditional "Lineman"-oriented topics to the high technology subjects that come into play with satellite antennas and fiber-optics links.

ALSO AVAILABLE

Covering the fundamental basics of circuits and their applications, *Industrial Electronics* gives students a grasp of the building blocks of the technology. Written by Paul B. Zbar and Richard L. Koelker, it includes an instructor's guide.

PREFACE

Basic Electricity: A Text-Lab Manual, Sixth Edition, is an introductory textbook in electrical technology for students of electricity and electronics. It provides a comprehensive laboratory program in basic electrical theory, electric circuits, and passive devices in both direct and alternating current. As in previous editions, the focus is on practical and analytical techniques essential to the modern technician. By emphasizing hands-on activities, the text helps students develop their troubleshooting and circuit-design skills in a systematic fashion.

The fifty-seven experiments are presented so that each new concept builds on the previous one. Topics range from an introduction to experimental methods, basic components, instruments and measurements, simple series, parallel, and series-parallel circuits to the more advanced circuit theorems, troubleshooting, and circuit design.

Many comments and suggestions from the students and instructors who have used previous editions of *Basic Electricity: A Text-Lab Manual* have contributed to the "fine tuning" of this sixth edition. The dynamic technological advances in the electricity/electronics field have required introducing new hardware and a new approach to fundamentals. The format and content of this edition carefully address the needs of students as well as those of the industries that will employ them. Topics no longer pertinent to a modern electronics curriculum have been discarded and new topics added in their place. However, the basic strengths of the previous editions have been retained.

The organization of each experiment consists of the following:

Objectives. Each experiment opens with a concise statement of the goals of the experiment. Thus, from the very beginning of each experiment, the student is aware of the direction and purpose of the laboratory procedures he or she will be following.

Basic Information. Before performing an experiment, the student must understand the underlying principles as well as the practical aspects of the concepts being investigated. This section focuses on those concepts essential to successfully completing the experimental assignments.

Summary. The purpose of the summary is to highlight the key elements of the basic information for quick reference.

Students can use this condensed presentation of the basic information as a convenient review before performing the experiment.

Self-Test. Following the summary is a self-test designed to measure the student's understanding of the basic theory and practices involved in the experiment. These tests also will prove extremely useful in reinforcing the student's knowledge of certain important concepts. Students are expected to understand the correct answers to all the questions in the test before proceeding to the experiment. Answers to the self-tests are given at the end of the procedure section.

Materials Required. As the initial step in the experiment, the student is provided with a list of the power supplies, instruments, and components necessary to conduct the experiment. In some cases, the instructor will need to provide special materials not specifically listed.

Procedure. Step-by-step instructions are given for each part of the experiment. During the experiment, the student is expected to wire circuits according to schematic diagrams, make electrical measurements using meters and instruments similar to those used in industry, tabulate data, and use formulas to calculate unknown quantities. By following the sequence of steps given in the procedure, the student will develop practical hands-on experience, learn safe laboratory practices, and apply analytical skills to the solution of practical problems.

Special performance sheets are provided at the end of the experiment with the necessary tables to be completed as well as space for calculations. The performance sheets also contain a series of stimulating questions and problems directly related to the student's experimental results. The student must report not only on the tabulated results but also on the theoretical and practical aspects underlying the experiment. In some cases students are required to plot graphs based on their experimental data. Thus, the student can demonstrate his or her technical communication and analytical skills directly on these performance sheets. The sheets are perforated so that they can be readily removed from the manual and submitted to the instructor. This leaves the rest of the experiment manual intact for review and possible reworking of the experiment.

As in previous editions, the text is fully illustrated with circuit diagrams, tables, and graphs designed to supplement and support the basic information. Detailed circuit diagrams are provided as required for performing the experiments. Sample problems and their step-by-step solutions are included in the basic information sections wherever necessary.

The authors would be remiss in not acknowledging the cooperation and support this text received from the electronics industry. The preceding Series Preface details the long-standing close working relationship with the Consumer Electronics Group of the Electronic Industries Association this series of text-lab manuals enjoys. Special thanks go to the members of the Service Education Sub-Committee, who reviewed the manuscript: Frank Steckel (EIA Consultant), Gerald Ganguzza (Sharp Electronics Corporation), and Edward Rosenthal (NEC Technologies, Inc.). Under their expert guidance this text combines the latest industry practices with sound educational and training principles.

The authors also wish to acknowledge, with thanks, Jack P. Moore (Hickok Teaching Systems, Inc.) and Barbara Fikaris for their guidance and help in their review of the manuscript.

Last but not least, the authors wish to thank their wives, May Zbar and Ellen Rockmaker, for their patience, encouragement, and inspiration. To Ellen go our special thanks for her many, many hours of manuscript preparation and proofreading of this edition.

Paul B. Zbar
Gordon Rockmaker

SAFETY

Electronics technicians work with electricity, electronic devices, motors, and other rotating machinery. They are often required to use hand and power tools in constructing prototypes of new devices or in setting up experiments. They use test instruments to measure the electrical characteristics of components, devices, and electronic systems. They are involved in any of a dozen different tasks.

These tasks are interesting and challenging, but they may also involve certain hazards if the technician is careless in his or her work habits. It is therefore essential that student technicians learn the principles of safety at the very start of their career and then practice these principles throughout their busy and exciting lives.

Safe work requires a careful and deliberate approach to each task. Before undertaking a job, the technician must understand what he or she is to do and how to do it. The technician must plan the job, setting out on the work bench in a neat and orderly fashion the tools, equipment, and instruments that will be needed. Extraneous items should be removed and cables should be secured as far as possible.

When working on or near rotating machinery, loose clothing should be anchored, neckwear firmly tucked away.

In the lab or shop, students should use caution when working with line voltages. Even voltages as low as 25 or 30 volts can be hazardous in certain situations. Line (power) voltages should be isolated from ground by means of an isolation transformer. Power-line voltages can kill, so these should not be contacted with the hands or any part of your bare skin.

Line cords should be checked before use. If the insulation on the cords is brittle or cracked, these cords must not be used. Measure voltages with one hand in your pocket. Be certain that your hands are dry and that you are not standing on wet floor when making tests and measurements in a live circuit. Shut off power before connecting test instruments in a live circuit.

Be certain that line cords of power tools and nonisolated equipment use safety plugs (polarized 3-post plugs). Do not defeat the safety feature of these plugs by using ungrounded adapters. Do not defeat any safety device, such as a fuse or circuit breaker, by shorting across it or by using a higher amperage fuse than that specified by the manufacturer. Safety devices are intended to protect you and your equipment.

Handle tools properly and with care. Don't indulge in horseplay or play practical jokes in the laboratory or shop. When using power tools, secure your work in a vise or jig. Wear protective clothing, gloves, and goggles when required.

FIRST AID

If an accident should occur, shut off power immediately. Report the accident at once to your instructor. It may be necessary for you to render first aid before a physician can come, so you should know the principles of first aid. A proper knowledge of these may be acquired by attendance at a Red Cross first aid course.

Some first aid suggestions are set forth here as a simple guide.

An injured person should be kept lying down until medical help arrives and should be kept warm to prevent shock. Do not attempt to give the victim water or other liquids if he or she is unconscious. Be sure nothing is done to cause further injury. Keep the injured person comfortable and cheerful until medical help arrives.

ARTIFICIAL RESPIRATION

Severe electrical shock may cause stoppage of breathing. Be prepared to start artificial respiration at once if breathing has stopped. The two recommended techniques are
1. Mouth-to-mouth breathing, considered the most effective
2. Schaeffer method

These techniques are described in first aid books. You should master one or the other so that if the need arises you will be able to save a life by applying artificial respiration.

These instructions are not intended to frighten you but to make you aware that there are dangers in the work of an electronics technician. But then there are hazards in every job.

Exercise good judgment and common sense and your life in the laboratory will be safe, interesting, and rewarding.

CONTENTS

SERIES PREFACE
PREFACE
SAFETY

Note on Experiment Content

Each of the experiments is set up like this:

OBJECTIVES The objectives are enumerated and clearly stated.
BASIC INFORMATION The theory and basic principles involved in the experiment are clearly stated.
SUMMARY A summary of the salient points is given.
SELF-TEST A self-test, based on the material included in Basic Information, helps students evaluate their understanding of the principles covered, prior to the experiment proper. The self-test should be taken before the experiment is undertaken. Answers to the self-test questions are given at the end of the Procedures section.
MATERIALS REQUIRED All the materials required to do the experiment—including test equipment and components—are listed.
PROCEDURE A detailed step-by-step procedure is given for performing the experiment.
ANSWERS TO SELF-TEST
PERFORMANCE SHEETS Special tear-out pages with fill-in tables and a series of questions related to the students' experimental results.

Experiments

4

CELLS IN SERIES AND PARALLEL

OBJECTIVES

1. To measure the voltage of cells in series
2. To measure the voltage of cells in parallel

BASIC INFORMATION

Electric cells are a source of direct current. They produce electric energy through chemical reactions. Two dissimilar materials separated by a chemical that allows the materials to exchange electrons produce a potential difference. The two dissimilar materials are the *electrodes*, or *poles*, of the cell; the material permitting the exchange of electrons is called an *electrolyte*. Different combinations of materials produce different potential differences, or voltages. The most common voltages in general use range from about 1.25 V to 1.5 V for the familiar C, D, AA, and AAA size cells. Although these cells are often called "dry" cells, this is a misnomer.

The electrolyte in all cells is liquid in one form or another. In some cells, the electrodes are immersed in a pool of liquid electrolyte; in others, the electrolyte is moist, powdery chemicals or a spongy, moist insulating material. In the past "wet" cells were actually glass jars of liquid electrolyte (usually acid or caustic soda), whereas dry cells were the cells encased in metal cans (usually zinc) covered with cardboard and totally sealed so that the interior chemicals would not dry out (or leak and ruin the equipment in which they were used). Modern manufacturing methods now permit totally sealed wet cells. The prime example of a wet cell is the one that is used in automobiles (see the discussion of series and parallel cell connection that follows).

Series Connection of Cells

Cells have two terminals: + and −. If the positive terminal of one cell is connected to the negative terminal of a second cell, as in Figure 4–1(a), the cells are said to be *series-connected*, *aiding*, or simply *series-aiding*. The voltage measured across AB in Figure 4–1(a) will be the sum of V_1 and V_2. If another cell is connected to the first two in the same fashion, as in Figure 4–1(b), the total voltage is $V_1 + V_2 + V_3$. In general, the total voltage across a group of series-aiding cells is the sum of the individual cell voltages.

If the two positive terminals or the two negative terminals are connected together, as in Figure 4–2(a) (p. 26), the cells are said to be *series-connected*, *opposing*, or simply *series-opposing*. In such a case, the voltage across AB will be the difference of the two voltages. If the two voltages are exactly equal, the difference is zero. That is, a voltmeter placed across the two cells would measure 0 V. If the two cells have unequal voltage, the polarity of the voltage across the series combination is that of the higher voltage, as shown in Figure 4–2(b). If this arrangement were connected to a circuit and current were produced, the cell with the lower voltage would be subjected to reverse current and probably would suffer permanent damage. Connecting cells in a series-opposing arrangement is not a suitable method of reducing the voltage across a combination of cells.

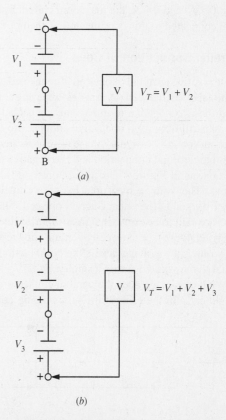

Figure 4–1. Cells connected in series-aiding. (a) Total voltage $V_T = V_1 + V_2$; (b) total voltage $V_T = V_1 + V_2 + V_3$.

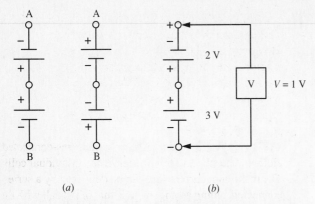

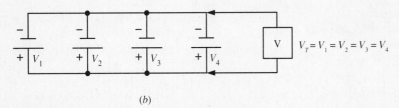

(a)   (b)

Figure 4–2. Cells connected in series-opposing. (a) If the cell voltages are equal, then $V_{AB} = 0$. (b) If the cell voltages are unequal, V_{AB} is the difference between the two voltages.

From a practical standpoint, cells connected in a series-aiding arrangement should all be of the same voltage and composition. In fact, they should all have had the same usage. This is because cells have internal resistance that affects their output voltage, the internal heat they produce, and the amount of current that can be delivered to an external load. The ideal arrangement of series cells is to have all cells of equal voltage and life. The maximum current that can be delivered to a load by a series-aiding arrangement of cells is equal to the lowest maximum current of any of the cells. That is, if three cells are series-aiding and their maximum currents are, respectively, 1 A, 0.5 A, and 0.1 A, the maximum current available to any load connected to this combination is, theoretically, 0.1 A.

Parallel Connection of Cells

If the positive terminal of one cell is connected to the positive terminal of another cell (as in series-opposing connection) but then the two negative terminals are connected together, no current will flow from one cell to another, provided they are of equal voltage. Now, if a voltage were measured between the positive terminal connection and the negative terminal connection, as in Figure 4–3(a), an interesting thing would be found. The voltage measured by the voltmeter would be exactly equal to the voltage across one cell. The cells in this case are said to be connected in *parallel*. If three or four [as in Figure 4–3(b)] or more cells of equal voltage are connected in parallel, the voltage across the combination will still be equal to the voltage across a single cell.

What advantage does the parallel connection have over a single cell? In the case of the series-aiding connection, the

overall voltage across the combination increases and is equal to the sum of the voltages of the individual cells. In the case of the parallel-connected cells, the voltage is unchanged, but the current that can be delivered to a load is equal to the sum of the currents that each individual parallel cell can deliver.

Suppose the positive terminal of one cell were connected to the negative terminal of the other cell, and the other negative and positive terminals were connected, as in Figure 4–3(c). The result would be a short circuit, high damaging current, and possibly an explosion. *Never* connect cells in this fashion.

Series-Parallel Connection of Cells

Figure 4–4 (p. 27) shows three parallel connections of cells connected in a series-aiding arrangement. Let us assume all cells have the same rating (and age), 1.5 V and 0.25 A maximum. What would the voltmeter read, and how much current would be available to a load connected across AH? The voltage across BC would be 1.5 V; the voltage across DE would be 1.5 V; and the voltage across FG would be 1.5 V, or three series-aiding voltages of 1.5 V each. The total voltage across AH would therefore equal 1.5 + 1.5 + 1.5, or 4.5 V. The parallel combination across BC is capable of delivering 0.5 A (0.25 + 0.25). But the 0.5 A has only one path to follow, so that the maximum current available to a load is 0.5 A.

The life of a cell is usually measured by its ability to deliver current. Because the required load current in a parallel connection is shared by each of the cells in parallel, individual cells are called upon to deliver less current. Thus the parallel connection not only increases the ability of the combination to deliver a higher current than a single cell, it also increases the life of each cell.

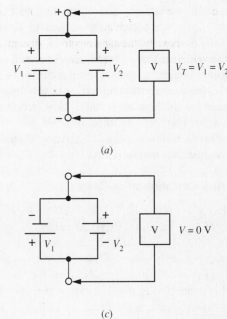

(a)

(c)

Figure 4–3. Cells connected in parallel.

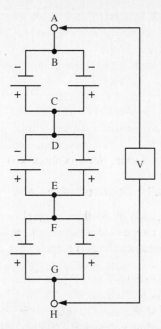

Figure 4–4. Cells connected in series-parallel. The series combination is shown aiding.

Cells and Batteries

In popular usage the terms *cell* and *battery* are used interchangeably. From a technical standpoint, cells and batteries differ. A *cell* is a single unit producing voltage from the chemical reaction of the electrode materials and an electrolyte. A *battery* is a combination of cells that are series, parallel, or series-parallel connected.

For example, a common automobile battery has a nominal voltage of 12 V. However, the cell that makes up the auto battery produces 2.1 V. The 12 V is obtained by connecting six cells in series (6 × 2.1 = 12.6 V). The 12.6 V across the terminals of the auto battery reduces to 12 V nominally as current is drawn by the car's electrical system. Auto batteries, however, are called upon to deliver extraordinary currents (in the hundreds of amperes) when starting. This capability is designed into the auto battery by connecting many pairs of electrodes in parallel, in effect, connecting many cells in parallel.

The popular 9-V rectangular battery used in portable cassettes, calculators, and radios consists of six cells of 1.5 V each (6 × 1.5 V = 9 V). The individual cells that make up this battery are relatively small. Because the current capacity of the 9 V battery is not increased by the series-aiding connection, these batteries are not used for applications requiring heavy current drains.

SUMMARY

1. Electric cells are a source of direct current.
2. Cells produce electric energy through a chemical reaction.

3. An electric cell consists of two electrodes made of dissimilar materials separated by an electrolyte.
4. The common C, D, AA, and AAA cells produce voltages ranging from 1.25 V to 1.5, depending upon the materials making up the electrodes and electrolyte.
5. If the positive terminal of one cell is connected to the negative cell of a second cell, the cells are series-connected, aiding.
6. The voltage across a combination of cells series-connected, aiding is the sum of the voltages of the individual cells.
7. The maximum current that can be delivered by a series-connected, aiding arrangement is that of the cell with the lowest maximum current.
8. If cells are connected so that the two positive terminals or two negative terminals are connected, the cells are series-connected, opposing. The total voltage is the difference of the individual cell voltages. If current is drawn, one of the cells will receive reverse current. This can lead to cell damage and possible hazardous conditions.
9. If the positive terminal of one cell is connected to the positive terminal of a second cell of equal voltage and the negative terminals of the cells are connected to each other, the voltage measured across the positive and negative terminals will be that of a single cell.
10. Cells connected as in item 9 are said to be parallel connected.
11. The current available from parallel-connected cells is equal to the sum of the currents of the individual cells.
12. Cells can be connected in series-parallel arrangement in order to increase both the voltage and current capacity of each single cell.
13. A combination of cells in a single package or housing is called a battery. The automobile battery consists of six series-connected cells, each producing 2.1 V. The overall voltage of the battery is 6 × 2.1 V = 12.6 V. A 9-V battery (used in portable radios, stereos, and calculators) contains six cells rated 1.5 V each.

SELF-TEST

Check your understanding by answering the following questions:

1. An electric cell is a source of _____ current.
2. (True/False) Whenever possible, cells should be connected in series-opposing to reduce the output voltage to a circuit. _____
3. The total voltage of six series-aiding cells rated 2.1 V each is _____ V.
4. The basic parts of an electric cell are the electrodes and a(n) _____ .
5. (True/False) The term dry cell is incorrect because internally no cell can be dry and still operate effectively.
6. In parallel-connected cells the _____ terminal of one cell is connected to the positive terminal of a second

cell while the _____ terminal of one cell is connected to the _____ terminal of the other.

7. Three cells series-aiding have maximum current ratings of 1 A, 1.2 A, and 1.5 A. The maximum current available to the circuit is _____ A.

8. The cells of question 7 are parallel-connected. The maximum current available to the circuit is (theoretically) _____ A.

MATERIALS REQUIRED

Power Supplies:
■ 4 1.5-V cells (D size) (the cells should be labeled 1 through 4)

Instruments:
■ DMM or VOM
■ DC ammeter or second DMM or VOM

Resistors:
■ 1 220-Ω, ½-W, 5%

Miscellaneous:
■ 2 SPST switches

PROCEDURE

1. The four cells you will use in this experiment should have the same voltage. Measure the voltage of each cell and record the value in Table 4–1 (p. 31).

2. Connect the three cells as in Figure 4–5. Measure the voltage across the combination of three cells and across the combination of cells 1 and 2, and across cells 2 and 3. Record the values in Table 4–1.

3. Reverse cell 1, as shown in Figure 4–6. Measure the voltage across the combination of three cells and across the combination of cells 1 and 2, and across cells 2 and 3. Record the values in Table 4–2 (p. 31).

4. With switch S_1 **open**, connect the circuit of Figure 4–7. The cells are connected series-aiding.

5. **Close** S_1. Measure the current in the circuit, the voltage across R, and the individual voltages across cell 1 and cell 2. Record the values in Table 4–3 (p. 31). Open S_1.

6. With S_1 **open**, reverse cell 2, as shown in Figure 4–8 (p. 29).

7. Do this step quickly. Connect the meters with the polarities as indicated. If analog meters are used, note if the pointer moves off scale to the left. Reverse the meter

leads and record any values as minus. If a digital meter is used record any negative readings. **Close** S_1. (*Do not allow S_1 to remain closed for any length of time—make your measurements quickly and then open the switch.*) Measure the current in the circuit. Measure the voltage across R and the individual voltages across cell 1 and cell 2. Record the values in Table 4–3. **Open** S_1. Disconnect the circuit.

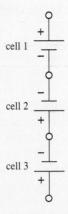

Figure 4–6. Connection of cells for procedure step 3.

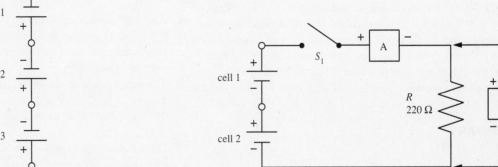

Figure 4–7. Circuit for procedure step 4.

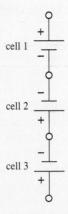

Figure 4–5. Connection of cells for procedure step 2.

8. With S_1 and S_2 **open**, connect the circuit of Figure 4–9.

9. **Close** S_1. Measure the current, the voltage across R, and the voltage across cell 1 and across cell 2. Record the values in Table 4–4 (p. 31). **Open** S_1.

10. **Close** S_2. Measure the current, the voltage across R, and the voltage across cell 3 and across cell 4. Record the values in Table 4–4. Switch S_2 should remain **closed**.

11. **Close** S_1. Measure the current and the voltage across R. Record the values in Table 4–4. **Open** S_1 and S_2.

12. With S_1 and S_2 **open**, reverse the connection of cell 3 in Figure 4–9, as shown in Figure 4–10. This step must be done quickly. If the pointer of any meter moves off scale to the left, **open** the switches and reverse the leads to the meter. **Close** S_1 and S_2. (*Do not allow* S_2 *to remain closed for any length of time —make your measurements quickly and open the switches.*) Measure the current through R and the voltage across R. Record the values in Table 4–4. **Open** S_1 and S_2. Disconnect the circuit.

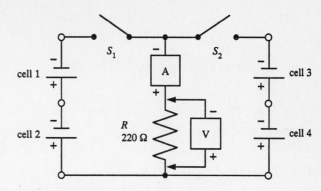

Figure 4–9. Circuit for procedure step 8.

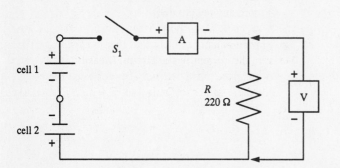

Figure 4–8. Circuit for procedure step 6.

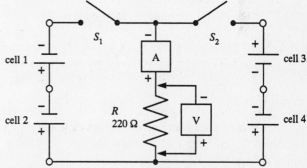

Figure 4–10. Circuit for procedure step 12.

ANSWERS TO SELF-TEST

1. direct
2. false
3. 12.6
4. electrolyte
5. true
6. positive; negative; negative
7. 1
8. 3.7

TABLE 4–1. Cells Series-Connected, Aiding

Cell Number	Nominal Cell Voltage, V	Measured Cell Voltage, V	Calculated Cell Voltage, V
1	1.5		
2	1.5		
3	1.5		
4	1.5		
1, 2, 3			
1, 2			
2, 3			

TABLE 4–2. Cells Series-Connected, Opposing

Cell Combination	Measured Voltage, V	Calculated Voltage, V
1, 2, 3		
1, 2		
2, 3		

TABLE 4–3. Series Cells in a Circuit

Circuit Element	Series-Aiding Cells Voltage, V	Series-Opposing Cells Voltage, V
Resistor, R		
Cell 1		
Cell 2		
Current (A)	$I =$	$I =$

TABLE 4–4. Cells Connected in Parallel

Step No.	Switch S_1	Switch S_2	Current Through Resistor, A	Voltage Across Resistor, V	Cell Voltage, V 1	2	3	4
9	Closed	Open						
10	Open	Closed						
11	Closed	Closed						
12	Closed	Closed						

QUESTIONS

1. Explain how the parallel connection of cells increases the life of each individual cell.

_____

_____

_____

_____

_____

_____

_____

_____

_____

_____

_____

2. List the combinations of electrode materials and electrolytes for various types of the common D-size cell. Indicate their rated output voltages.

_____

_____

_____

_____

_____

_____

_____

_____

3. If three 1.5-V cells are connected in series-opposing, what is the maximum output voltage possible?

_____

_____

_____

_____

_____

_____

_____

4. Why is the name dry-cell battery a misnomer?

_____

_____

_____

_____

_____

_____

5. Design a battery using 1.5-V, 40-mA cells that will have a rated voltage of 15 V at 120 mA. Draw a schematic representation of the battery.

EXPERIMENT

5

CONDUCTORS AND INSULATORS

OBJECTIVES

1. To identify conductors and insulators
2. To measure the resistance of a variety of conductors and insulators.

BASIC INFORMATION

Conductors and Insulators

The useful effects of electricity result from the movement of electric charges in a circuit. This movement of electric charges is called *current*. Electric charges move easily through paths called *conductors*, but it is very difficult for these charges to move through *insulators*. Conductors are materials that permit current to flow easily with little electrical pressure (voltage) applied, whereas insulators are materials that permit very little or no current to flow. Copper is an example of a good conductor; rubber is a good insulator.

A complete (closed) circuit or path for direct current consists of a voltage source such as a battery or power supply, a

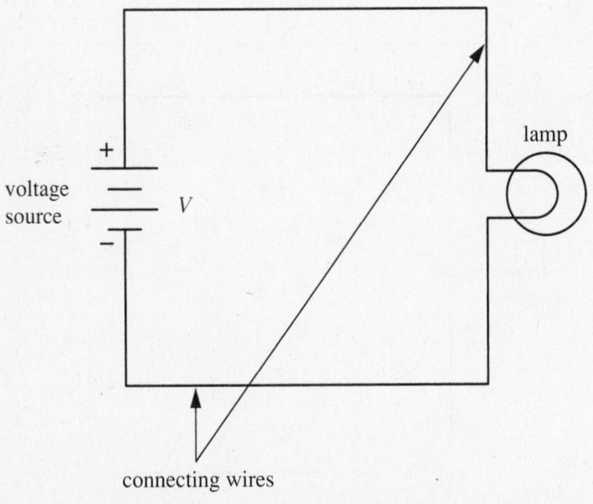

Figure 5–1. A complete (closed) electric circuit showing a voltage source, *V*, connecting wires, and a lamp (the electric load).

load such as an electric lamp, and connecting conductors (copper wire) (Figure 5–1). In such a closed circuit there is electric current, and if sufficient current flows, the lamp will light. If rubber cords instead of copper wires were used to tie the battery to the lamp, the lamp would not light, because rubber is an insulator and does not permit current to flow through it in the circuit.

The tungsten filament in the electric lamp is a conductor, but it is not as good a conductor as the connecting copper wires. Tungsten is used for the filament because its properties cause it to glow and give off light when it is heated by an electric current.

It is apparent that not all materials are equally good conductors of electricity. Specific materials are used for electrical components because of their unique characteristics as conductors and nonconductors.

Wire Conductors

The ability of a conductor to allow current flow is called *conductance*, denoted by the symbol *G*. The unit of conductance is the *siemen*. The opposition to current flow is called *resistance*, denoted by the symbol *R*. The unit of resistance is the *ohm*.

The degree of resistance of a material is called its *resistivity*. Different materials have different values of resistivity depending on their molecular structure.

The resistance of a material depends not only on its resistivity but also on its dimensions and temperature. For example, a round copper wire of a given diameter and length will have a certain value of resistance measured in ohms. If the length of this wire is doubled, its resistance will double. Its resistivity will remain the same as long as the temperature is constant. Resistance also depends on the cross-sectional area of a conductor. The resistance of a copper wire of the same length as the original wire but with double the cross-sectional area has one-half the resistance of the original wire. That is, resistance decreases as the cross-sectional area of a conductor increases; resistance increases as the length of the conductor increases.

For a given length and diameter of wire, the higher the resistivity, the higher the resistance. Stated another way, good conductors have low resistivities.

A formula that relates length, cross-sectional area, and resistivity to resistance is

$$R = \frac{\rho l}{A} \qquad (5\text{--}1)$$

where

R = resistance in ohms
l = length in feet
A = cross-sectional area in circular mils
ρ = resistivity at a given temperature

The resistivity of copper at 20°C (usually considered room temperature) is 10.37. Other materials often used as low-resistance conductors are silver, gold, and aluminum. The resistivity of silver is 9.85 and of aluminum is 17.00. In some applications, high-resistance conductors are necessary. High-resistance materials include nickel, iron, tungsten, a combination of nickel and chrome known as nichrome, and carbon.

An example will show how the resistance formula is used.

Problem. Find the resistance of 100 ft of copper wire having a diameter of 0.1 in.

Solution. A diameter of 0.1 in. is equal to 100 mils (1 mil = 0.001 in.). An area in circular mils is found by squaring the diameter in mils. Thus 100 mils in diameter yields a cross-sectional area d^2 of 10,000 circular mils. Substituting this value in formula (5-1), we have

$$R = \frac{\rho l}{A}$$

$$R = \frac{10.37 \times 100}{10,000}$$

$$R = 0.1037 \ \Omega$$

Wire Size

Round wire used as electrical conductors is specified by gage sizes. In the United States, the American Wire Gage (AWG) is used as the standard measure. The higher the AWG number, the smaller the diameter of the wire. For example, no. 12 copper wire (a common size for house wiring) has a much larger diameter than no. 22 wire (a common hookup wire for electronic circuits). The largest gage size is 0000, or, as it is often written, 4/0. Wire sizes larger than 4/0 are usually specified by giving their cross-sectional area in circular mils.

Electrical reference books often contain tables of wire sizes and properties. Included in these tables are AWG sizes, diameters, cross-sectional areas, and resistance per 1000 ft of wire.

The amount of current a conductor is able to carry before being damaged is usually much more than the conductor is *permitted* to carry by code regulations or law for reasons of safety to personnel and equipment. For example, a no. 12 copper wire with plastic insulation is limited by the National Electrical Code® to carrying 20 A of current when used as building wire under certain conditions. Yet the same no. 12 wire, uninsulated, when properly used in circuits in the laboratory, can safely carry considerably more than 20 A without melting or creating a safety hazard.

Wire Components

Some components are made by winding a length of wire around a form. For example, ceramic high-power resistors consist of a wire winding around a ceramic core. Iron-core inductors (chokes) are wound around a form through which an iron core is placed. Transformers consist of several windings around some specified core. If an ohmmeter is placed across each winding of these components, it will measure the resistance of that winding. Therefore, one way of determining if a wire winding is continuous, that is, if it is not broken, is to measure its resistance across the two end terminals. The normal resistance is usually specified by the manufacturer. If a winding measures infinite resistance, it is open. Because the resistance of these windings is often very low, the idea of the *continuity test* is to see that a complete electrical path exists rather than to measure resistance. A wide swing of the meter needle or a reading of very low resistance indicates continuity of the electrical path.

Carbon Resistors as Conductors

In a previous experiment we learned how to determine the resistance of carbon resistors using a color code. Carbon is also considered a conductor of electricity. A complete circuit that contains a dc power source, connecting wires, and a resistor (Figure 5–2) will permit current to flow.

The physical construction of a resistor will determine its resistance. Here, as in circular wire, the resistance of the resistor varies inversely with the cross-sectional area of the carbon element in the resistor and directly with the length of the element. By controlling the diameter and length of the inner carbon element, we can control the resistance of R.

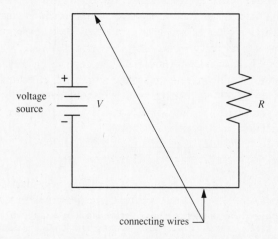

Figure 5–2. A complete electric circuit containing V, a dc power source, connecting wires, and R, a resistor.

Comparing a resistor with a copper wire, we note that the resistance of the wire is *distributed* throughout its length. In a small carbon resistor, the resistance is lumped between the two terminals.

In the circuit of Figure 5–2 the current can be controlled by controlling the value of R, in ohms. With V held constant, the higher the resistance of R, the lower the current will be. That is, as the resistance of R increases, the opposition to current increases and the current decreases.

Insulators

There are some materials whose resistivity is so high that they permit *very* little current to flow. These materials are called *insulators*, or *nonconductors*. Rubber was previously mentioned as an insulator. Wood, glass, paper, mica, bakelite, most plastics, and air are other examples of insulators.

If the leads of an ohmmeter are connected across an insulator, say a rubber cord, the meter will register infinite resistance. An infinite resistance will not permit direct current to flow in a circuit. Thus, Figure 5–3 is an open circuit, because direct current will not flow through the insulator.

Insulating materials are used to cover copper wires in electricity and electronics. The purpose of insulating electrical wires is to prevent them from making electrical contact (shorting) with any other conductor or component that they may touch accidentally. Wires on which the insulation has melted, cracked, or become frayed are safety hazards and should not be used.

Insulators may break down and become conductors if the voltage across them is high enough. Thus an air gap between two voltage terminals that are 1 cm apart will break down if the voltage across these terminals equals or exceeds 30,000 V. A descriptive term frequently used for this phenomenon is *arc-over*.

Semiconductors

The basic active materials of many modern electronic components are silicon and germanium. These two elements, in their pure state, are insulators. However, when certain impurities, such as gallium or antimony, are added, they can be made somewhat conductive, and they are then called *semiconductors*. The resistivity of semiconductors depends on the nature and amount of impurity injected. In general, however, the resistivity of semiconductors is said to be between that of insulators and conductors. Semiconductor devices are also referred to as *solid-state* devices.

The Human Body as a Conductor

The human body is a conductor of electricity, as evidenced by the electric shocks that people have experienced when touching the positive and negative terminals of a voltage source. Electric current will flow through the body, and the higher the voltage, the higher the current. Electric shock is *dangerous*

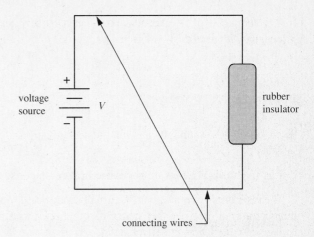

Figure 5–3. An open circuit. The rubber insulator does not permit direct current to flow through the circuit.

and can be *fatal*. That is why the student of electricity and electronics should follow the safety precautions taught in order to avoid shock.

The resistance of the body is not a constant value. Its value at a specific point in time can be measured with an ohmmeter. The method to follow is given in the Procedure section.

SUMMARY

1. Conductors are materials that permit current to flow through them with ease. Silver, copper, and aluminum are examples of good conductors.
2. Insulators are materials that do not permit current to flow through them. Rubber, glass, mica, and most plastics are examples of insulators.
3. Conductors vary in their conductivity. Thus, copper is a better conductor than aluminum.
4. The resistance of wire varies directly with its length and inversely with its cross-sectional area. Resistance also depends on the material from which the conductor is made.
5. The American Wire Gage (AWG) is used to measure wire size.
6. Wire tables list AWG wire sizes, cross-sectional area of wire, resistance of 1000 ft of wire, and other electrical and physical properties of wire.
7. Insulators can withstand voltages up to a critical voltage. At the critical voltage or higher, insulators break down and permit current flow by arc-over.
8. The breakdown voltage of an insulator depends on its material.
9. The resistance of a good conductor is very low and that of an insulator very high.
10. Semiconductors are materials with resistivity that lies between that of a conductor and insulator. Silicon and germanium to which impurities have been added are semiconductors.

11. The human body is a conductor of electricity. Body contact with electrical wires must be avoided to prevent shock.

SELF-TEST

Check your understanding by answering the following questions:

1. List five materials that are good electrical conductors:
   _____ ; _____ ; _____ ;
   _____ ; _____ .
2. List five materials that are electrical insulators:
   _____ ; _____ ; _____ ;
   _____ ; _____ .
3. For current to flow in a circuit, it must have
   (a) _____ ; (b) _____ ; and
   (c) _____ .
4. A circuit in which current flows is called a _____ circuit.
5. Current will not flow in a(n) _____ circuit.
6. No. 22 AWG round copper wire has a _____ (larger/smaller) diameter than no. 12 wire.
7. The resistance of no. 10 wire is _____ (higher/lower) than that of no. 12.
8. A 50-ft piece of no. 22 wire has _____ the resistance of a 100-ft piece of the same wire.
9. An insulator may _____ if it is subjected to a voltage higher than its breakdown voltage.
10. (True/False) Semiconductors will, under the proper conditions, permit electric current to flow. _____
11. The diameter of a round copper wire is 0.025 in. The circular mil area of this wire is _____ .
12. The diameter of a round copper wire is 45 mils. Its resistivity is 10.37. The resistance of 1000 ft of this wire is _____ .

MATERIALS REQUIRED

Instruments:
■ Digital multimeter or analog VOM

Resistors:
■ 1 carbon resistor (resistance value not important)

Capacitors:
■ 1   0.01- μF, 25 V dc

Semiconductors:
■ 1 silicon rectifier, 1N2615 or equivalent

Miscellaneous:
■ 12-in. length of #40 AWG solid copper wire
■ 6-in. length of #40 nichrome wire
■ 12-in. length of #40 nichrome wire
■ 2-in. square of solid rubber, $\frac{1}{4}$-in. thick
■ 2-in. square of wood, $\frac{1}{4}$-in. thick
■ 2-in. square of acrylic plastic, $\frac{1}{4}$-in. thick
■ Coil of wire

PROCEDURE

CAUTION: Be sure to hold the insulated probes on the meter leads when you measure resistance. This becomes *particularly important* when you are using the *higher* ranges of the meter ($R \times 1000$, etc.). The reason is that the ohmmeter will read your body resistance if you hold the metal tips of the meter leads. This measurement of body resistance will affect the resistance reading of the object you are measuring, particularly if the object has high resistance.

1. In this experiment you will measure the resistance of each of the 10 objects listed in Table 5–1 (p. 39). Check your ohmmeter first by zeroing it as necessary. The test leads of your meter should be color coded with the common lead, or minus, black, and the plus lead red.
2. Connect the red test lead of the meter to one lead of the resistor and the black test lead to the other lead of the resistor. Measure the resistance and record it in column A of row 1 (Resistor).
3. Reverse the red and black test leads of the meter so that they are connected across the resistor again. Measure the resistance again and record the value in column B of row 1.
4. Indicate in column C, row 1, whether the resistor behaved like a conductor or an insulator.
5. Repeat steps 2 through 4 for the capacitor and record in row 2 the resistance values and whether the capacitor behaved as a conductor or an insulator.
6. Repeat steps 2 through 4 for the other eight objects listed in Table 5–1.

NOTE: Measure the resistance across the 2-in. dimension for the wood, rubber, and acrylic samples.

Body Resistance

7. Set the ohmmeter on the $R \times 10,000$ or $R \times 100,000$ range and check that it is properly zeroed.
8. Measure your body resistance by grasping the metal tip of one lead with one hand and the metal tip of the other lead with your other hand. Record the value: Body resistance = _____ Ω.

ANSWERS TO SELF-TEST

1. silver; copper; gold; aluminum; tungsten; and the like
2. wood; glass; paper; mica; air; and the like
3. (a) voltage source; (b) conductive load; (c) connecting conductive wiring
4. closed
5. open
6. smaller
7. lower
8. one-half
9. arc-over (or breakdown)
10. true
11. 625
12. 5.12 Ω

TABLE 5–1. Resistance of Conductors and Insulators

Object	Resistance, Ω				Object	Resistance, Ω		
	A	B	C			A	B	C
1. Resistor					6. Coil			
2. Capacitor					7. Wood			
3. 12-in. length #40 copper wire					8. Plastic			
4. 12-in. length #40 nichrome wire					9. Rubber			
5. 6-in. length #40 nichrome wire					10. Silicon rectifier 1N2615			

QUESTIONS

1. How were you able to determine whether a material was a conductor of direct current? (Refer to your actual test results.)

_____

_____

_____

2. Compare columns A and B of Table 5–1. Did the polarity of the test leads affect the resistance measurements of the materials? If so, explain the effect; refer to actual test results.

_____

_____

_____

_____

3. Is the silicon rectifier a good conductor or a good insulator? How do your test results confirm your answer?

_____

_____

_____

_____

4. Formula (5–1) shows that the resistance of a wire varies directly with length for a given cross-sectional area of wire. Which measurement or measurements in this experiment, if any, confirm this relationship?

_____

_____

_____

_____

Conductors and Insulators   **39**

5. Calculate the resistivity of nichrome wire using Formula (5–1) and the test results in Table 5–1. Use standard wire tables to determine the cross-sectional area of the #40 nichrome wire. Refer to a standard engineering handbook to compare your answer with standard published values. Explain any possible difference between your calculated value and the standard value.

_____

_____

_____

_____

6. Explain how to zero an analog ohmmeter at both ends of its scale.

_____

_____

_____

_____

7. Was the ohmmeter used in this experiment able to measure accurately the resistance of the insulating materials being tested? Explain your answer by referring to actual test results in Table 5–1.

_____

_____

_____

_____

8. Is the ohmic resistance of your body higher or lower than that of the surrounding air? What evidence from this experiment proves your answer?

_____

_____

_____

_____

9. The ends of bare copper wire are often scraped with a knife or rubbed with fine sandpaper before being connected to a circuit or being soldered. What is a possible reason for this practice?

_____

_____

_____

_____

11

CURRENT IN A PARALLEL CIRCUIT

OBJECTIVES

1. To verify experimentally that the total current in a parallel circuit is greater than the current in any branch
2. To verify experimentally that the total current in a parallel circuit is equal to the sum of the currents in each of the parallel branches.

BASIC INFORMATION

Branch Currents

In considering the series circuit in Experiment 6, it was found that a closed circuit is required for current, that current stops when the circuit is open, and that current in a series circuit is the same everywhere. How does a parallel circuit differ from a series circuit?

Figure 11–1 shows three resistors connected in parallel and a voltage V applied across the combination. If the line connecting the battery to the parallel network is broken at X or at Y and an ammeter is inserted in the circuit at X or Y (Figure 11–2), the ammeter will measure the total current delivered by the voltage source. This line current is drawn by the three resistors from the source.

A simple experiment suggests an important characteristic of a parallel circuit. If, in Figure 11–2, resistor R_1 is removed from the circuit, the line current measured by the ammeter decreases. If R_2 is then removed from the circuit, the line current decreases further. What remains is a simple series circuit consisting of V, R_3, and the ammeter. The line current is now the current drawn by R_3 from V; it may be computed directly by Ohm's law.

The results of this experiment indicate that in Figures 11–1 and 11–2 there are actually three conductive paths for current—namely, R_1, R_2, and R_3. When all three paths are closed, there is maximum line current. When path R_1 is broken, there is less line current because only two paths remain, R_2 and R_3, and as was shown before, when both R_1 and R_2 are open, only path R_3 remains. The individual paths are called *branches,* or *legs,* of the parallel circuit.

One characteristic of a parallel circuit containing only resistors is that the total current I_T in the circuit is greater than the current in any branch. It follows that each branch current in a parallel circuit containing only resistors is less than the total or line current I_T.

An example will illustrate this characteristic. Assume in Figure 11–3 (p. 74) that the voltage source V is 6.6 V. Note that 6.6 V appears across each branch resistor in the circuit—that is, the voltage across R_1 is 6.6 V, that across R_2 is 6.6 V, and

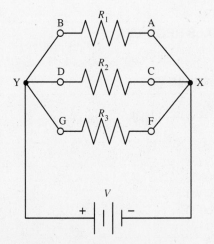

Figure 11–1. Voltage V applied across three resistors connected in parallel.

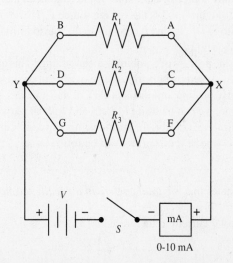

Figure 11–2. Measuring total current in a parallel circuit.

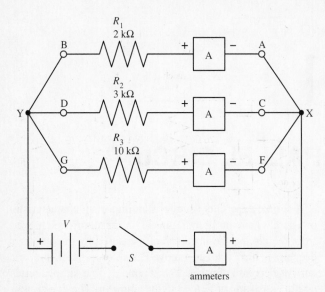

Figure 11–3. Circuit used to find the relationship between branch currents and total current in a parallel circuit.

that across R_3 is 6.6 V also. The individual currents in each branch, I_1 in R_1, I_2 in R_2, and I_3 in R_3, can be calculated by Ohm's law. Thus,

$$I_1 = \frac{V}{R_1} = \frac{6.6}{2000} = 0.0033 \text{ A} = 3.3 \text{ mA}$$

$$I_2 = \frac{V}{R_2} = \frac{6.6}{3000} = 0.0022 \text{ A} = 2.2 \text{ mA}$$

$$I_3 = \frac{V}{R_3} = \frac{6.6}{10,000} = 0.00066 \text{ A} = 0.66 \text{ mA}$$

In a parallel circuit, the total current is greater than the current in any branch. Therefore the total current I_T in this circuit must be greater than 3.3 mA.

Total Current in a Parallel Circuit

The only source of current for R_1, R_2, and R_3 in the parallel circuit of Figure 11–3 is the voltage source V. Therefore, each of the currents I_1, I_2, and I_3 combine in the line to form the line, or total, current I_T.

Does the process of combining mean adding? This question can be answered by performing an experiment on the parallel circuit. By placing ammeters in each of the branches and one in the line, we can measure I_1, I_2, I_3, and I_T. If the experiment is performed carefully, the relationship between branch currents and line or total current will be found to be

$$I_T = I_1 + I_2 + I_3$$

which states that *the total current in a parallel circuit is the sum of the branch currents*.

The total current in the preceding example is

$$I_T = 3.3 + 2.2 + 0.66$$

$$= 6.16 \text{ mA}$$

The total current, 6.16 mA, is also seen to be greater than any of the branch currents, 3.3 mA, 2.2 mA, or 0.66 mA.

SUMMARY

1. In a parallel circuit containing only resistors, the individual current in each branch is less than the total line current.
2. The total line current is greater than each individual branch current.
3. The total line current is equal to the sum of all the branch currents.
4. The voltage across each branch is the same.

SELF-TEST

Check your understanding by answering the following questions:

1. The voltage across each branch of a parallel network must be _____ .
2. In Figure 11–3, the voltage V_1 across R_1, the voltage V_2 across R_2, and the voltage V_3 across R_3 must all be _____ V.
3. In Figure 11–3, if $V = 10$ V, find the branch currents I_1, I_2, and I_3. $I_1 = $ _____ A, $I_2 = $ _____ A, and $I_3 = $ _____ A.
4. The total current in the circuit of question 3 must be greater than the _____ branch current.
5. In Figure 11–3, for the branch currents listed in question 3, $I_T = $ _____ A.

MATERIALS REQUIRED

Power Supply:
■ Variable 0–15 V dc, regulated

Instruments:
■ DMM or VOM (with at least 100 mA dc range)

Resistors (5% ¹/₂-W):
■ 1   820-Ω
■ 1   1000-Ω
■ 1   2200-Ω
■ 1   3300-Ω
■ 1   4700-Ω

Miscellaneous:
■ SPST switch

PROCEDURE

1. Measure the resistance of each of the five resistors supplied and record values in Table 11–1 (p. 77).

2. With the power supply **off** and switch S_1 **open,** connect the circuit of Figure 11–4(a). If you are using a single DMM to measure all quantities in this experiment, do not connect the ammeter to measure I_T at this point.

3. With power **on, close** S_1 and adjust the power supply V_{PS} to 10 V. Branches 1, 2, and 3 must be connected across the power supply and drawing current when voltage V_{PS} is adjusted. **Open** S_1.

4. Connect the ammeter in the circuit to measure I_T as shown. Do not change the adjustment on the power supply.

5. In the next steps it will be necessary to measure the current in each of the branches as well as the total current in the circuit. If only a single ammeter is available, it will be necessary to break the branch circuit to insert the meter in each case to measure I_1, I_2, and I_3. The main circuit will need to be broken to measure I_T. **Open** S_1 in each case before changing the meter and circuit connections. The following steps assume all necessary meter and circuit connections have been made to enable the currents and voltages to be measured.

6. **Close** S_1. Measure V_{PS}, I_T, I_1, I_2, and I_3. Record these values in Table 11–2 (p. 77). Calculate I_T (the sum of all the branch currents) and record your answer in Table 11–2.

7. Remove the 820-Ω resistor from branch 1. Adjust V_{PS} = 10 V. Measure I_T, I_2, and I_3. Record the values in Table 11–2. Calculate I_T as in step 6 and record your answer in Table 11–2.

8. Remove the 1000-Ω resistor from branch 2 so that only branch 3 remains in the circuit. Adjust V_{PS} to equal 10 V. Measure I_T and I_3. Record the values in Table 11–2. Calculate I_T as in step 6 and record your answer in Table 11–2. **Open** S_1 and turn the power **off.**

9. Connect the circuit in Figure 11–4(b). Reread the suggestions in steps 2 through 5. Voltage V_{PS} should remain 10 V with the three branches connected.

10. With power **on** and S_1 **closed,** measure V_{PS}, I_T, I_1, I_2, and I_3. Record all values in Table 11–2. Calculate I_T (the sum of the branch currents) and record your answer in Table 11–2.

11. Remove the 1000-Ω resistor from branch 1. Adjust V_{PS} = 10 V. Measure I_T, I_2, and I_3. Record the values in Table 11–2. Calculate I_T as in step 10 and record your answer in Table 11–2.

12. Remove the 2200-Ω resistor from branch 2 so that only branch 3 remains in the circuit. Adjust V_{PS} = 10 V. Measure I_T and I_3. Record the values in Table 11–2. Calculate I_T as in step 10 and record your answer in Table 11–2. **Open** S_1 and turn power **off.**

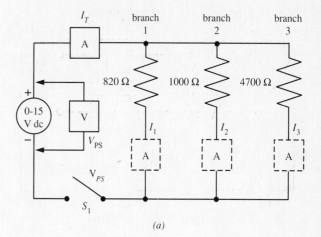

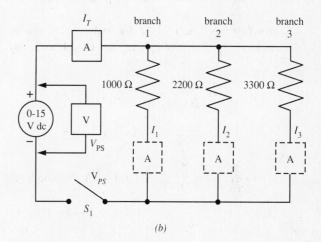

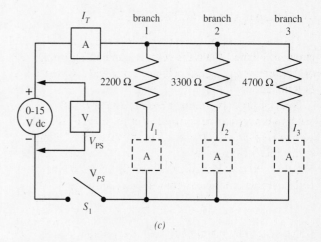

Figure 11–4. (a) Circuit for procedure step 2. (b) Circuit for procedure step 9. (c) Circuit for procedure step 13.

13. Connect the circuit in Figure 11–4(c). Reread the suggestions in steps 2 through 5. Voltage V_{PS} should remain 10 V with the three branches connected.

14. With power **on** and S_1 **closed,** measure V_{PS}, I_T, I_1, I_2, and I_3. Record the values in Table 11–2. Calculate I_T (the sum of the branch currents) and record your answer in Table 11–2.

15. Remove the 2200-Ω resistor from branch 1. Adjust $V_{PS} = 10$ V. Measure I_T, I_2, and I_3 and record the values in Table 11–2. Calculate I_T (as in step 14) and record your answer in Table 11–2.

16. Remove the 3300-Ω resistor from branch 2 so that only branch 3 remains in the circuit. Adjust $V_{PS} = 10$ V. Measure I_T and I_3 and record the values in Table 11–2. Calculate I_T (as in step 14) and record your answer in Table 11–2. **Open** S_1 and turn the power **off.**

ANSWERS TO SELF-TEST

1. equal
2. equal to
3. 0.005; 0.0033; 0.001
4. highest
5. 0.0093

Name _____ Date _____

TABLE 11–1. Measured Values, Experimental Resistors

Resistor	R_1	R_2	R_3	R_4	R_5
Rated value, Ω	820	1000	2200	3300	4700
Measured value, Ω					

TABLE 11–2. Measured and Computed Values in Parallel Circuit

Step	Rated Value of Branch Resistors, Ω					Measured Values					I_T Calculated (sum of branch I) A
						V	A				
	R_1	R_2	R_3	R_4	R_5	V	I_T	I_1	I_2	I_3	A
6	820	1000			4700						
7		1000			4700						
8					4700						
10		1000	2200	3300							
11			2200	3300							
12				3300							
14			2200	3300	4700						
15				3300	4700						
16					4700						

QUESTIONS

1. Explain, in your own words, how your experimental results confirmed the two objectives of this experiment. Refer to the data in Tables 11–1 and 11–2 to support your discussion.

_____

_____

_____

_____

_____

2. Why was it important to measure the values of the resistors (step 1)?

_____

_____

_____

_____

_____

_____

Current in a Parallel Circuit

3. Discuss the effect on the total current of parallel-connected resistors if:
   (a) the number of resistors in parallel is increased.
   (b) the resistance of each resistor is increased.
   Support your answers in (a) and (b) by referring to your experimental data in Table 11–2.

   _____

   _____

   _____

   _____

   _____

   _____

   _____

   _____

   _____

   _____

   _____

   _____

   _____

4. Examine your data in Tables 11–1 and 11–2. What general relationship is indicated
   between the branch current and the total circuit current? State this relationship in your
   own words; then write the relationship as a mathematical formula.

   _____

   _____

   _____

   _____

   _____

   _____

   _____

   _____

   _____

   _____

   _____

   _____

   _____

   _____

   _____

   _____

   _____

12

RESISTANCE OF A PARALLEL CIRCUIT

OBJECTIVE

To verify experimentally the relationship between branch resistances and the total resistance of a parallel circuit.

BASIC INFORMATION

Total Resistance in a Parallel Circuit

The resistance R_T to which voltage V is connected in Figure 12–1 limits the current in the circuit to I_T. If a single resistor could be found that would draw the same current I_T when connected across V, then the value of this resistor would be equivalent to the three parallel resistors. This equivalent resistor would also represent the total resistance R_T of the three parallel resistors.

Measuring Total Resistance

With V removed, an ohmmeter placed across the end points X and Y of a parallel circuit such as in Figure 12–2 would measure the total resistance of the three resistors.

CAUTION: Always disconnect resistors from their power source before making resistance measurements with an ohmmeter.

The Ohm's law formula for finding resistance is

$$R = \frac{V}{I}$$

If R is the total resistance across V, then I is the total current delivered by V to the resistance. This provides another method for measuring R_T.

If the circuit in series with V were broken and an ammeter inserted in the break, as in Figure 12–3 (p. 80), the ammeter would read the total current delivered by V. The value of V can be measured by connecting a voltmeter directly across the voltage source. Two of the three factors of Ohm's law, V and I, would then be known and the third factor, R, could be calculated using the formula

$$R_T = \frac{V}{I_T} \qquad \textbf{(12–1)}$$

In Experiment 11 it was found that the total current drawn by a parallel circuit is greater than the current in any branch. From Ohm's law and the fact that resistance is inversely proportional to current (that is, if voltage is held constant, current will decrease as resistance increases) a similar characteristic is true for parallel resistors.

In a parallel circuit the total resistance of the circuit is *less* than the *lowest* resistance in the parallel combination. For example, if three resistors with values of 47 Ω, 68 Ω, and 100 Ω were connected in parallel as in Figure 12–2, the total resistance would be *less* than 47 Ω. (The exact value and a method for finding it are discussed next.)

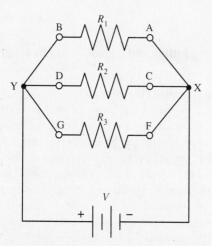

Figure 12–1. Voltage *V* applied across three resistors connected in parallel.

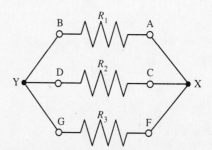

Figure 12–2. Finding the total resistance of three parallel resistors.

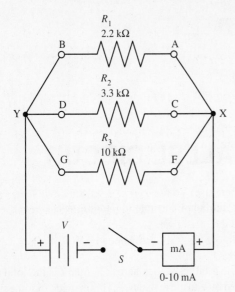

Figure 12–3. Measuring total current in a parallel circuit.

The Relationship Between Branch Resistance and Total Resistance

It seems logical that there should be some definite relationship between parallel resistances $R_1, R_2, R_3, \ldots$ and their total or equivalent resistance. And it should be possible to express this relationship in a formula.

In the discussion that follows, we make a very important (but valid) assumption. We will assume that the resistance of the conductors in the circuits is zero and that the only resistance in the circuit is that of the resistors themselves.

In Figure 12–1 this assumption allows us to say that electrically points B, D, and G are the same as point Y and points A, C, and F are the same as point X.

This fact makes one condition of the circuit very obvious: *The voltage across each branch resistor is exactly the same,* and in the circuit of Figure 12–1, it is equal to V.

Using Ohm's law we can find the current in each branch of the circuit in Figure 12–1:

$$I_1 = \frac{V}{R_1}$$

$$I_2 = \frac{V}{R_2}$$

$$I_3 = \frac{V}{R_3}$$

The total current delivered to this circuit by V is

$$I_T = I_1 + I_2 + I_3$$

$$= \frac{V}{R_1} + \frac{V}{R_2} + \frac{V}{R_3}$$

We can rewrite formula (12–1) as

$$I_T = \frac{V}{R_T}$$

Substituting this in the previous formula we get

$$\frac{V}{R_T} = \frac{V}{R_1} + \frac{V}{R_2} + \frac{V}{R_3}$$

Canceling the Vs,—that is, dividing both sides of the formula by V—results in

$$\frac{1}{R_T} = \frac{1}{R_1} + \frac{1}{R_2} + \frac{1}{R_3} \qquad \textbf{(12–2)}$$

Formula (12–2) states the relationship between branch resistance and total resistance of a parallel circuit: *The reciprocal of the total resistance of a parallel circuit is equal to the sum of the reciprocals of the individual branch resistances.*

To find R_T we need to take the reciprocals of both sides of formula (12–2)

$$R_T = \frac{1}{\dfrac{1}{R_1} + \dfrac{1}{R_2} + \dfrac{1}{R_3}} \qquad \textbf{(12–3)}$$

Notice that it is necessary to take the reciprocal of the *entire right side of the formula,* not merely the reciprocals of the individual terms.

Formulas (12–2) and (12–3) apply to any parallel circuit no matter how many branches are involved. The general form is

$$\frac{1}{R_T} = \frac{1}{R_1} + \frac{1}{R_2} + \frac{1}{R_3} + \ldots$$

where the dots indicate that any number of reciprocal resistances can be added in a particular case.

Although formulas (12–2) and (12–3) were obtained using Ohm's law, their validity can be checked experimentally. Using the methods discussed before, the total resistance of a parallel circuit can be measured directly using an ohmmeter or else calculated using measured values of V and I_T and Ohm's law.

Measuring Individual Resistances in a Parallel Circuit

To verify formulas (12–2) and (12–3) using the circuit in Figure 12–3, we need to measure the individual branch resistances $R_1, R_2,$ and R_3 in the parallel network. How can this be done? Obviously it *cannot* be done by placing the ohmmeter across each resistor in the network, because this procedure would give the measurement of R_T, not the branch resistor. We can measure R_1 by disconnecting it from the parallel network and measuring it outside of the circuit. Or we can disconnect one lead of R_1, say at point A, thus removing the effect of the network. We can then measure R_1 by placing an ohmmeter across it. A similar procedure can be followed to measure the resistances of R_2 and R_3.

SUMMARY

1. The voltage V across each branch (i.e., each resistor in Figures 12–1 and 12–3) of a parallel circuit is the same.
2. The total or equivalent resistance R_T of two or more resistors connected in parallel, as in Figure 12–3, can be found experimentally by measuring the total current I_T, measuring the voltage V across the parallel network, and substituting the measured values in the formula

$$R_T = \frac{V}{I_T}$$

3. Another method of finding the total resistance R_T of two or more parallel-connected resistors, as in Figure 12–2, is to place an ohmmeter across the parallel circuit. The meter will measure R_T.
4. Resistance should never be measured when there is power applied to the circuit. If the parallel resistance of R_1, R_2, and R_3 in Figure 12–3 is required, power must first be disconnected.
5. A formula that expresses the relationship between R_T and R_1, R_2, R_3, etc., of parallel-connected resistors is:

$$R_T = \frac{1}{\dfrac{1}{R_1} + \dfrac{1}{R_2} + \dfrac{1}{R_3} + \cdots}$$

6. Another way of writing the formula for R_T is

$$\frac{1}{R_T} = \frac{1}{R_1} + \frac{1}{R_2} + \frac{1}{R_3} + \cdots$$

7. To measure one of two or more resistors connected in parallel, say R_1 in Figure 12–2, disconnect one lead of R_1 from the circuit. Then measure the resistance of R_1.

SELF-TEST

Check your understanding by answering the following questions:

1. In the circuit of Figure 12–1, $I_T = 0.02$ A. $V = 5$ V. The total resistance R_T equals _____ Ω.
2. (True/False) For the conditions in question 1, it is possible for R_1 to equal 100 Ω. _____
3. For the conditions in question 1, the voltage V across $R_2 = $ _____ V.
4. (True/False) To measure the resistance of R_3 in Figure 12–2, simply place the ohmmeter leads across GF and read the resistance. _____
5. In Figure 12–2 the resistance of $R_1 = 25$ Ω, $R_2 = 33$ Ω, $R_3 = 75$ Ω. $R_T = $ _____ Ω.

MATERIALS REQUIRED

Power Supplies:
■ Variable, 0-15 V dc, regulated

Instruments:
■ DMM or VOM

Resistors (5%, ½-W):
■ 1   820-Ω
■ 1   1000-Ω
■ 1   2200-Ω
■ 1   3300-Ω
■ 1   4700-Ω

Miscellaneous:
■ SPST switch

PROCEDURE

A. Finding R_T by Formula

A1. Measure the resistance of each of the resistors supplied and record the value in Table 12–1 (p. 85).

A2. Connect the two resistors shown in parallel in Figure 12–4(a) (p. 82). Using an ohmmeter, measure the resistance of the parallel combination. Record the value in Table 12–2 (p. 85).

A3. Connect a third resistor to the parallel combination as shown in Figure 12–4(b). Using an ohmmeter, measure the resistance of the parallel combination. Record the value in Table 12–2.

A4. Connect a fourth resistor to the parallel combination as shown in Figure 12–4(c). Using an ohmmeter, measure the resistance of the parallel combination. Record the value in Table 12–2.

A5. Connect a fifth resistor to the parallel combination as shown in Figure 12–4(d). Using an ohmmeter, measure the resistance of the parallel combination. Record the value in Table 12–2. Do not disconnect this combination.

A6. For each of the parallel combinations in steps 2 through 5, calculate the value of R_T using measured values of resistance from Table 12–1 and the formulas discussed in the Basic Information section. Record your answers in Table 12–2.

B. Finding R_T Using the Voltage-Current Method

B1. With power **off** and switch S_1 **open,** using the parallel combination of resistors in step 5 of part A, connect the circuit of Figure 12–5(a) (p. 83). Power **on,** S_1 **closed.** A constant voltage, $V_{PS} = 10$ V, will be applied to all circuits in part B. Measure V_{PS} (it should be 10 V) and I_T. Record values in Table 12–3 (p. 85).

B2. Remove the 4700-Ω resistor to obtain the circuit of Figure 12–5(*b*). Adjust V_{PS} to 10 V. Measure I_T and record values in Table 12–3.

B3. Remove the 3300-Ω resistor from the circuit of step B2 as in Figure 12–5(*c*). Adjust V_{PS} to 10 V. Measure I_T and record values in Table 12–3.

B4. Remove the 2200-Ω resistor from the circuit of step B3, leaving two resistors in parallel as in Figure 12–5(*d*). Adjust V_{PS} to 10 V. Measure I_T and record values in Table 12–3. Power **off,** S_1 **open.**

B5. For each circuit in steps B1 through B4, calculate R_T by using the Ohm's law formula discussed in the Basic Information section. Record your answers in Table 12–3.

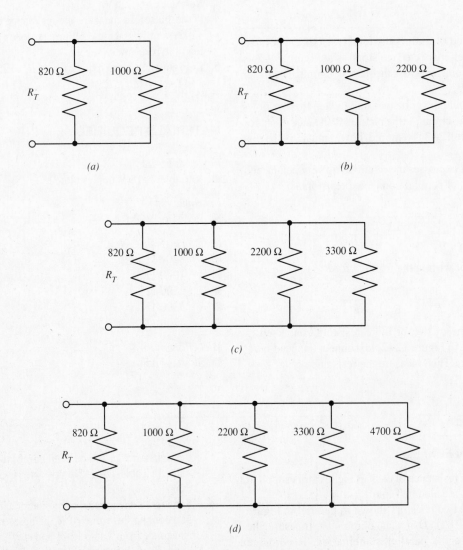

Figure 12–4. (*a*) Circuit for procedure step A2. (*b*) Circuit for procedure step A3. (*c*) Circuit for procedure step A4. (*d*) Circuit for procedure step A5.

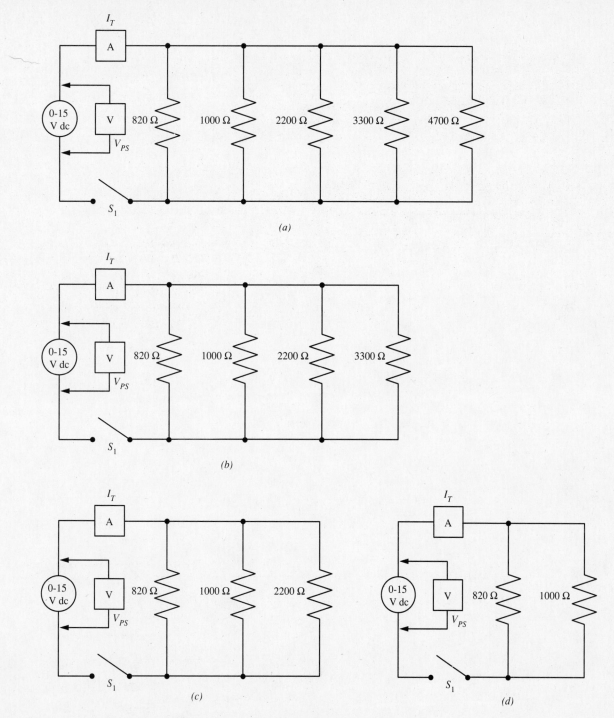

Figure 12–5. (*a*) Circuit for procedure step B1. (*b*) Circuit for procedure step B2. (*c*) Circuit for procedure step B3. (*d*) Circuit for procedure step B4.

ANSWERS TO SELF-TEST

1. 250
2. false
3. 5
4. false
5. 12

Experiment 12　　　　Name _____ Date _____

TABLE 12–1. Measured Values, Experimental Resistors

Resistor	R_1	R_2	R_3	R_4	R_5
Rated value, Ω	820	1000	2200	3300	4700
Measured value, Ω					

TABLE 12–2. Part A: Finding R_T of Parallel-Connected Resistors by Formula

Step	Rated Value, Ω					Measured Value of R_T, Ω	Calculated Value of R_T, Ω
	R_1	R_2	R_3	R_4	R_5		
A2	820	1000					
A3	820	1000	2200				
A4	820	1000	2200	3300			
A5	820	1000	2200	3300	4700		

TABLE 12–3. Part B: Finding R_T by the Voltage-Current Method

Step	Measured Values		Calculated Values R_T, Ω
	V_{PS}, V	I_T, A	
B1			
B2			
B3			
B4			

QUESTIONS

1. Explain, in your own words, the relationship between branch resistances and the total resistance of a parallel circuit.

_____

_____

_____

_____

_____

2. Write the relationship discussed in Question 1 as a mathematical formula.

_____

_____

　　Resistance of a Parallel Circuit　　**85**

3. Discuss the effect on total resistance of a parallel circuit, if:
   (a) the number of parallel resistors is increased.
   (b) the resistance of each resistor is increased.
   Support your answers by referring to your experimental data.

_____

_____

_____

_____

_____

_____

_____

_____

4. Discuss three methods used in finding the total resistance of parallel-connected resistors.

_____

_____

_____

_____

_____

_____

_____

_____

_____

5. Parts A and B use similar circuits. For each comparable combination of resistors
   in parts A and B, compare the calculated values. Discuss the possible reasons for
   differences, if any.

_____

_____

_____

_____

_____

_____

_____

_____

_____

_____

13

DESIGNING PARALLEL CIRCUITS

OBJECTIVES

1. To design a parallel circuit that will meet specified voltage, current, and resistance requirements
2. To construct and test the circuit to see that it meets the design requirements

BASIC INFORMATION

Designing a Parallel Circuit to Meet Specified Resistance Requirements

The formulas for total resistance of parallel-connected resistors can be applied to the solution of simple design problems. An example will indicate the techniques to be used.

Problem 1. A technician has a stock of the following color-coded resistors: four 68-Ω, five 82-Ω, two 120-Ω, three 180-Ω, two 330-Ω , and one each of 470-Ω, 560-Ω, 680-Ω, and 820-Ω. A circuit being designed needs a 37-Ω resistance. Find a combination of resistors, using the least possible number of components, that will satisfy the design requirement. Assume the measured values of the resistors are the same as the color-coded values.

Solution. Because the value of 37 Ω is less than the resistance of the smallest resistor in stock, a parallel arrangement will be required. (Recall from a previous experiment that the total resistance of a parallel circuit is less than the resistance value of the smallest resistor.) The formula for total resistance R_T of a parallel circuit is

$$\frac{1}{R_T} = \frac{1}{R_1} + \frac{1}{R_2} + \frac{1}{R_3} + \cdots \qquad \textbf{(13–1)}$$

If two resistors in parallel satisfy the design requirements, this formula can be rewritten as

$$\frac{1}{R_T} = \frac{1}{R_1} + \frac{1}{R_2}$$

Upon simplification this becomes

$$\frac{1}{R_T} = \frac{R_1 + R_2}{R_1 R_2}$$

Solving for R_T yields

$$R_T = \frac{R_1 \times R_2}{R_1 + R_2} \qquad \textbf{(13–2)}$$

That is, the total resistance of two parallel resistors is equal to the product of their resistances divided by the sum of their resistances. Formula (13–2) is a general formula that can be used to find the total resistance of all two-resistor parallel combinations. Using this formula where it applies can save considerable calculating time.

Assume that two resistors will meet the design requirements of this problem and that the resistors are 68 Ω and 82 Ω. Substituting these values in formula (13–2) gives

$$R_T = \frac{68 \times 82}{68 + 82} = \frac{68 \times 82}{150} = 37.2 \ \Omega$$

It is evident then that the two values selected meet the problem requirements, because when connected in parallel, their total resistance is very close to 37 Ω. Trial and error will show that no other combination of two resistors in stock will produce a value closer to 37 Ω.

It is not always possible to select the two required resistors so easily. Another method can be used. Suppose we assume that 68 Ω is one of the resistors. We wish to find another resistor R_X that, when connected in parallel with 68 Ω, will produce an equivalent or total resistance of 37 Ω. Substitute the known values of R_T and R_1 in formula (13–2):

$$37 = \frac{68 \times R_X}{68 + R_X}$$

This formula can be rewritten to solve for R_X

$$R_X = \frac{37 \times 68}{68 - 37}$$

$$= 81.2 \ \Omega$$

The 82-Ω resistor in stock could therefore satisfy (within limits) the requirements of the circuit.

Problem 2. Assume the same stock as in Problem 1 is available. The technician must design a circuit requiring a 60-Ω resistor.

Solution. The 60-Ω requirement is exactly half the value of the 120-Ω resistor in stock. Assume that two resistors will satisfy the design requirements, and one of the resistors is 120 Ω. Substitute the known values in formula (13–2):

$$R_T = \frac{120 \times R_2}{120 + R_2} = 60$$

Solving for R_2, we have

$$R_2 = \frac{60 \times 120}{120 - 60} = \frac{7200}{60}$$

$$= 120 \ \Omega$$

Thus, the two 120-Ω resistors in parallel will produce an equivalent resistance of one-half the value of one of the resistors; two 68-Ω resistors in parallel are equivalent to 34 Ω; two 1000-Ω resistors in parallel are equivalent to 500 Ω, and so on. What would be the equivalent resistances of three, four, or more equal resistors in parallel? Formula (13–1) will be used to test the condition of three equal resistors connected in parallel.

$$\frac{1}{R_T} = \frac{1}{R} + \frac{1}{R} + \frac{1}{R} = \frac{3}{R}$$

$$R_T = \frac{R}{3}$$

Similarly, applying formula (13–1) to four, five, and six equal resistors results in total resistances equal to R/4, R/5, and R/6, respectively. We can draw a general rule and formula from these calculations: *The total resistance of n equal resistors connected in parallel is the value of a single resistance R divided by the number of equal resistors,* or

$$R_T = \frac{R}{n} \qquad \textbf{(13–3)}$$

The technician could also use three 180-Ω resistors connected in parallel, because their total resistance is also 60 Ω.

$$R_T = \frac{180}{3} = 60 \ \Omega$$

However, here three components are used rather than the two 120-Ω resistors.

Problem 3. In the circuit of Figure 13–1 a technician measures the resistance between points A and B and finds it is 180 Ω. The circuit requires a resistance between A and B of 45 Ω. How can the technician modify the circuit to meet the required value, assuming the same stock of resistors as in problem 1 is available?

Solution. Assume that there is a resistor R_X that, when connected in parallel with 180 Ω, will bring the resistance R_{AB} down to 45 Ω. Substitute the known values in formula (13–2).

$$45 = \frac{180 \times R_X}{180 + R_X}$$

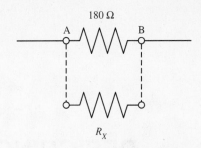

Figure 13–1. When resistor R_X is connected across the 180-Ω resistor, the total resistance across A–B must be 45 Ω.

Solving for R_X, we have

$$R_X = \frac{45 \times 180}{180 - 45} = \frac{8100}{135}$$

or

$$R_X = 60 \ \Omega$$

What is required, then, is a 60-Ω resistor. From problem 2 it is evident that two 120-Ω resistors, connected in parallel with the 180-Ω resistor, will yield the required result:

$$R_{AB} = 45 \ \Omega$$

To verify the design we will find the total resistance of a parallel circuit consisting of three branch resistors, two 120-Ω and one 180-Ω.

$$\frac{1}{R_T} = \frac{1}{120} + \frac{1}{120} + \frac{1}{180}$$

$$\frac{1}{R_T} = \frac{3 + 3 + 2}{360} = \frac{8}{360}$$

$$R_T = \frac{360}{8} = 45 \ \Omega$$

The procedure, then, for designing a parallel circuit having a specified resistance value R_T from a group of known resistors is to apply formulas (13–1), (13–2), and (13–3) to the resistors at hand and find a combination that equals or is closest to the required value. After this is done the technician should connect the resistors in parallel and measure their total resistance with an ohmmeter to confirm the solution.

Designing a Parallel Circuit to Meet Specified Resistance and Current Requirements

Ohm's law and the formulas for total resistance in a parallel circuit are applied in the solution of this type of problem. Again, an example illustrates the procedure.

Problem 4. In the circuit of Figure 13–2(a), find the value of V to which the variable dc power supply must be set to obtain a current of 20 mA.

Solution. In order to find the applied voltage V, it is first necessary to determine the total resistance R_T in the circuit. When that is known, V can be found by using the Ohm's law formula $V = I \times R_T$.

1. *Procedure for finding R_T:* The 680-Ω resistor between A and B is connected in parallel with the two series-connected resistors, 1200 and 820 Ω. The first step is to replace the 1200- and 820-Ω resistors with a single resistor R_1 whose resistance value is equal to the sum of the two resistors. Thus,

$$R_1 = 1200 + 820 = 2020 \ \Omega$$

The resulting equivalent circuit is then Figure 13–2(b), whose resistance and current characteristics are the same as those of Figure 13–2(a).

The total resistance can now be calculated by use of formula (13–2).

$$R_T = \frac{680 \times 2020}{680 + 2020} = 509 \ \Omega$$

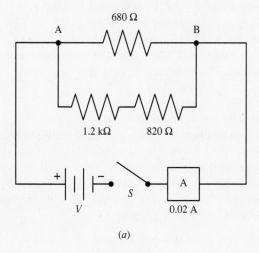

(a)

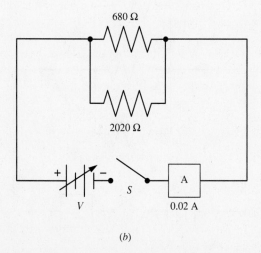

(b)

Figure 13–2. Finding the voltage that will draw 0.02 A from the supply.

2. *Procedure for finding V:*

$$V = I_T \times R_T$$
$$= 0.02 \times 509 = 10.2 \ V$$

The required voltage is approximately 10.2 V.

The technician must now connect the circuit of Figure 13–2(a) and apply a measured 10.2 V from the dc supply. The measured current should be 20 mA.

In actual practice the technician would have connected the circuit of Figure 13–2 and adjusted the output of the dc supply until the ammeter measured 20 mA. The technician could read the dc voltage, which should be 10.2 V, on the power supply voltmeter.

Designing a Parallel Circuit to Meet Specified Voltage and Current Requirements

Ohm's law and the formulas for total current and total resistance are applied in the solution of this type of problem. The following examples illustrate the procedures to use.

Problem 5. In the circuit of Figure 13–3, a current-divider network is required that will permit 0.02 A in resistor R_1 and 0.03 A in resistor R_2. A dc voltage source of 15 V powers the circuit. What values of R_1 and R_2 must be used to meet the circuit requirements?

Solution. The voltage across each branch of a parallel network is the same. Therefore, the voltage V_1 across R_1 equals the voltage V_2 across R_2. In this case,

$$V_1 = V_2 = 15 \ V$$

By Ohm's law,

$$R_1 = \frac{V_1}{I_1} = \frac{15}{0.02} = 750 \ \Omega$$

$$R_2 = \frac{V_2}{I_2} = \frac{15}{0.03} = 500 \ \Omega$$

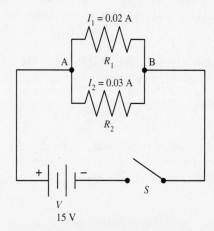

Figure 13–3. Finding the values of R_1 and R_2 to meet the given voltage and current requirements.

Designing Parallel Circuits **89**

The required resistors are 750 and 500 Ω. The technician should therefore connect 750- and 500-Ω resistors in the circuit of Figure 13–3 and measure the branch currents I_1 and I_2 to verify the solution.

Problem 6. A technician must design a circuit that will deliver 0.25 A from a 5-V supply. Using the appropriate formula, determine what combination of resistors, using the least number of components, will satisfy the design requirements. The stock of resistors that the technician has on hand includes two 40-, two 60-, three 150-, three 180-, three 680-, and one each of 330- and 470-Ω resistors.

Solution. The total resistance that will satisfy the requirements of the problem is

$$R_T = \frac{5}{0.25} = 20 \ \Omega$$

A series combination will not give the required solution, but a parallel arrangement may. By inspection, it is apparent from the stock of resistors that two 40-Ω resistors in parallel will give the required value of 20 Ω. Although other combinations are possible (Can you find them?), this is the only solution requiring only two resistors.

SUMMARY

1. If a circuit requires a resistance value R_T that is less than the lowest resistance value of a stock of resistors, it may be possible to approximate the required value by paralleling two or more of the available resistors. The procedure is to choose a resistor R_1 whose resistance is slightly higher than the required R_T, then find R_2 from the formula for total resistance

$$R_T = \frac{R_1 \times R_2}{R_1 + R_2}$$

If R_2 is in stock, the two resistors have been found. Otherwise it may be necessary to repeat the process. Finally, if two resistors whose parallel combination is R_T cannot be found, it may be necessary to use three or four resistors in parallel.

2. The equivalent resistance R_T of n equal-valued resistors R_1, connected in parallel, is

$$R_T = \frac{R_1}{n}$$

3. If it is necessary to calculate the voltage V that will produce I amperes in a circuit containing R ohms, the Ohm's law formula can be used:

$$V = I \times R$$

4. It is possible to find experimentally the voltage V necessary to produce I amperes in a circuit containing R ohms.

A circuit is connected containing the variable dc source in series with an ammeter and the resistance R. The voltage is varied until the required I is measured. The voltage V, measured at the terminals of the dc source, is the required voltage.

5. If it is required to calculate the resistance R that will draw I amperes from a voltage source V, use the formula

$$R = \frac{V}{I}$$

If the value of R is available in stock, that solves the problem. Otherwise, use series or parallel resistors in a combination that will satisfy the required R.

SELF-TEST

Check your understanding by answering the following questions:

1. What two resistors in parallel will yield a total resistance R_T of 30 Ω? Give at least two solutions.
   (a) _____ Ω in parallel with _____ Ω;
   (b) _____ Ω in parallel with _____ Ω.
2. Three resistors, 22-, 33-, and 47-Ω, are connected in parallel. Their total resistance will be (more/less) _____ than 22 Ω.
3. A 10-Ω resistance can be constructed from _____ 50-Ω resistors connected in parallel.
4. In the circuit of Figure 13–4, the voltage V that will deliver 0.15 A to the resistors is _____ V.
5. In a circuit with two parallel resistors similar to Figure 13–3, there is a total current of 0.003 A. The applied voltage, V, is 15 V. The total resistance R_{AB} between points A and B is _____ Ω.

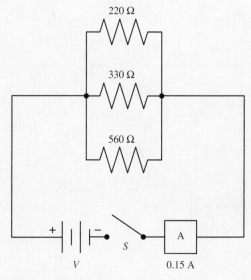

Figure 13–4. Circuit for question 4.

MATERIALS REQUIRED

Power Supply:
- Variable 0-15 V dc, regulated

Instruments:
- DMM or VOM,
- 0–100 mA milliammeter

Resistors ($\frac{1}{2}$-W, 5%)
- 1   820-Ω
- 1   1000-Ω
- 1   2200-Ω
- 1   3300-Ω
- 1   4700-Ω
- 1   5600-Ω

Miscellaneous:
- SPST switch

PROCEDURE

1. Measure the resistance of each of the resistors supplied for this experiment and record its value in Table 13–1 (p. 93).
2. Using the formulas for finding the R_T of a parallel circuit, find the parallel combinations of two or three resistors that will produce the values of R_T in Table 13–2 (p. 93). Calculate R_T using the rated values of the supplied resistors only. Record the rated values in Table 13–2.
3. Connect each of the parallel combinations in Table 13–2 and measure the R_T with an ohmmeter. Record the measured values in Table 13–2.
4. Using the resistors supplied, design a two-resistor parallel circuit that will draw approximately 20 mA when 15 V is applied across the parallel combination. If more than one combination appears to be suitable, choose the circuit closest to drawing the 20 mA. In Table 13–3 (p. 93), record the rated values of the resistors used.
5. With power **off** and switch S_1 **open,** connect the circuit in step 4 as in Figure 13–5. Turn **on** power; close S_1. Adjust V_{PS} to 15 V and measure I_T. Record the value in Table 13–3.

6. Using the formulas in the Basic Information section, calculate the voltage required to deliver 0.02 A to a circuit consisting of a 1000-Ω resistor in parallel with a 2200-Ω resistor. Record your answer in Table 13–3.
7. Connect the circuit of step 6 to a variable voltage supply as in Figure 13–6. Adjust V_{PS} until the ammeter reads 0.02 A. Measure V_{PS} and record the value in Table 13–3.
8. Design a three-branch parallel circuit with branch currents as shown in Figure 13–7 (p. 92). Choose the resistors from the six supplied for this experiment. Calculate the voltage required to deliver the branch currents specified. Record your answers in Table 13–4 (p. 93).
9. With power **off** and S_1 **open,** connect the circuit of Figure 13–7 using your designed resistor values. Turn **on** the power; **close** S_1. Adjust V_{PS} until the specified I_T is being delivered. Measure I_1, I_2, I_3, I_T, and V_{PS} and record the values in Table 13–4.
10. Design a parallel circuit containing four branches for which $I_T = 10$ mA (approximately). The individual branch currents are not critical. The voltage applied to the circuit cannot exceed 15 V. Draw a diagram of the circuit showing a meter for measuring I_T and a switch

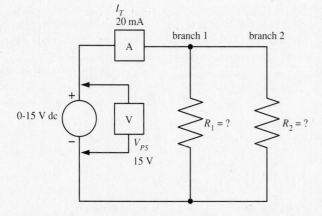

Figure 13–5. Circuit for procedure step 5. Values of R_1 and R_2 are determined in step 4.

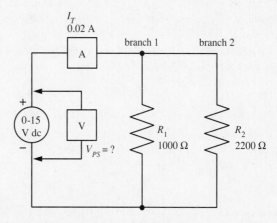

Figure 13–6. Circuit for procedure step 7.

Designing Parallel Circuits   **91**

for opening and closing the circuit. Show all calculations for finding the resistance value of each branch, and the value of V_{PS}. Record your answers in Table 13–5 (p. 93).

11. With power **off** and S_1 **open**, connect your circuit of step 10. Adjust V_{PS} to your design value. Measure I_T and record the values of V_{PS} and I_T in Table 13–5. Turn power **off**; **open** S_1.

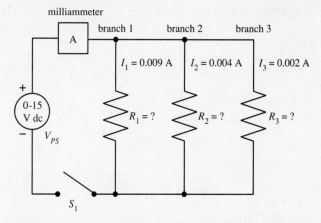

Figure 13–7. Circuit for procedure steps 8 and 9. (Values of R_1 and R_2 and R_3 are determined in step 8.)

ANSWERS TO SELF-TEST

1. (a) 60, 60; (b) 50, 75
2. less
3. 5
4. 16.1
5. 5000

Experiment 13   Name _____   Date _____

TABLE 13–1. Measured Value of Experimental Resistors

Resistor	R_1	R_2	R_3	R_4	R_5	R_6
Rated value, Ω	820	1000	2200	3300	4700	5600
Measured value, Ω						

TABLE 13–2. Designing a Parallel Resistive Network for a Required R_T

R_T Required, Ω	Combination of Parallel Resistors (Rated value, Ω)			Measured Value R_T, Ω
	Branch 1 R	Branch 2 R	Branch 3 R	
374				
417				
530				
825				
1068				
1320				
1440				
2555				

TABLE 13–3. Designing a Circuit That Will Yield I_T with V Known

Steps	V_{PS}, V			I_T, A		Parallel Resistors, Ω	
	Given Value	Calculated Value	Measured Value	Required Value	Measured Value	Branch 1 R	Branch 2 R
4, 5	15			0.02			
6, 7				0.02		1000	2200

TABLE 13–4. Designing a Current-Divider Circuit

Branch Current, A						Rated Value of Resistor, Ω			Voltage, V	
Required			Measured			Branch 1 R	Branch 2 R	Branch 3 R	Calculated	Measured
I_1	I_2	I_3	I_1	I_2	I_3					
0.009	0.004	0.002								

TABLE 13–5. Design Circuit: Given I_T and R, Find V

Steps	Rated Value of Resistors, Ω				V_{PS}, V (Design Value)	I_T Measured, A
	Branch 1 R	Branch 2 R	Branch 3 R	Branch 4 R		
10, 11						

Designing Parallel Circuits   **93**

1. Did all the resistors used in your design fall within their tolerance ratings? Refer to your measurements in Table 13–1 and the resistor color code to support your answer.

_____

_____

_____

_____

2. Why was it necessary to measure the resistance of the resistors used in your design?

_____

_____

_____

_____

3. Did the measurement of I_T in step 5 confirm that the values of the design resistors found in step 4 would satisfy the specifications of the problem? If not, explain the difference.

_____

_____

_____

_____

_____

4. Was the measured voltage, V_{PS}, in step 9 the same as that calculated in step 8? If not, discuss possible reasons for the difference.

_____

_____

_____

_____

_____

_____

_____

_____

5. On a separate sheet of $8\frac{1}{2} \times 11$ paper, design a three-branch parallel circuit that divides the current such that the current in the first branch is double the current in the second branch and three times more than the current in the third branch. The total current in the circuit, I_T is 110 mA. The total resistance of the parallel circuit, R_T, is approximately 305 Ω. Find the value of each parallel resistor. (The resistors can have any whole-number values.) Also find the value of the applied voltage. Show all formulas and calculations. Draw and label the circuit diagram.

EXPERIMENT
14

RESISTANCE OF SERIES-PARALLEL CIRCUITS

OBJECTIVES

1. To verify experimentally the rules for finding the total resistance R_T of a series-parallel circuit
2. To design a series-parallel network that will meet specified current requirements

BASIC INFORMATION

Total Resistance of a Series-Parallel Circuit

Figure 14–1 shows a series-parallel arrangement of resistors. In this circuit, R_1 is in series with the parallel circuit between points B and C, which is then in series with R_3. What is the total resistance between points A and D? Obviously we can measure R_T with an ohmmeter, or R_T can be found by the voltage-current method described in Experiment 8. Is it possible, however, to write a formula by which R_T of a series-parallel circuit can be computed without measurement?

Again refer to Figure 14–1. If R_1 and R_3 were removed and if the parallel circuit between points B and C were left standing alone, we could compute the total resistance R_{T2} between B and C by the formula for parallel resistors. We can, therefore, replace the parallel circuit in Figure 14–1 with its equivalent resistance R_{T2}, and the circuit will now take the form of Figure 14–2. The total resistance R_T of the series circuit in Figure 14–2 is the same as the total resistance R_T between points A and D in Figure 14–1.

This suggests that to find the total resistance of a series-parallel network, replace the parallel circuits first by their equivalent resistances and treat the resultant network like a simple series circuit. In the case of Figure 14–1, the total resistance R_T between points A and D is

$$R_T = R_1 + R_{T2} + R_3$$

The circuit in Figure 14–3 (p. 96) is similar to the circuit in Figure 14–1 except that there are two parallel circuits in Figure 14–3. To find the total resistance of the circuit, first find the equivalent resistance of each of the parallel sections. This results in a series circuit having five components, as in Figure 14–4 (p. 96).

The equivalent resistance R_{T1} of the parallel circuit between points B and C can be found using the formula from Experiment 13.

$$R_{T1} = \frac{R_2 R_4}{R_2 + R_4}$$

The same formula can be used to find the equivalent resistance R_{T2} between points D and F. Note, however, that one branch of this parallel section has two resistors in series. The total resistance of the branch is therefore $R_6 + R_7$. Applying the formula for two branch parallel circuits (as before), we have

$$R_{T2} = \frac{R_5 (R_6 + R_7)}{R_5 + (R_6 + R_7)}$$

The total resistance R_T of the resulting series circuit (Figure 14–4) can be found by adding all resistances in series, including the two equivalent resistances

$$R_T = R_1 + R_{T1} + R_3 + R_{T2} + R_8$$

The method just discussed is applicable to all series-parallel networks. However, in solving more complex series-parallel circuits, the initial step usually involves identifying series and parallel combinations of components. This is not always easy because the circuit diagrams are rarely as straightforward as those in Figures 14–1 through

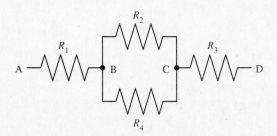

Figure 14–1. A series-parallel resistor circuit.

Figure 14–2. Equivalent circuit for Figure 14–1.

14–4. Often a choice must be made in isolating one particular section of an overall complex network and then concentrating on solving that section.

An example is used to demonstrate this method of solving a complex circuit. Other methods for solving complex networks are discussed elsewhere in this book.

Problem 1. Find the current I_T drawn by the circuit in Figure 14–5 (p. 97).

Solution. The circuit in Figure 14–5 can be redrawn as in Figure 14–6(*a*). These two circuits are exactly the same electrically. Notice that R_2 and R_3 are in series and that the $R_2 + R_3$ combination is across—that is, in parallel with—R_5. The circuit in Figure 14–6(*b*) shows the result of finding the equivalent resistance of R_5 in parallel with $(R_2 + R_3)$. Again, a series circuit consisting of R_4 in series with R_{T1} can be simplified by finding the total resistance $R_{T2} = R_4 + R_{T1}$.

Figure 14–6(*c*) shows the circuit after that simplification. Finally, the parallel branches R_{T2} and R_1 can be combined using the formula for two parallel resistances,

$$R_T = \frac{R_1 R_{T2}}{R_1 + R_{T2}}$$

We are left with the equivalent circuit in Figure 14–6(*d*). As far as the voltage source is concerned, this is the circuit it is feeding, and I_T is the current drawn by R_T. Therefore,

$$I_T = \frac{V}{R_T}$$

A numerical example shows how the simplifying process works step by step.

Problem 2. Assume that the following values apply to the circuit of Figure 14–5: $V = 12$ V, $R_1 = 500\ \Omega$, $R_2 = 680\ \Omega$, $R_3 = 320\ \Omega$, $R_4 = 1000\ \Omega$, and $R_5 = 1000\ \Omega$. Find I_T.

Solution. The steps described in problem 1 are followed in this problem, substituting the known values.

$$R_2 + R_3 = 680 + 320 = 1000\ \Omega$$

$$R_{T1} = \frac{R_5\ (R_2 + R_3)}{R_5 + (R_2 + R_3)} = \frac{1000 \times 1000}{1000 + (680 + 320)} = 500\ \Omega$$

The equivalent resistance R_{T1} is in series with R_4; therefore,

$$R_{T2}\ =\ R_4\ +\ R_{T1}\ =\ 1000\ +\ 500\ =\ 1500\ \Omega$$

Finally,

$$R_T\ =\ \frac{R_1\ R_{T2}}{R_1\ +\ R_{T2}}\ =\ \frac{500 \times 1500}{500 + 1500}\ =\ 375\ \Omega$$

$$I_T\ =\ \frac{12}{375}\ =\ 0.032\ \text{A, or 32 mA}$$

Verifying the Simplification Process

The process of combining series resistances into a single equivalent resistance and, similarly, of combining parallel branches into a single equivalent resistance, can be tested experimentally.

Again, the circuit of Figure 14–5 is used to describe the method. First, measure the resistance of R_1 through R_5. Next, connect R_2 in series with R_3. Substitute a resistor whose value is equal to $R_2 + R_3$. Connect this new resistor across R_5 and measure the resistance of the parallel circuit. This is R_{T1}. Again, use a resistor whose value is R_{T1} and connect it in series with R_4. Measure the series combination to find R_{T2}. Substitute a resistor whose value is R_{T2} and connect it across R_1. Measure the parallel combination to find R_T. Finally, connect R_T across V in series with an ammeter, adjust V to 12 V, and measure I_T.

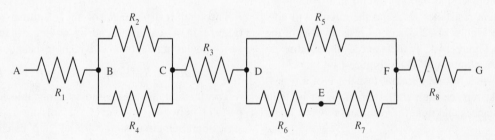

Figure 14–3. Two parallel circuits in a series-parallel combination.

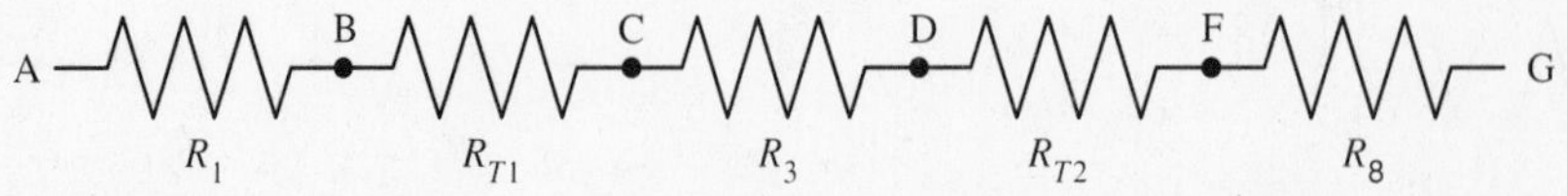

Figure 14–4. Equivalent resistance for Figure 14-3.

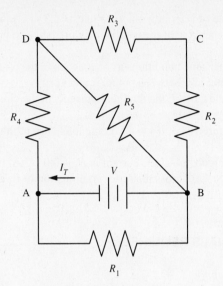

Figure 14–5. Circuit for problem 1.

SUMMARY

1. In a series-parallel network like that in Figure 14–1, the total resistance R_T of the network measured across the end terminals A and D can be found by replacing each parallel combination by its equivalent resistance R_{T2}, leaving an equivalent series circuit (Figure 14–2). The total resistance R_T can then be calculated by using the series-resistance formula:

$$R_T = R_1 + R_{T2} + R_3 + \cdots$$

2. In determining the equivalent resistance R_{T2} of one of the parallel combinations, such as R_2, R_4 in Figure 14–1, the formula for parallel resistors is used.

3. A parallel branch may sometimes have two or more series resistors, such as branch R_6–R_7 in Figure 14–4. In that case R_6 and R_7 are combined like series resistors, and their equivalent series resistance $R_{6\text{-}7} = R_6 + R_7$ is substituted in the circuit.

4. In a parallel network the voltage across each branch is the same.

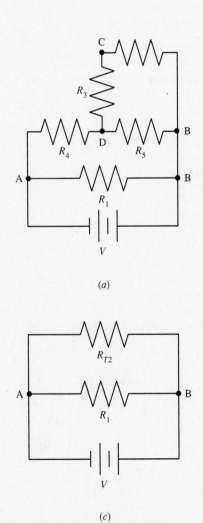

(a)

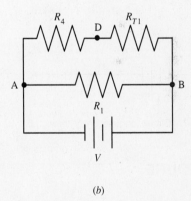

(b)

(c)

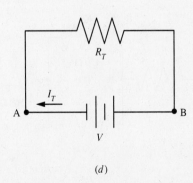

(d)

Figure 14–6. Equivalent circuit simplification for Figure 14–5.

SELF-TEST

Check your understanding by answering the following questions:

1. In Figure 14–1, $R_1 = 280\ \Omega$, $R_2 = 120\ \Omega$, $R_3 = 330\ \Omega$, and $R_4 = 470\ \Omega$. The value of R_{T2} measured from B to C is _____ Ω.
2. For the same conditions as in question 1, the value of R_T measured from A to D is _____ Ω.
3. In Figure 14–3, $R_1 = 470\ \Omega$, $R_2 = 56\ \Omega$, $R_3 = 33\ \Omega$, $R_4 = 68\ \Omega$, $R_5 = 120\ \Omega$, $R_6 = 20\ \Omega$, $R_7 = 100\ \Omega$, and $R_8 = 100\ \Omega$. The equivalent resistance R_{T2} between points B and C is _____ Ω.
4. For the same conditions as in question 3, the equivalent resistance R_{6-7} in the lower branch of the parallel network between points D and F is _____ Ω.
5. For the same conditions as in question 3, the equivalent resistance R_{T3} of the parallel network between points D and F is _____ Ω.
6. For the same conditions as in question 3, the total resistance R_T between points A and G is _____ Ω.
7. In Figure 14–7, the total resistance R_T in the circuit is _____ Ω.
8. In Figure 14–7, if $V = 25$ V, the total current in the circuit is _____ A.
9. If in Figure 14–7 the current in R_2 is 0.0675 A, the voltage V_{AB} across the series combination of R_4 and R_5 is _____ V.

MATERIALS REQUIRED

Power Supply:
- Variable 0–15 V dc, regulated

Instruments:
- DMM or VOM
- 0–10 mA milliammeter

Resistors ($\frac{1}{2}$-W, 5%):
- 1   330-Ω
- 1   470-Ω
- 1   560-Ω
- 1   1200-Ω
- 1   2200-Ω
- 1   3300-Ω
- 1   4700-Ω
- 1   10,000-Ω

Miscellaneous:
- SPST switch

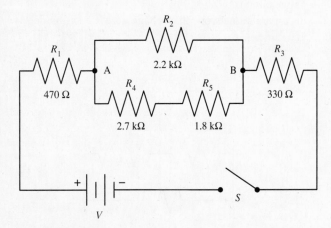

Figure 14–7. Circuit for questions 7 through 9.

PROCEDURE

1. Measure the resistance of each of the resistors supplied and record its value in Table 14–1 (p. 101).

2. Connect the resistors as in Figure 14–8(*a*). Measure the resistance between A and D (R_T) and the resistance between B and C (R_{BC}). Record the values in Table 14–2 (p. 101).

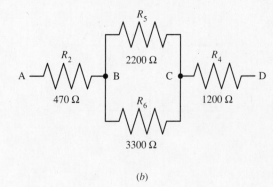

(*a*)   (*b*)

Figure 14–8. (*a*) Resistor combination for procedure step 2. (*b*) Resistor combination for procedure step 4.

3. Complete the first row of Table 14−2 under "Calculated Value" in the following manner:

R_T (a)   Calculate the total resistance between A and D by adding the measured values of R_1, R_{BC}, and R_3.

R_{BC} Use the formulas from the Basic Information section to calculate the resistance of the parallel combination of R_2 and R_4. Use measured values from Table 14−1 in your calculations.

R_T (b)   Calculate the total resistance between A and D by adding the measured values of R_1 and R_3 to the calculated value of R_{BC}.

4. Connect the resistors as in Figure 14−8(b). Complete the second row of Table 14−2 using the procedures in steps 2 and 3.

5. Connect the resistors as in Figure 14−9(a). Measure the resistance across A and G (R_T), across B and C (R_{BC}), and across D and F (R_{DF}). Record the values in the third row of Table 14−2.

6. Complete the third row of Table 14−2 under "Calculated Value" in the following manner:

R_T (a)   Calculate the total resistance between A and D by adding the measured values of R_1, R_{BC}, R_3, R_{DF}, and R_8.

R_{BC} Use the formulas from the Basic Information section to calculate the resistance of the parallel combination of R_2 and R_4. Use measured values from Table 14−1 in your calculations.

R_T (b)   Calculate the total resistance between A and G by adding the measured value of R_1, the calculated value of R_{BC}, the measured value of R_3, the calculated value of R_{DF}, and the measured value of R_8.

7. Connect the resistors as in Figure 14−9(b). Complete the fourth row of Table 14−2 using the procedures in steps 5 and 6. To calculate R_{DF}, calculate the resistance of the series combination of R_3 and R_7 in parallel with resistance R_8.

8. With power **off** and switch S_1 **open,** connect the circuit of Figure 14−10 (p. 100). Power **on.** Adjust V_{PS} to 15 V and maintain this voltage for the next step.

9. **Close** S_1. Measure the voltage across R_1, R_2, branch 1 (points B and E) (V_{BE}), branch 2 (V_{CD}), points A and E (V_{AE}), and points B and F (V_{BF}). Record the values in Table 14−3 (p. 101).

10. Using the rated values of the resistors supplied, design a four-resistor series-parallel circuit as in Figure 14−1, such that the circuit will draw a total current I_T of approximately 5 mA when connected across a 10-V dc supply. Draw the circuit showing all resistor values, an ammeter to measure I_T, and a switch S_1 to **open** and **close** the circuit. Show all calculations used to find the circuit resistances.

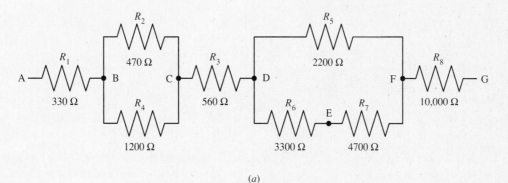

(a)

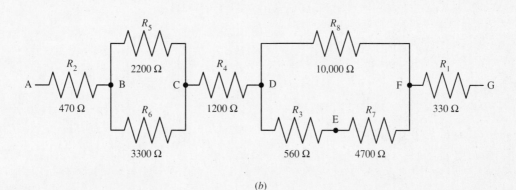

(b)

Figure 14−9. (a) Resistor combination for procedure step 5. (b) Resistor combination for procedure step 7.

11. With power **off,** and S_1 **open,** connect the circuit you designed in step 10. Measure R_T.

CAUTION: Do not attempt to measure R_T with power **on.**

After approval by the instructor, turn power **on** and **close** S_1. Measure V_{PS} and I_T. Record all values in Table 14–4 (p. 101). Power **off;** S_1 **open.**

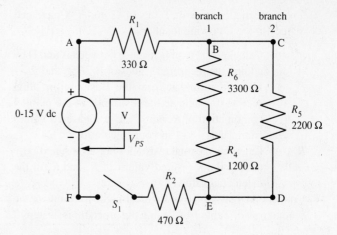

Figure 14–10. Circuit for procedure step 8.

ANSWERS TO SELF-TEST

1. 96
2. 706
3. 30.7
4. 120
5. 60
6. 694
7. 2278
8. 0.011
9. 148.5

Experiment 14 Name _____ Date _____

TABLE 14-1. Measured Values of Resistors

Resistor	R_1	R_2	R_3	R_4	R_5	R_6	R_7	R_8
Rated value, Ω	330	470	560	1200	2200	3300	4700	10,000
Measured value, Ω								

TABLE 14-2. Ohmmeter Method for Determining R_T in a Series-Parallel Network

Steps	Measured Value			Calculated Value			
	R_T	R_{BC}	R_{DF}	R_T (a)	R_{BC}	R_{DF}	R_T (b)
2, 3 [Figure 14–8(a)]							
4 [Figure 14–8(b)]							
5, 6 [Figure 14–9(a)]							
7 [Figure 14–9(b)]							

TABLE 14-3. Branch Voltage in a Series-Parallel Network

V Applied	V_1 (across R_1)	V_2 (across R_2)	Branch 1 V_{BE}	Branch 2 V_{CD}	V_{AE}	V_{BF}

TABLE 14-4. Design Problem

Design Values			Measured Values		
V_{PS}	I_T	R_T	V_{PS}	I_T	R_T
10 V	5 mA				

QUESTIONS

1. Explain, in your own words, the rules for finding the total resistance of a series-parallel circuit.

_____

_____

_____

_____

2. Explain why it is essential to disconnect power from a circuit before measuring resistance in the circuit with an ohmmeter.

_____

_____

_____

Resistance of Series-Parallel Circuits

3. When measuring the value of a resistor in a series-parallel circuit, one lead of the resistor should be disconnected from the circuit. Explain why.

_____

_____

_____

_____

_____

4. What does your data in Tables 14–1 and 14–2 confirm about the total resistance of a series-parallel circuit? Refer to specific measurements in the tables to support your conclusions.

_____

_____

_____

_____

_____

5. What measurements will you need to make to find the current in each resistor of a series-parallel circuit?

_____

_____

_____

_____

6. Refer to Figure 14–7 (neglect the resistor values). If I_2 is the current through R_2, $I_{4,5}$ is the current through resistors R_4 and R_5, and I_T is the total current supplied by V, what is the relationship between I_2R_2, I_3R_3, $I_{4,5}(R_4 + R_5)$, and I_TR_T, where R_T is the total resistance of the series-parallel circuit? (State the relationship as a mathematical formula.)

_____

_____

_____

_____

7. Refer to Question 6. Do your data in Tables 14–2 and 14–3 support your answer? If so, cite specific measurements.

_____

_____

_____

_____

_____

15

KIRCHHOFF'S VOLTAGE LAW (ONE SOURCE)

OBJECTIVES

1. To find a relationship between the sum of the voltage drops across series-connected resistors, and the applied voltage
2. To verify experimentally the relationship found in objective 1

BASIC INFORMATION

Kirchhoff's voltage law is used to solve complex electric circuits. The law, named for Gustav Robert Kirchhoff (1824–1887), the physicist who formulated it, is the basis for modern circuit analysis.

Voltage Law

In the circuit of Figure 15–1, the series resistors R_1, R_2, R_3, and R_4 can be replaced by their total or equivalent resistance R_T, where

$$R_T = R_1 + R_2 + R_3 + R_4 \qquad (15–1)$$

Use of R_T will not affect the total current I_T. The relationship between I_T, R_T, and the voltage source V is given by Ohm's law,

$$V = I_T \times R_T \qquad (15–2)$$

Substituting formula (15–1) in (15–2) gives

$$V = I_T (R_1 + R_2 + R_3 + R_4)$$

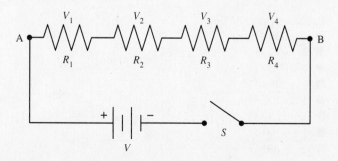

Figure 15–1. Voltages across resistors in a series circuit.

which, upon multiplication, becomes

$$V = I_T R_1 + I_T R_2 + I_T R_3 + I_T R_4 \qquad (15–3)$$

Because Ohm's law applies to any part of a circuit as well as the entire circuit, formula (15–3) shows that

$$I_T R_1 = \text{the voltage drop across } R_1 = V_1$$
$$I_T R_2 = \text{the voltage drop across } R_2 = V_2$$
$$I_T R_3 = \text{the voltage drop across } R_3 = V_3$$
$$I_T R_4 = \text{the voltage drop across } R_4 = V_4$$

Formula (15–3) can now be rewritten as

$$V = V_1 + V_2 + V_3 + V_4 \qquad (15–4)$$

Formula (15–4) is the mathematical expression of Kirchhoff's voltage law.

Formula (15–4) may be generalized for circuits containing one or more series-connected resistors in a closed circuit. The law also applies to series-parallel circuits [Figure 15–2 (p. 104)]. Here, $V = V_1 + V_2 + V_3 + V_4 + V_5$, where V_1, V_3, and V_5 are the voltage drops across R_1, R_4, and R_8, respectively. Voltages V_2 and V_4 are across the parallel circuits between A and B and between C and D, respectively.

Expressed in words, formula (15–4) states that in a closed circuit or loop, the applied voltage equals the sum of the voltage drops in the circuit.

The use of algebraic signs or polarity is helpful in solving electric circuit problems. The circuit of Figure 15–3 (p. 104) illustrates the convention used in assigning a + or a – sign to a voltage in a circuit. In the case of electron-flow current, electrons move from a negative to positive potential. The arrow in Figure 15–3 shows the direction of current, and the – and + signs indicate the following: Point A is negative with respect to B; point B is negative with respect to C; point C is negative with respect to point D; and D is negative with respect to point E. This is consistent with our assumption of electron-flow current in this circuit. With regard to the voltage source, point E is positive with respect to point A, indicating a voltage rise.

To establish the algebraic sign for the voltages in the closed circuit, move in the direction of assumed current. Consider as positive any voltage source or voltage drop whose + (positive) terminal is reached first, and negative

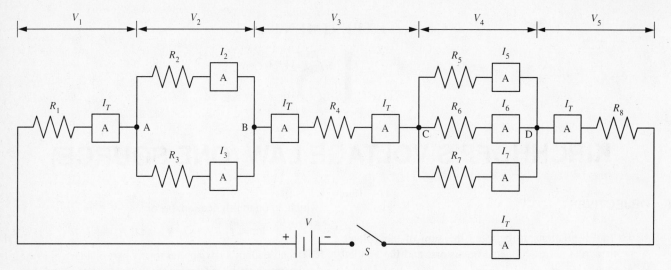

Figure 15–2. Application of Kirchhoff's law to a series-parallel circuit.

any voltage source or drop whose – (negative) terminal is reached first. Starting at point A in Figure 15–3 and moving in the direction of current, we have $-V_1$, $-V_2$, $-V_3$, $-V_4$, and $+V$. With this convention in mind, Kirchhoff's voltage law may now be generalized as follows.

The algebraic sum of the voltages in a closed circuit equals zero.

Applying the convention on signs and Kirchhoff's law to the closed circuit of Figure 15–3 and starting at point A, we may write the following:

$$-V_1 - V_2 - V_3 - V_4 + V = 0 \qquad \textbf{(15–5)}$$

Is this formula consistent with formula (15–4)? Yes, for by transposing the terms on the right side of formula (15–4) to the left side, we get $V - V_1 - V_2 - V_3 - V_4 = 0$, a result identical with that in formula (15–5).

SUMMARY

Kirchhoff's voltage law can be expressed in two ways.

1. The sum of the voltage drops in a closed circuit equals the applied voltage.
2. The algebraic sum of the voltages in a closed circuit equals zero.

SELF-TEST

Check your understanding by answering the following questions:

1. In Figure 15–1 $V_1 = 3$ V, $V_2 = 5.5$ V, $V_3 = 6$ V, and $V_4 = 12$ V. The applied voltage V must then equal _____ V.
2. In Figure 15–2 $V_1 = 1.5$ V, $V_2 = 2.0$ V, $V_4 = 2.7$ V, $V_5 = 6$ V, and $V = 15$ V. The voltage $V_3 = $ _____ V.

MATERIALS REQUIRED

Power Supply:
■ Variable 0–15 V dc, regulated

Instruments:
■ DMM or VOM

Resistors (½-W, 5%):
■ 1 330-Ω
■ 1 470-Ω
■ 1 820-Ω
■ 1 1000-Ω
■ 1 1200-Ω
■ 1 2200-Ω
■ 1 3300-Ω
■ 1 4700-Ω

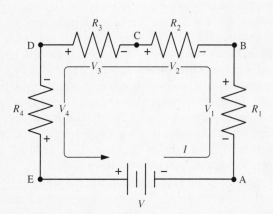

Figure 15–3. Convention for assigning polarity to voltages in a closed circuit.

Miscellaneous:
■ SPST switch

PROCEDURE

1. Measure each of the resistors supplied and record its value in Table 15–1 (p. 107).
2. With power **off** and switch S_1 **open,** connect the circuit of Figure 15–4. Turn the power **on.** Adjust the power supply so that V_{PS} = 15 V.
3. **Close** S_1. Measure the voltage across R_1 (V_1), R_2 (V_2), R_3 (V_3), and R_4 (V_4) and record the values in Table 15–2 (p. 107). Calculate the sum of the voltages V_1, V_2, V_3, and V_4 and record your answer in Table 15–2. S_1 **open;** power **off.**

4. Connect the circuit in Figure 15–5. Power **on.** Adjust the power supply so that V_{PS} = 15 V.
5. **Close** S_1. Measure voltages V_1, V_2, V_3, V_4, and V_5 as shown in Figure 15–5. Record the values in Table 15–2. Calculate the sum of the voltages V_1, V_2, V_3, V_4, and V_5 and record your answer in Table 15–2. S_1 **open;** power **off.**

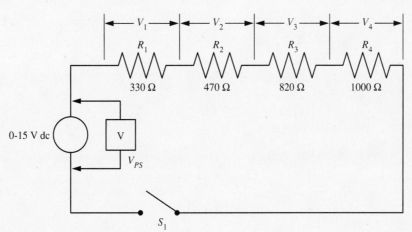

Figure 15–4. Circuit for procedure step 2.

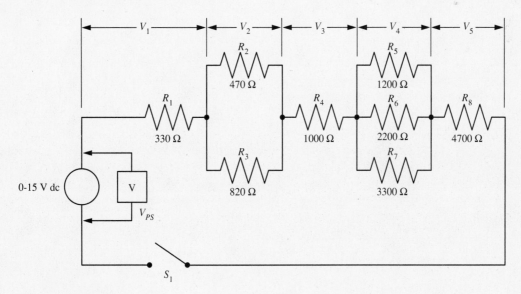

Figure 15–5. Circuit for procedure step 4.

ANSWERS TO SELF-TEST

1. 26.5
2. 2.8

Kirchhoff's Voltage Law (One Source)

Experiment 15   Name _____   Date _____

TABLE 15–1.   Color-Coded Values, Experimental Resistors

	R_1	R_2	R_3	R_4	R_5	R_6	R_7	R_8
Rated value, Ω	330	470	820	1000	1200	2200	3300	4700
Measured value, Ω								

TABLE 15–2.   Verifying Kirchhoff's Voltage Law

Step	V	V_1	V_2	V_3	V_4	V_5	Sum of Vs
3						✕	
5							

QUESTIONS

1. State, in your own words, the relationship between the voltage drops across series-connected resistors and the voltage applied to the entire series circuit.

_____
_____
_____
_____
_____
_____
_____
_____
_____
_____
_____
_____
_____
_____

2. Express your answer for Question 1 as a mathematical formula.

_____
_____
_____
_____
_____
_____

     Kirchhoff's Voltage Law (One Source)

3. Refer to Table 15–2. Do your experimental data support your answers to Questions 1 and
   2? (Refer to actual data in the table.) If not, explain the discrepancy.

   _____

   _____

   _____

   _____

   _____

   _____

   _____

   _____

   _____

4. On a separate sheet of 8½ × 11 paper, design a series-parallel circuit resembling
   Figure 15–2. The applied voltage is 35 V. The current supplied by the power supply is
   5 mA. Use only the eight resistors listed in the Materials Required list (use the rated
   values shown). Draw a fully labeled circuit diagram; show all design calculations and
   formulas used. The design specification allows a deviation of current of ±1%. The
   supply voltage cannot vary.

EXPERIMENT

16

KIRCHHOFF'S CURRENT LAW

OBJECTIVES

1. To find a relationship between the sum of the currents entering any junction of an electric circuit and the current leaving that junction
2. To verify experimentally the relationship found in objective 1

BASIC INFORMATION

Current Law

Experiment 11 verified that the total current I_T in a circuit containing resistors connected in parallel is equal to the sum of the currents in each of the parallel branches. This was one demonstration of Kirchhoff's current law, limited to a parallel network. The law is general, however, and applies to any circuit. Kirchhoff's current law states that

The current entering any junction of an electric circuit is equal to the current leaving that junction.

In the series-parallel circuit of Figure 16–1, the total current is I_T. It enters the junction at A in the direction

indicated by the arrow. The currents leaving the junction at A are I_1, I_2, and I_3, as shown. The currents I_1, I_2, and I_3 then enter the junction at B, and I_T leaves the junction at B. What is the relationship between I_T, I_1, I_2, and I_3?

The voltage across the parallel circuit can be found using Ohm's law:

$$V_{AB} = I_1 \times R_1 = I_2 \times R_2 = I_3 \times R_3$$

The parallel network may be replaced by its equivalent resistance R_T, in which case Figure 16–1 is transformed into a simple series circuit and $V_{AB} = I_T \times R_T$. It follows, therefore, that

$$I_T \times R_T = I_1 \times R_1 = I_2 \times R_2 = I_3 \times R_3 \quad \textbf{(16–1)}$$

Formula (16–1) may be rewritten as

$$I_1 = I_T \times \frac{R_T}{R_1}$$

$$I_2 = I_T \times \frac{R_T}{R_2} \qquad \textbf{(16–2)}$$

$$I_3 = I_T \times \frac{R_T}{R_3}$$

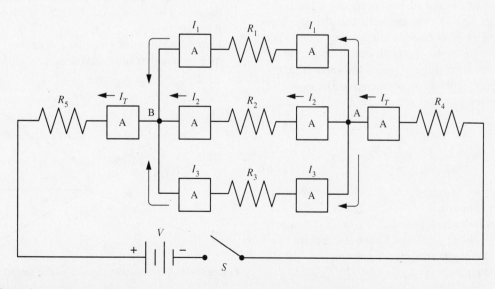

Figure 16–1. The total current through the supply is the sum of the currents in each of the branches.

Adding I_1, I_2, and I_3 gives

$$I_1 + I_2 + I_3 = I_T \times \frac{R_T}{R_1} + I_T \times \frac{R_T}{R_2} + I_T \times \frac{R_T}{R_3}$$

$$I_1 + I_2 + I_3 = I_T \times R_T \left(\frac{1}{R_1} + \frac{1}{R_2} + \frac{1}{R_3} \right)$$

But

$$\frac{1}{R_1} + \frac{1}{R_2} + \frac{1}{R_3} = \frac{1}{R_T}$$

Therefore,

$$I_1 + I_2 + I_3 = I_T \times R_T \times \frac{1}{R_T} = I_T$$

That is,

$$I_T = I_1 + I_2 + I_3 \qquad \textbf{(16–3)}$$

Formula (16–3) is a mathematical statement of Kirchhoff's law, applied to the circuit of Figure 16–1. In general, if I_T is the current entering a junction of an electric circuit, and I_1, I_2, I_3, ..., I_n are the currents leaving that junction, then

$$I_T = I_1 + I_2 + I_3 + \cdots + I_n \qquad \textbf{(16–4)}$$

This applies equally well if I_T is a current leaving a junction and I_1, I_2, I_3, ..., I_n are the currents entering that junction.

Kirchhoff's current law is often stated in another way: *The algebraic sum of the currents entering and leaving a junction is zero.*

Recall that this is similar to the formulation of Kirchhoff's voltage law: *The algebraic sum of the voltages in a closed path or loop is zero.*

Just as it was necessary to agree on a polarity convention for voltages in a loop, so it is necessary to agree on a current convention at a junction. If the current entering a junction is considered positive (+) and the current leaving a junction is considered negative (−), then the statement that the algebraic sum of the currents entering and leaving a junction is zero can be shown to be identical with formula (16–4). Consider the circuit of Figure 16–2. The total current I_T enters the junction at A and is considered +. The currents I_1 and I_2 leave the junction at A and are designated −. Then,

$$+I_T - I_1 - I_2 = 0 \qquad \textbf{(16–5)}$$

and

$$I_T = I_1 + I_2 \qquad \textbf{(16–6)}$$

Obviously, the two statements of Kirchhoff's current law lead to the same formula.

An example shows how Kirchhoff's current law may be applied to solve circuit problems. Suppose in Figure 16–3 that I_1 and I_2 are currents entering the junction at A and are, respectively, +5 A and +3 A. Currents I_3, I_4, and I_5 are

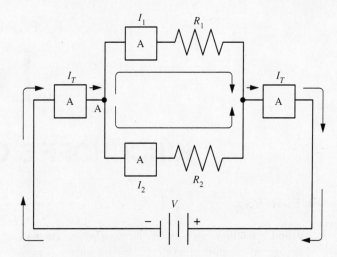

Figure 16–2. The algebraic sum of the currents entering and leaving a junction is equal to zero.

leaving A. Currents I_3 and I_4 are, respectively, 2 A and 1 A. What is the value of I_5? Applying Kirchhoff's current law,

$$I_1 + I_2 - I_3 - I_4 - I_5 = 0$$

and substituting the known values of current, we get

$$5 + 3 - 2 - 1 - I_5 = 0$$
$$5 - I_5 = 0$$
$$I_5 = 5 \text{ A}$$

SUMMARY

1. Kirchhoff's current law states that the current entering any junction of an electric circuit is equal to the current leaving that junction.
2. To use Kirchhoff's current law in solving circuit problems, polarity is assigned to current entering a junction (assume it is +) and to current leaving a junction (assume it is −).
3. Using the polarities given in 2, Kirchhoff's current law may be stated as follows: The sum of the currents enter-

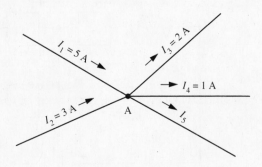

Figure 16–3. Currents entering and leaving junction A.

ing and leaving a junction is zero. Thus, for Figure 16–1, at junction A

$$I_T - I_1 - I_2 - I_3 = 0$$

SELF-TEST

Check your understanding by answering the following questions:

1. In Figure 16–1 the current entering the junction at A is 0.5 A. $I_1 = 0.25$ A, $I_2 = 0.1$ A. The current I_3 must therefore equal _____ A.
2. In Figure 16–1 the current leaving the junction at B is 1.5 A. The sum of currents I_1, I_2, and I_3 must be _____ A.
3. In applying Kirchhoff's current law to junction B in Figure 16–1, the polarity given to each current by convention is as follows:
   (a) I_1 _____
   (b) I_2 _____
   (c) I_3 _____
   (d) I_T _____
4. The equation that describes the relationship among the currents at junction A in Figure 16–3 is _____ .

5. In Figure 16–3, $I_2 = 4$ A, $I_3 = 4$ A, $I_4 = 3$ A, $I_5 = 1$ A. $I_1 = $ _____ A.

MATERIALS REQUIRED

Power Supply:
■ Variable 0–15 V dc, regulated

Instruments:
■ DMM or VOM
■ 0–10-mA milliammeter

Resistors ($\frac{1}{2}$-W, 5%):
■ 1   330-Ω
■ 1   470-Ω
■ 1   820-Ω
■ 1   1000-Ω
■ 1   1200-Ω
■ 1   2200-Ω
■ 1   3300-Ω
■ 1   4700-Ω

Miscellaneous:
■ SPST switch

PROCEDURE

NOTE: This experiment requires numerous measurements of current in series-parallel circuits. If only one ammeter is available, it will be necessary to break the line for which a current reading must be taken. Disconnect power to the circuit by opening S_1 each time the position of the ammeter is changed.

1. Measure the resistance of each of the resistors supplied and record its value in Table 16–1 (p. 113).
2. With power **off** and S_1 **open,** connect the circuit of Figure 16–4 (p. 112). Power **on.** Adjust the power supply so that $V_{PS} = 15$ V.
3. **Close** S_1. Measure currents I_{TA}, I_2, I_3, I_{TB}, I_{TC}, I_5, I_6, I_7, I_{TD}, and I_{TE}. Record the values in Table 16–2 (p. 113). Calculate the sum of I_2 and I_3 and the sum of I_5, I_6, and I_7. Record your answers in Table 16–2. S_1 **open**; power **off**.
4. Design a series-parallel circuit consisting of three parallel branches and two series resistors similar to the circuit of Figure 16–1. The currents in the three parallel branches should be such that the current in the second branch is

approximately twice the current in the first branch, and the current in the third branch is approximately three times the current in the first branch. (Stated another way, the currents in the three parallel branches are in the ratio 1:2:3, approximately.) Use only the resistors supplied for this experiment. The total current in the circuit is 6 mA. The maximum voltage available is 15 V. Show the position of the meters used to measure current through each resistor and I_T. Include a switch to disconnect power to your circuit. Draw a complete diagram of the circuit showing the rated values of the resistors chosen, the calculated current in each line, and the applied voltage. Show all calculations used to find the values of the resistors.

5. With power **off** and the switch **open,** connect the circuit you designed in step 4. After approval of the circuit by your instructor, turn power **on.** Adjust the power supply to your design voltage. Read all currents in the circuit and record the values in Table 16–3 (p. 113). S_1 **open**; power **off.**

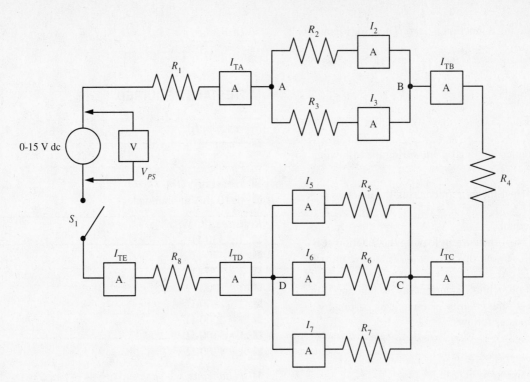

Figure 16–4. Circuit for procedure step 2.

ANSWERS TO SELF-TEST

1. 0.15
2. 1.5
3. (a) +; (b) +; (c) +; (d) −
4. $I_1 + I_2 - I_3 - I_4 - I_5 = 0$
5. 4

Experiment 16   Name _____ Date _____

TABLE 16–1. Measured Value of Resistors

	R_1	R_2	R_3	R_4	R_5	R_6	R_7	R_8
Rated value, Ω	330	470	820	1000	1200	2200	3300	4700
Measured value, Ω								

TABLE 16–2. Verifying Kirchhoff's Current Law

	I_{TA}	I_2	I_3	I_{TB}	I_{TC}	I_5
current, mA						

	I_6	I_7	I_{TD}	I_{TE}	$I_2 + I_3$	$I_5 + I_6 + I_7$
current, mA						

TABLE 16–3. Design Problem Data

Calculated Value, A				*Measured Value*, A			
Branch 1 I_1	Branch 2 I_2	Branch 3 I_3	I_T	Branch 1 I_1	Branch 2 I_2	Branch 3 I_3	I_T

QUESTIONS

1. Explain, in your own words, the relationship between the currents entering and leaving a junction point in a circuit.

_____

_____

_____

_____

_____

_____

2. Write the relationship explained in Question 1 as a mathematical formula.

_____

_____

_____

_____

_____

_____

Kirchhoff's Current Law   **113**

3. Refer to Figure 16–4. What information would you need to find I_2 and I_3 in this circuit?

_____

_____

_____

_____

_____

_____

_____

_____

4. Steps 4 and 5 in the procedure required you to design a series-parallel circuit with V_{PS} max = 15 V and I_T = 6 mA. Resistors R_1, R_2, and R_3 were in the ratio 1:2:3. Discuss your results, referring to your experimental data in Table 16–3. Your report should indicate whether you were able to meet the design specifications exactly. If not, explain the discrepancy. Also, explain how you went about choosing (or eliminating) each of the given resistors. (Submit your design circuit diagram with this report.)

_____

_____

_____

_____

_____

_____

_____

_____

_____

_____

_____

_____

_____

_____

_____

_____

_____

_____

17

VOLTAGE-DIVIDER CIRCUITS (LOADED)

OBJECTIVES

1. To find what effect load has on the voltage relationships in a voltage-divider circuit
2. To verify the results of objective 1 by experiment

BASIC INFORMATION

In the simple dc voltage-divider circuits studied in Experiment 10, no load current was drawn. The only current was the "bleeder" current in the divider network itself. In electronics, divider networks are frequently used as a source to supply voltage to a load, which draws current. When this is the case, the divider-voltage relationships that were obtained for no-load conditions no longer hold. The actual changes depend on the amount of current drawn and on the circuit connections. In the circuit of Figure 17–1, V is a constant-voltage source that maintains 15 V across the divider network with and without load. Without load, points A, B, and C are 5, 10, and 15 V, respectively, with respect to G. Moreover, a 5-mA bleeder current I_1 is drawn. Now if we add a load resistor R_L (as shown in Figure 17–2 at point B) that draws 2 mA of load current I_L, the voltages at A and B will be different from the condition of no load. We can calculate the voltages in the loaded circuit from the information given.

Assume that there is a bleeder current I_1 when a load current of I_L is drawn by R_L. By Kirchhoff's law we can set up the formula

$$I_1(R_2 + R_3) + (I_1 + I_L)(R_1) = 15 \quad \text{(17–1)}$$

Solving for I_1 (since this is the only unknown), we get

$$I_1(R_1 + R_2 + R_3) = 15 - (R_1)I_L \quad \text{(17–2)}$$

$$I_1 = \frac{15 - (I_L \times R_1)}{R_1 + R_2 + R_3} \quad \text{(17–3)}$$

Substituting 1000 Ω for each of R_1, R_2, and R_3, and I_L = 0.002 A in formula (17–3), we get

$$I_1 = 0.00433 \text{ A, or } 4.33 \text{ mA} \quad \text{(17–4)}$$

The voltage from A to G is

$$V_{AG} = I_1 \times R_3$$
$$V_{AG} = 0.00433 \times 1000 = 4.33 \text{ V} \quad \text{(17–5)}$$

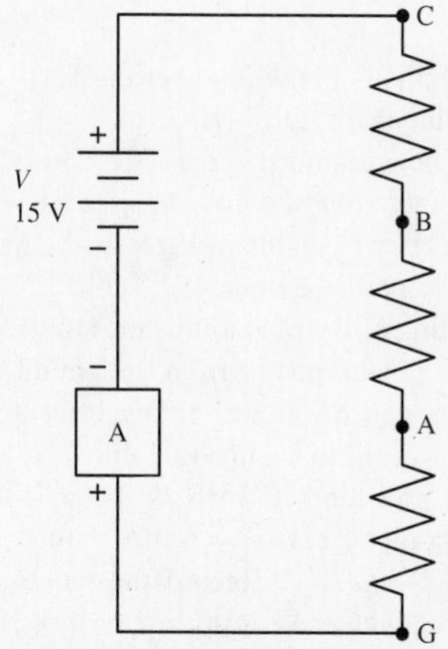

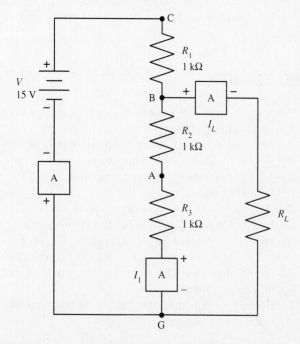

Figure 17–1. Voltage divider without load.

Figure 17–2. Voltage divider with fixed load R_L.

and the voltage from B to G is

$$V_{BG} = 0.00433 \times 2000 = 8.66 \text{ V} \qquad (17\text{--}6)$$

This shows that both the voltages and the bleeder current are affected when a load is added to a voltage-divider circuit.

To verify this effect experimentally, it is necessary to have a voltage source across the divider that will remain constant with and without load. In Figure 17–2 the divider network consists of R_1, R_2, and R_3 connected to a voltage source. When a load resistor (R_L) is added as shown, the voltage source is readjusted if necessary to maintain the same voltage across the network as under no-load conditions. The required measurements are made under no-load and load conditions.

SUMMARY

1. A series circuit such as in Figure 17–1 is a simple unloaded voltage divider.
2. In an unloaded voltage divider such as the circuit of Figure 17–1 the voltage V across R_1 can be found by using the formula

$$V_1 = V \times \frac{R_1}{R_T}$$

where V is the source voltage and R_T is the sum of the series resistances. The voltages at points A, B, and C are given with respect to G. They can be calculated using the preceding formula.

3. If a load is added to a voltage divider, as in Figure 17–2, the bleeder current and the voltages across the resistors in the voltage divider will change.
4. To solve for the voltages and bleeder current in a loaded voltage divider, Kirchhoff's voltage and current laws can be used.

SELF-TEST

Check your understanding by answering the following questions:

1. In the circuit of Figure 17–1 let $R_1 = 1000 \ \Omega$, $R_2 = 1800 \ \Omega$, $R_3 = 2200 \ \Omega$, and $V = 15$ V.
   (a) What is the voltage V_1 across R_1? $V_1 = $ _____ V.
   (b) What is the voltage across the combination of R_2 and R_3? $V_{BG} = $ _____ V.
   (c) What is the voltage across R_3? $V_{AG} = $ _____ V.
2. The bleeder current in question 1 is _____ A, or _____ mA.
3. In the voltage divider of question 1, a load resistor is placed across R_2 and R_3, as shown in Figure 17–2. If the load current I_L is 5 mA, what is the bleeder current? $I_1 = $ _____ A, or _____ mA.
4. For the conditions of question 3, what are the following voltages? $V_1 = $ _____ V, $V_{BG} = $ _____ V, and $V_{AG} = $ _____ V.

MATERIALS REQUIRED

Power Supply:
- Variable 0–15 V dc, regulated

Instruments:
- DMM or VOM
- 0–10-mA milliammeter

Resistors:
- 3   1200-Ω ($\frac{1}{2}$-W, 5%)
- 1   10-kΩ, 2-W potentiometer

Miscellaneous:
- SPST switch

PROCEDURE

1. With power **off** and S_1 **open,** connect the circuit in Figure 17–3(*a*) (p. 117).
2. Power **on** and S_1 **closed.** Adjust the power supply so that $V_{PS} = 10$ V. Measure I_1 (called the *bleeder current*), voltage V_{BD}, and voltage V_{CD}. Record your answers in Table 17–1 (p. 119). **Open** S_1.
3. Connect the 10-kΩ potentiometer across points B and D as shown in Figure 17–3(*b*). With the voltmeter across the power supply, adjust the potentiometer until $I_L = 2$ mA with $V_{PS} = 10$ V. Measure I_1, V_{BD}, and V_{CD}. Record the values in Table 17–1. Do not change the setting of the potentiometer. **Open** S_1.
4. Disconnect the potentiometer from the circuit and use an ohmmeter to measure the load resistance R_L across

points E and F. This is the load setting for $I_L = 2$ mA. Record this value in Table 17–1.
5. Reconnect the potentiometer across BD. **Close** S_1. Adjust the potentiometer until $I_L = 4$ mA. Again adjust the power supply as necessary to maintain $V_{PS} = 10$ V. Measure I_1, V_{BD}, and V_{CD}. Record the values in Table 17–1. Do not change the setting of the potentiometer. **Open** S_1.
6. Disconnect the potentiometer from the circuit and measure the load resistance as in step 4. This is the load setting for $I_L = 4$ mA. Record this value in Table 17–1.
7. Reconnect the potentiometer across BD. **Close** S_1. Adjust the potentiometer until $I_L = 6$ mA with $V_{PS} = 10$ V. Measure I_1, V_{BD}, and V_{CD}. Record the values in Table 17-1. Do not change the setting of the potentiometer. **Open** S_1.

8. Disconnect the potentiometer from the circuit and measure the load resistance, as in step 4. This is the load setting for $I_L = 6$ mA. Record this value in Table 17–1. **Open** S_1; turn the power **off.**

9. Using the methods discussed in the Basic Information section, calculate the bleeder current I_1, voltages V_{BD} and V_{CD}, and the load resistance R_L for each load condition (0 mA, 2 mA, 4 mA, and 6 mA) in this experiment. Record your answers in Table 17–1.

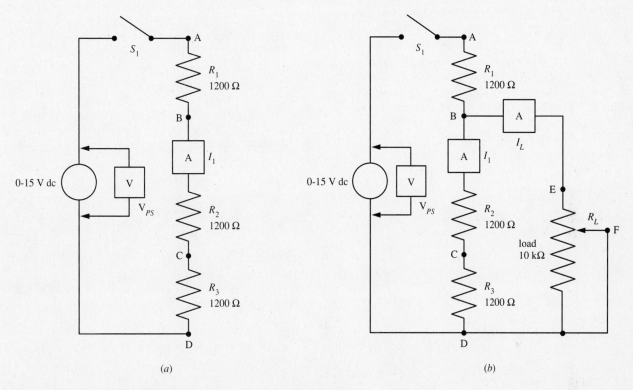

Figure 17–3. (a) Circuit for procedure step 1. (b) Circuit for procedure step 3.

ANSWERS TO SELF-TEST

1   (a) 3; (b) 12; (c) 6.6
2. 0.003; 3
3. 0.002; 2
4. 7; 8; 4.4

Voltage-Divider Circuits (Loaded)   **117**

TABLE 17–1. Effect of Load on a Voltage Divider

			Measured Values				Calculated Values			
Steps	V	I_L (load current), mA	I_1, mA	V_{BD}, V	V_{CD}, V	R_L, Ω	I_1, mA	V_{BD}, V	V_{CD}, V	R_L, Ω
2	10	0				✕				
3, 4	10	2								
5, 6	10	4								
7	10	6								

QUESTIONS

1. Explain, in your own words, the effect load has on the voltage relationships in a voltage-divider circuit.

_____

_____

_____

_____

_____

_____

_____

_____

_____

_____

_____

2. Refer to your data in Table 17–1. What effect does changing the load resistance have on load current? Give specific examples from your data. Explain the reason for this effect.

_____

_____

_____

_____

_____

_____

_____

_____

_____

_____

Voltage-Divider Circuits (Loaded)   **119**

3. Refer to your data in Table 17–1. How is the bleeder current, I_1, affected by changes in the load current, I_L? Give specific examples from your data. Explain the reason for this effect.

_____

_____

_____

_____

_____

_____

_____

_____

_____

_____

_____

4. Refer to Figure 17–3(b) and Table 17–1. What effect do changes in the load current, I_L, have on the divider tap voltages, V_{CD} and V_{BD}? Give specific examples from your data. Explain the reason for this effect.

_____

_____

_____

_____

_____

_____

_____

_____

_____

_____

5. Compare the measured values of I_1, V_{BD}, V_{CD}, and R_L with the calculated values. Explain any differences.

_____

_____

_____

_____

_____

_____

_____

_____

18

DESIGNING VOLTAGE- AND CURRENT-DIVIDER CIRCUITS

OBJECTIVES

1. To design a voltage divider that will meet specified voltage and current requirements
2. To design a current divider that will meet specified current and voltage requirements
3. To construct and test the circuits to see that they meet the design requirements

BASIC INFORMATION

Designing a Voltage Divider for Specific Loads

The electronics technician may be called upon to design a voltage divider that will meet specified requirements of voltage, load current, and bleeder current. The process to follow is illustrated by a sample problem.

Problem 1. Design a voltage-divider circuit for a 30-V power supply. The loads that must be served are 0.05 A at 30 V, and 0.04 A at 25 V.

A "rule-of-thumb" value of bleeder current is 10 percent (approximately) of load current. The bleeder current under these conditions is 0.01 A.

Solution. Draw a diagram, as in Figure 18–1. Label the known load currents and voltages. Designate the *unknown resistors* as R_1 and R_2. (The loads are shown as resistors drawing the required currents.)

Apply Kirchhoff's current law and show the currents entering and leaving the junctions A, B, and C. The total current I_T that must be supplied is $I_T = 0.05 + 0.04 + 0.01 = 0.1$ A. Current I_T is shown entering the junction at C and leaving the junction at A.

$$I_1 \times R_1 = 25 \text{ V} \qquad (18-1)$$

or

$$R_1 = \frac{25}{I_1} \qquad (18-2)$$

Substituting the bleeder current value (0.01 A) for I_1 gives

$$R_1 = \frac{25}{0.01} = 2500 \ \Omega \qquad (18-3)$$

The current in R_2 is 0.05 A, the sum of the bleeder current and load 1 current. Moreover, as is evident from Figure 18–1, the voltage from A to B is 5 V. Therefore,

$$(0.05)(R_2) = 5 \text{ V} \qquad (18-4)$$

and

$$R_2 = \frac{5}{0.05} = 100 \ \Omega \qquad (18-5)$$

We have thus found the values of R_1 and R_2 that will deliver 0.04 A at 25 V and 0.05 A at 30 V with a bleeder current of 0.01 A, from a 30-V power supply.

Designing Current-Divider Circuits

It is sometimes necessary for the technician to design current-divider circuits. To avoid trial-and-error methods, study the conditions of the problem and apply the appropriate circuit formulas. The exact steps to follow in this procedure will depend on the nature of the design problem, but in general, the following steps should be taken to find the solution:

1. Draw a circuit diagram labeling all the known values and the unknowns.

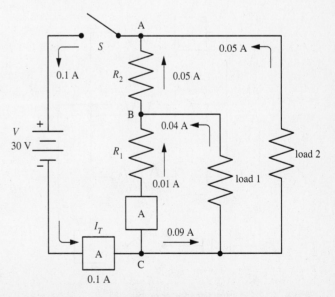

Figure 18–1. Voltage-divider circuit for problem 1.

2. Write the Ohm's and Kirchhoff's laws formulas that describe the electrical relationships involved.
3. Solve these formulas for the unknown component values.
4. Construct the circuit from the calculated values and test the circuit to see if the design requirements have been met.

Two problems illustrate this process.

Problem 2. Design the three-branch parallel circuit shown in Figure 18–2 so that a 15-V source V will deliver a total current I_T of 1.2 A. A further requirement is that the current I_T be distributed among R_1, R_2, and R_3 so that $I_1:I_2:I_3$ = 2:4:6. What values of R_1, R_2, and R_3 will satisfy these conditions?

Solution. We know V, I_T, and the ratios $I_1:I_2:I_3$. If we could find the actual values of I_1, I_2, and I_3, we would then know two of the Ohm's law values for each of the branches—namely, V and I. From V and I, R can be found, since

$$R = \frac{V}{I} \qquad (18\text{–}6)$$

We can set up two equations for I from the current ratios given.

$$\frac{I_1}{I_2} = \frac{2}{4}$$

$$\frac{I_1}{I_3} = \frac{2}{6} \qquad (18\text{–}7)$$

Solving for I_2 and I_3 in terms of I_1 yields

$$I_2 = \frac{4}{2}I_1 = 2I_1$$

$$I_3 = \frac{6}{2}I_1 = 3I_1 \qquad (18\text{–}8)$$

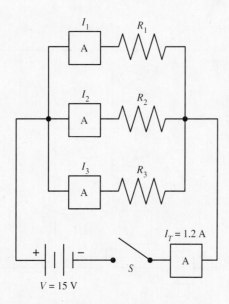

Figure 18–2. Currrent divider for problem 2.

Now, applying Kirchhoff's current law,

$$I_T = I_1 + I_2 + I_3 = 1.2 \qquad (18\text{–}9)$$

Substituting in formula (18–9) the values of I_2 and I_3 from (18–8) gives

$$I_1 + 2I_1 + 3I_1 = 1.2 \qquad (18\text{–}10)$$

Solving,

$$6I_1 = 1.2$$

$$I_1 = 0.2 \text{ A} \qquad (18\text{–}11)$$

Substituting 0.2 for I_1 in (18–8) gives

$$I_2 = 0.4 \text{ A}$$

$$I_3 = 0.6 \text{ A} \qquad (18\text{–}12)$$

Now we can find the values of R_1, R_2, and R_3.

$$R_1 = \frac{V}{I_1} = \frac{15}{0.2} = 75 \text{ }\Omega$$

$$R_2 = \frac{V}{I_2} = \frac{15}{0.4} = 37.5 \text{ }\Omega \qquad (18\text{–}13)$$

$$R_3 = \frac{V}{I_3} = \frac{15}{0.6} = 25 \text{ }\Omega$$

Problem 3. Design a divider network that will supply three resistive loads from a 24-V source such that loads 1, 2, and 3 receive 0.02 A, 0.03 A, and 0.05 A, respectively. All three loads are in parallel, and the resistance of load 1 is 300 Ω.

Solution. Draw the three parallel loads R_1, R_2, and R_3. Since the current and resistance of load 1 are given, we can find the voltage across load 1 and also across each of the other two parallel loads. Thus,

$$V_1 = I_1 \times R_1 = 0.02(300) = 6 \text{ V} \qquad (18\text{–}14)$$

We have a 24-V source. Therefore, we will have to drop 18 V before we reach the three parallel loads. A series-parallel circuit, such as that in Figure 18–3, will serve our requirements. We can now find the values of R_4 and loads R_2 and R_3.

First, solving for R_2 and R_3,

$$R_2 = \frac{V_{AB}}{I_2} = \frac{6}{0.03} = 200 \text{ }\Omega$$

$$R_3 = \frac{V_{AB}}{I_3} = \frac{6}{0.05} = 120 \text{ }\Omega \qquad (18\text{–}15)$$

Finally, the total current I_T in R_4 can be found by Kirchhoff's current law:

$$I_T = I_1 + I_2 + I_3 = 0.02 + 0.03 + 0.05$$

$$= 0.1 \text{ A}$$

Therefore,

$$R_4 = \frac{18}{0.1} = 180 \text{ }\Omega \qquad (18\text{–}16)$$

SUMMARY

1. In designing either voltage- or current-divider circuits, a mathematical solution is found first. Ohm's and Kirchhoff's laws are applied in the process.
2. The general steps in solving a design problem are as follows:
   (a) Draw a diagram of the proposed circuit, labeling all the knowns and unknowns.
   (b) Set up one or more formulas that reflect the electrical relationships involved.
   (c) Solve these formulas for the values of the unknown components.
   (d) Construct the circuit from the calculated values and test the circuit to see if the design requirements have been met.
3. In designing voltage-divider circuits for specified load currents, a rule-of-thumb value for bleeder current is 10 percent (approximately) of load current.

SELF-TEST

Check your understanding by answering the following questions:

1. It is required to design a voltage-divider circuit like that in Figure 18–4 so that a 20-V source V can supply two loads: load 1 is 0.1 A at 5 V; Load 2 is 0.4 A at 20 V. Assume a bleeder current I_1 of 0.05 A. The values of R_1 and R_2 that will accomplish this result are:
   (a) $R_1 =$ _____ Ω
   (b) $R_2 =$ _____ Ω

2. For the same conditions as in question 1, the resistances of the two loads are:
   (a) Resistance of load 1 is _____ Ω
   (b) Resistance of load 2 is _____ Ω
3. In a divider circuit similar to that in Figure 18–5 (p. 124) it is required to supply three equal-load resistors, drawing a total current of 0.15 A, at a load voltage of 5 V. The supply source V is 50 V. The values of $R_1, R_2, R_3,$ and R_4 that will achieve this are:
   (a) $R_1 =$ _____ Ω
   (b) $R_2 =$ _____ Ω
   (c) $R_3 =$ _____ Ω
   (d) $R_4 =$ _____ Ω

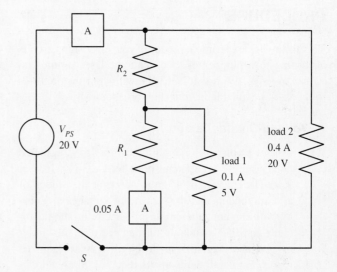

Figure 18–4. Circuit for questions 1 and 2.

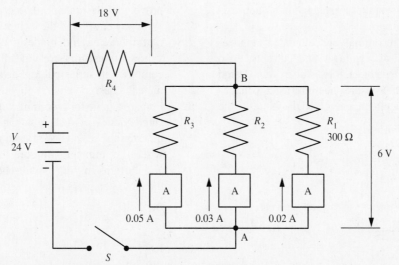

Figure 18–3. A voltage- and current-divider circuit that supplies three resistors, R_1, R_2, and R_3.

MATERIALS REQUIRED

Power Supply:
■ Variable 0–15 V dc, regulated

Instruments:
■ DMM or VOM
■ 0–10-mA milliammeter

Resistors:
■ Specify ½-W, 5% resistors as required by your design

Miscellaneous:
■ SPST switch

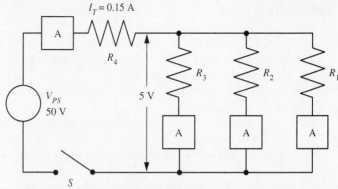

Figure 18–5. Circuit for question 3.

PROCEDURE

The following experiments require you to design circuits according to a number of specifications. Use commercial values of ½-W, 5% resistors. Check with your instructor as to the values available in your shop or laboratory.

A. Voltage-Divider Design

A1. Design a voltage-divider circuit for a 15-V regulated power supply. The circuit must feed a 3-mA load at 9 V. Draw a circuit diagram showing the values of all components and the significant voltages and currents throughout the circuit. Your circuit should include an SPST switch for turning power to the circuit **on** and **off.** Present your design in a neat organized report, listing all components required. Show all calculations leading to your design components.
Have your instructor approve your design before proceeding to the next step.

A2. With power **off** and the switch **open,** connect the voltage-divider circuit you designed in step A1. If certain values of resistors are unavailable, resistors can be combined in series-parallel arrangements to approximate the design value.

A3. Prepare a table for recording your calculated and measured values of design voltages and currents. Also measure the resistance of each resistor in your design with an ohmmeter. Measure the significant voltages and currents in your circuit and record the values in the table you prepared.

B. Current-Divider Circuit

B1. Design a current-divider circuit that will feed three parallel branch loads from a 15-V power supply. The total current to be delivered by the power supply is 5 mA at 10 V. The branch currents are to divide as follows:
(a) The current in branch 2 must be 1½ times more than the current in branch 1.
(b) The current in branch 3 must be 2½ times more than the current in branch 1.
(In other words, the currents in the three branches are to divide in the ratio 1:1.5:2.5.)
Draw a circuit diagram of your design showing the values of all components and the significant voltages and currents throughout the circuit. Your circuit should include an SPST switch for turning power to the circuit **on** and **off.**
Present your design in a neat organized report, listing all components required. Show all calculations leading to your design components. Have your instructor approve your design before proceeding to the next step.

B2. With power **off,** and the switch **open,** connect the current-divider circuit you designed in step B1. If certain values of resistors are unavailable, resistors can be combined in series-parallel arrangements to approximate the design value.

B3. Prepare a table for recording your calculated and measured values of design voltages and currents. Also measure the resistance of each resistor in your design with an ohmmeter. Measure the significant voltages and currents in your circuit and record the values in the table you prepared.

ANSWERS TO SELF-TEST

1. (a) 100; (b) 100
2. (a) 50; (b) 50
3. (a) 100; (b) 100; (c) 100; (d) 300

QUESTIONS

1. (a) Circuit diagram and design calculations for part A:

   (b) Table of measured and calculated values of design voltages and currents (step A3).

2. (a)   Circuit diagram and design calculations for part B:

(b)   Table of measured and calculated values of design voltages and currents (step B3).

3. Compare your measured values of current with the design values in part A. Refer to your data in the table of Question 1(b). Explain any differences.

_____
_____
_____
_____
_____
_____
_____
_____
_____
_____
_____
_____
_____
_____
_____
_____

4. Compare your measured values of current with the design values in part B. Refer to your data in the table of Question 2(b). Explain any differences.

_____
_____
_____
_____
_____
_____
_____
_____
_____
_____
_____
_____
_____
_____
_____

EXPERIMENT

19

TROUBLESHOOTING ELECTRIC CIRCUITS USING VOLTAGE, CURRENT, AND RESISTANCE MEASUREMENTS

OBJECTIVES

1. To study methods of troubleshooting circuits using voltage, current, and resistance measurements
2. To apply these methods in finding defective parts in common types of circuits

BASIC INFORMATION

Electronic technicians must be competent in many areas. They must be able to choose and use the tools and instruments required of a particular job. They must also know the physical as well as the electrical characteristics of the electronic equipment and components with which they must deal.

One of the most interesting and challenging jobs of the technician is troubleshooting and repairing defective equipment. *Troubleshooting* is the process of examining and testing a piece of defective equipment to find the source of its trouble. The technician may simply use his or her eyes and ears to locate the problem. If necessary, one or more instruments can be used in the process. In this experiment three instruments, or meters, will be used—the ammeter, the voltmeter, and the ohmmeter.

To introduce the idea of troubleshooting, this experiment uses circuits consisting of a network of resistors fed by a dc power source. Somewhere in the network a defective resistor is changing the expected behavior of the circuit.

If this circuit was part of a complex piece of equipment, the first job would be to find the circuit whose behavior was not as specified by the manufacturer or designer. Once that circuit was discovered and isolated, the next job would be to find the defective component in that circuit. By sight, touch, or even smell a suspected defective component can be located. After this component is replaced with a known good component, the circuit should operate normally, thus confirming that the defective component had been found. If sight, touch, and smell do not lead to the defective component, more systematic methods must be used. In this experiment you will apply your knowledge of dc circuits and the

use of basic dc instruments to find the defective resistors in a number of different circuits.

Defects of a Circuit

Changes in the operation of a circuit can be caused by changes in the characteristics of individual parts of the circuit.

Resistors. The resistance of a resistor may change. If the circuit path through the resistor opens, the result will be equivalent to an open switch. The effect would be that of having a tremendously high resistance in the circuit. A resistor can also be short-circuited, in which case the resistance across it will be very close to zero.

Switches. Mechanical switches can introduce very high resistance in a circuit due to corrosion of their parts or poor pressure on their contacts. Defective switches may also fail to close, which causes the circuit to remain open at all times. Switches may also be short-circuited, in which case the circuit will always remain closed no matter what position the switch is in.

Circuit conductors. Conductors may break, causing that part of the circuit to open. Faulty insulation or improper placement of components may cause adjacent conductors to touch, producing a short circuit.

Power source. As a battery ages, its internal resistance increases. When current is drawn from the battery, the internal voltage drop will decrease the terminal voltage below its rated value.

The output voltage of an unregulated power supply will also drop as current is drawn from the supply. This is a normal condition, but it must be taken into account when testing the behavior of a circuit under load. A defective power supply will produce little or no voltage.

Troubleshooting Circuit Components

Resistors, switches, and circuit conductors can be tested with an ohmmeter when out of a circuit. Such a test is called a *static test,* or *measurement.*

When measuring the resistance of a resistor against its rated value, consideration must be given to the tolerance rating of resistor. For example, a 1200-Ω, 10 percent resistor is considered good if its resistance measures anywhere between 1080 Ω (1200 − 120) and 1320 Ω (1200 + 120).

The operation of a switch can be tested by connecting an ohmmeter across its terminals. When the switch is **open** (or **off**), the meter should have an extremely high (or infinite) reading in ohms. When the switch is **closed** (or **on**), the meter should have an extremely low (or zero) reading in ohms.

Circuit conductors—whether individual wire, multiwire cable, heavy busses, or printed circuit traces—can be given a *continuity check* using an ohmmeter. This type of check is used primarily to verify that a complete path exists between the ends of the conductors. It can also be used to discover whether a conductor is making improper contact with the equipment frame or chassis (a condition called *grounding*) or whether improper contact is being made with other conductors (a condition called *shorting*).

CAUTION: Ohmmeter tests should never be performed in live circuits—that is, circuits to which voltage has been applied.

Often, intermittent defects in circuit components can be discovered by tapping, vibrating, or flexing the part while it is connected to the ohmmeter. Any sudden changes in the reading of the meter during this test may indicate a defect that could become permanent over time or else require tightening, cleaning, or other processes to correct the intermittent behavior.

The first, and probably most important, test is that given to the power supply before the load or circuit is connected. Is the source delivering the correct no-load voltage? Is the source regulated (varying very little if at all under load) or unregulated (varying with load)? Is the source capable of delivering the current likely to be required by the load or circuit? The self-contained meter of the source or an external voltmeter can be used to measure the no-load output or terminal voltage of the power supply. The nameplate on the supply will give its current rating. This will indicate whether it has the capacity to deliver enough current to the circuit.

Troubleshooting with Dynamic Measurements

Resistors, switches, conductors, and other circuit elements may test "good" when measured outside of a circuit or when the circuit is not on. However, these same elements may behave differently when tested in a circuit with a voltage supplied. Readings taken under operating conditions are known as *dynamic measurements*. Dynamic measurements may be necessary because it is difficult or even impossible to remove a particular component from a circuit.

Comparisons of calculated values of voltage and current with the values obtained through dynamic measurements give clues to the nature of defects and help pinpoint the part that is likely to be defective.

Basic Troubleshooting Rules

The conclusions drawn from the results of dynamic measurements are based on two related principles:

1. In a circuit that is operating normally and that contains only resistors, the voltage across each resistor and the current in every part of the circuit must follow Ohm's and Kirchhoff's laws.
2. If the voltage across any resistor or the current in any part of the circuit does not follow Ohm's and Kirchhoff's laws, then the circuit is not operating normally, and it can be assumed there is a defect in the circuit.

In applying the two preceding principles it should be understood that the value of resistance used in the Ohm's law and Kirchhoff's law calculations must be the actual measured resistance rather than the rated or color-coded value.

Behavior of DC Series Circuits

The simple series circuit in Figure 19–1 contains a 100-V dc power source V; a control device, switch S; and three resistors: R_1 = 2000 Ω, R_2 = 3000 Ω, and R_3 = 5000 Ω. An ammeter is placed in the circuit to measure current. When S is open, as shown, there is no complete path for current and the meter will measure 0 A. Since there is no current, the voltage drop across each resistor is zero, as given by Ohm's law.

$$V_1 = I \times R_1 = 0 \times 2000 = 0 \text{ V}$$
$$V_2 = I \times R_2 = 0 \times 3000 = 0 \text{ V}$$
$$V_3 = I \times R_3 = 0 \times 5000 = 0 \text{ V}$$

When S is closed, the circuit is completed and current will flow. Applying Ohm's law, with I as the circuit current, V the applied voltage, and R_T the total resistance of the series circuit, we have

$$I = \frac{V}{R_T} = \frac{100}{(2000 + 3000 + 5000)} = \frac{100}{10{,}000}$$
$$= 0.010 \text{ A, or } 10 \text{ mA}$$

Similarly, using Ohm's law we can calculate the voltage drop across each resistor:

$$V_1 = 0.01 \times 2000 = 20 \text{ V}$$
$$V_2 = 0.01 \times 3000 = 30 \text{ V}$$
$$V_3 = 0.01 \times 5000 = 50 \text{ V}$$

If any of the circuit parts in Figure 19–1 change in value or become defective, the measured values of current and voltage will be different from those calculated. This condition will affect the operation of any load dependent on the rated (or calculated) voltage or current. For example, if one of the resistors is actually a small motor that requires a particular voltage across it to operate, a change in the circuit may reduce (or increase) the voltage across the motor, and it will not operate as intended.

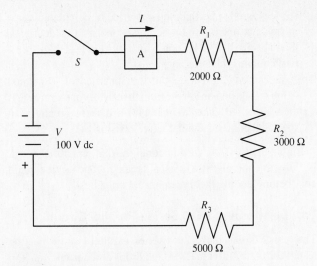

Figure 19–1. Troubleshooting a series dc circuit.

Troubleshooting a Series Circuit

Again, the circuit of Figure 19–1 is used to demonstrate the troubleshooting process. Because the circuit is not delivering 10 mA, it is suspected of being defective. Voltage and current measurements are taken around the circuit with the following results: $V = 100$ V, $I = 4$ mA, $V_1 = 8$ V, $V_2 = 12$ V, $V_3 = 80$ V. Because 4 mA is everywhere in the circuit, the values of the three resistors can be calculated using Ohm's law and the measured values of V and I.

$$R_1 = \frac{V_1}{I} = \frac{8}{0.004} = 2000 \ \Omega$$

$$R_2 = \frac{V_2}{I} = \frac{12}{0.004} = 3000 \ \Omega$$

$$R_3 = \frac{V_3}{I} = \frac{80}{0.004} = 20,000 \ \Omega$$

Because the circuit specified the value of R_3 to be 5000 Ω to obtain 10 mA, replacing R_3 with a known 5000-Ω resistor will restore the circuit to its specified value of current.

The measured values can be interpreted in another way. Since current has *decreased* from 10 mA to 4 mA, the total resistance must have *increased*. It is usually correct to make an initial assumption that just one part of a circuit is defective. If we make that assumption in this problem, we can compare some measured values of voltage with the calculated values in a good circuit. Here we notice that with a decrease to 40 percent of the specified current in the circuit, the voltages across R_1 and R_2 are 40 percent of their calculated value, but the voltage across R_3 is 160 percent of its calculated value. Thus, the voltage drops across R_1 and R_2 decreased as expected, but the voltage across R_3 rose, and that behavior was not expected. Therefore, R_3 is suspected of being defective (or at least not the specified value of resistance). All the foregoing is valid only if the voltage applied to the circuit is as specified. Thus, the first voltage to verify is the voltage supplying the circuit.

Characteristics of a Parallel Circuit

The circuit of Figure 19–2 is used to explain the important characteristics of a parallel circuit. We can make the usual assumptions that the voltage supply is constant at 100 V and the circuit conductors have zero resistance. The three resistors will draw current according to Ohm's law.

$$I_1 = \frac{V}{R_1} = \frac{100}{1000} = 0.100 \ \text{or} \ 100 \ \text{mA}$$

$$I_2 = \frac{V}{R_2} = \frac{100}{3000} = 0.033 \ \text{or} \ 33 \ \text{mA}$$

$$I_3 = \frac{V}{R_3} = \frac{100}{6000} = 0.017 \ \text{or} \ 17 \ \text{mA}$$

By Kirchhoff's law,

$$\begin{aligned} I_T &= I_1 + I_2 + I_3 \\ &= 100 \ \text{mA} + 33 \ \text{mA} + 17 \ \text{mA} \\ &= 150 \ \text{mA} \end{aligned}$$

The total resistance of the circuit can be found using Ohm's law

$$R_1 = \frac{V}{I_T} = \frac{100}{0.150} = 667 \ \Omega$$

The total resistance can also be found from the formula:

$$R_T = \frac{1}{\dfrac{1}{R_1} + \dfrac{1}{R_2} + \dfrac{1}{R_3}}$$

$$R_T = \frac{1}{\dfrac{1}{1000} + \dfrac{1}{3000} + \dfrac{1}{6000}}$$

$$= \frac{1}{10 \times 10^{-4} + 3.33 \times 10^{-4} + 1.67 \times 10^{-4}}$$

$$= \frac{1}{15 \times 10^{-4}} = 667 \ \Omega$$

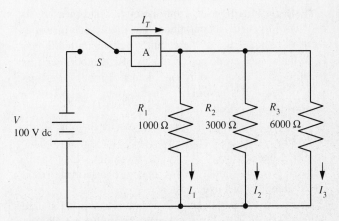

Figure 19–2. Troubleshooting a parallel dc circuit.

Troubleshooting DC Parallel Circuits

Again, using the circuit of Figure 19–2, if the values of voltage, current, or resistance differ from the calculated values given in the previous section, we may assume the circuit has a defect.

First we measure the supply voltage. It measures 100 V, so we know the circuit is getting its proper voltage. However, the total current measures only 105 mA. It is therefore likely that one of the resistors is defective and is not drawing its rated current. (Remember, always assume in the beginning that only one part is defective.)

We could place an ammeter in each branch and measure the branch currents, or we could disconnect one lead of each resistor from the circuit and measure its resistance with an ohmmeter.

In the preceding example I_1 was measured and found to be 55 mA. Resistor R_1 was therefore more than 1000 Ω. Its calculated value was

$$R_1 = \frac{V}{I_1} = \frac{100}{0.055} = 1818 \ \Omega$$

An ohmmeter check was able to verify this value. Replacement with a known 1000-Ω resistor restored the circuit to its specified values.

An examination of the specified values of current would have led to another clue to the defective resistor. Note that the specified currents I_1 and I_2 combined equal 133 mA. The other combinations are $I_2 + I_3 = 117$ mA and $I_1 + I_3 = 50$ mA.

Since we know one of the resistors is drawing less than its specified current, the total current may be inconsistent with these two-resistor currents. Obviously $I_1 + I_2$ and $I_1 + I_3$ as specified (133 mA and 117 mA, respectively) exceed the actual total current of 105 mA. The combination $I_2 + I_3$ has a specified current of 50 mA, but its actual value is not known. But the combinations $I_1 + I_2$ and $I_1 + I_3$ do tell us something. If R_2 is higher than specified, R_1 and R_3 would be good, and that would result in at least 117 mA in the circuit. If R_3 is higher than specified, then R_1 and R_2 would be good, and at least 133 mA would be drawn by the circuit.

Thus, the evidence points to R_1 as being defective. The next step is to remove R_1 from the circuit and measure its resistance to verify its condition. Thus, only one measurement and knowledge of the circuit specification led directly to the suspected defective resistor. Remember, our first premise is to assume only one defect. Further investigation may in fact reveal that more than one defect is responsible for a circuit malfunction.

As another example, a value of I_T greater than the specified value would have led to the conclusion that one of the branches had a resistor whose value was less than specified. A process similar to the one just described would have led to the defective resistor.

Therefore, in a parallel circuit with the value of the source voltage V as specified, we can make the following conclusions:

1. A measured I_T *less* than the specified or calculated value means that one of the branch resistors has a value *higher* than specified. The branch with a current less than specified contains the defective resistor.
2. A measured I_T *greater* than the specified or calculated value means that one of the branch resistors has a value less than specified. The branch with a current greater than specified contains the defective resistor.

Finally, if $I_T = 0$ A, we can conclude that either the switch is defective or else there is a break in one of the main conductors feeding the three parallel branches.

Troubleshooting a DC Series-Parallel Circuit

The resistors that make up a series-parallel circuit must be tested for their individual voltage drops and current much in the manner of the series and parallel circuits previously discussed. The total current I_T must also be measured. Figure 19–3 is a series-parallel circuit with R_2 in parallel with R_3 and R_1 in series with R_4. The resistance of the parallel combination R_2 and R_3 is

$$R_{2,3} = \frac{R_2 \times R_3}{R_2 + R_3} = \frac{3000 \times 6000}{3000 + 6000} = 2000 \ \Omega$$

For the series combination, $R_1 + R_4 = 2000 + 5000 = 7000 \ \Omega$. Therefore, $R_T = 2000 + 7000 = 9000 \ \Omega$. The total current I_T is

$$I_T = \frac{V}{R_T} = \frac{90}{9000} = 0.010 \ \text{A, or } 10 \ \text{mA}$$

Using this current, we can calculate the voltage drop across each resistor.

$$V_1 = 2000 \times 0.01 = 20 \ \text{V}$$
$$V_4 = 5000 \times 0.01 = 50 \ \text{V}$$
$$V_{2,3} = 2000 \times 0.01 = 20 \ \text{V}$$

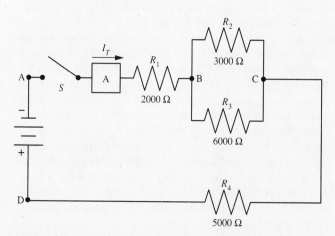

Figure 19–3. Troubleshooting a series-parallel dc circuit.

132   *Experiment 19*

The current through R_2 and R_3 can be found using $V_{2,3}$.

$$I_2 = \frac{V_{2,3}}{R_2} = \frac{20}{3000} = 0.0067 \text{ A, or } 6.7 \text{ mA}$$

$$I_3 = \frac{V_{2,3}}{R_3} = \frac{20}{6000} = 0.0033 \text{ A, or } 3.3 \text{ mA}$$

Troubleshooting can begin with a measurement of I_T (after the value of V has been verified, of course). If I_T is greater than the specified or calculated value, then R_T is less than specified. Similarly, if I_T is less than specified, R_T is greater than specified. In each case, measurements of voltage drops and currents must be made.

For example, in the circuit of Figure 19–3, current and voltage measurements were made indicating $I_T = 8$ mA and $V = 90$ V. Since I_T is less than the calculated value, R_T must be higher than specified. Voltage $V_{2,3}$ was then measured and found to be 34 V. Based on the specified $R_{2,3} = 2000$ Ω and the measured current, $I_T = 8$ mA, the expected voltage across the parallel combination of R_2 and R_3 should have been

$$V_{2,3} = R_{2,3} \times I_T = 2000 \times 0.008 = 16 \text{ V}$$

Thus, either R_2 or R_3 is higher than specified. A check of the current I_2 or I_3 would reveal which is lower than specified and, consequently, which resistor is higher than specified. If R_2 and R_3 were separated and disconnected from the circuit, an ohmmeter would indicate immediately which resistor was defective (or out of specification). Remember, just to take current readings, the branches containing R_2 and R_3 would need to be broken, an ammeter inserted, and the circuit reconnected.

The method of troubleshooting a series-parallel circuit involves the following steps:

1. Knowing the specified values of resistors, current, and voltage. The specified values are those given in the schematic diagram of the circuit or calculated from known values of resistance, voltage, and current in a normally operating circuit.
2. Measuring the actual values of current and voltage around the circuit.
3. Comparing the actual with the specified values to determine any differences.
4. Examining the circuit in terms of its individual series and parallel circuits and applying the rules of troubleshooting that these types of circuits require.

SUMMARY

1. The basic elements of a dc circuit are resistors, switches, conductors, and power sources. Any of these may be defective or have intermittent troubles.
2. Each element of a circuit can be tested for resistance and/or continuity with an ohmmeter.

3. The first step in troubleshooting is measurement of the voltage source under no load and verification of its capacity to deliver the necessary current to the load.
4. Dynamic measurements of voltage and current are made with the specified voltage source connected to the circuit.
5. In troubleshooting circuits, the initial assumption is that only one element of the circuit is defective.
6. Troubleshooting generally involves comparing dynamic measurements of voltage and current with specified values of voltage and current as given by a schematic diagram or calculated using Ohm's and Kirchhoff's laws.
7. In a series circuit, a lower-than-specified current means a higher-than-specified resistance. Similarly, a higher-than-specified current means a lower-than-specified resistance.
8. In a parallel circuit, a lower-than-specified total current means that one of the branch resistors has a higher-than-specified resistance. Similarly, a higher-than-specified total current means that one of the branch resistors has a lower-than-specified resistance.
9. In troubleshooting a series-parallel circuit, the procedures for troubleshooting individual series and parallel circuits are followed.

SELF-TEST

Check your understanding by answering the following questions:

1. In the circuit of Figure 19–1, the rated value of R_1 is 2200 Ω, R_2 is 4700 Ω , and R_3 is 8200 Ω. The applied voltage is $V = 150$ V. When the switch is closed, the ammeter reads 4.5 mA. The voltage V_1 (measured across R_1) is 10.0 V, $V_2 = 103$ V, and $V_3 = 37.0$ V.
   (a) If the circuit were normal, the values of current and voltage would be
      (i) $I =$ _____ mA
      (ii) $V_1 =$ _____ V
      (iii) $V_2 =$ _____ V
      (iv) $V_3 =$ _____ V
   (b) The total resistance in the circuit has _____ (increased/decreased).
   (c) The defective component is _____ .
2. In the circuit of Figure 19–1 the resistors are the same as in question 1 and $V = 150$ V. When the switch is closed, an ammeter in good working condition indicates there is no current in the circuit. The measured voltages are $V = 150$ V, $V_1 = 0$ V, $V_2 = 0$ V, $V_3 = 0$ V. The component that is possibly defective (open) is _____ .
3. In the circuit of Figure 19–2, the rated values of the resistors are as follows: $R_1 = 1200$ Ω, $R_2 = 1800$ Ω, $R_3 = 3000$ Ω, and $V = 300$ V. When the switch S is closed, the measured currents are as follows: $I_T = 0.4$ A,

$I_1 = 0.133$ A, $I_2 = 0.167$ A, and $I_3 = 0.1$ A.

(a) If the circuit were normal, the total current I_T would measure _____ A.

(b) The total resistance in the circuit has therefore _____ (increased/decreased).

(c) The normal currents should be

    (i) $I_1 =$ _____ A

    (ii) $I_2 =$ _____ A

    (iii) $I_3 =$ _____ A

(d) The defective resistor is _____, whose resistance has _____ (increased/decreased).

4. In the circuit of Figure 19–2, when switch S is closed, $I_T = 0$, $I_1 = 0$, $I_2 = 0$, $I_3 = 0$, $V = 100$ V. With switch S open, an ohmmeter check from the positive terminal of V to the right-hand terminal of S shows continuity. Assuming all the wiring is good, the defective component must be (a) _____ , which is (b) _____ (open/shorted).

5. In the circuit of Figure 19–3 the measured voltages are as follows: $V_{AB} = 18$ V, $V_{BC} = 27$ V, $V_{CD} = 45$ V. The defective component is (a) _____ , which is (b) _____ (open/shorted).

MATERIALS REQUIRED

Power Supply:

■ Variable 0–15 V dc, regulated

Instruments:

■ DMM or VOM

■ 0–10-mA ammeter

■ 0–100-mA ammeter

(A second multimeter with the required ammeter ranges can be substituted for the ammeters listed.)

Resistors ($\frac{1}{2}$-W, 5%):

(The resistors used in this experiment will be selected from the following group.)

■ 1	680-Ω	■ 1	2200-Ω
■ 1	820-Ω	■ 1	2700-Ω
■ 1	1000-Ω	■ 1	3300-Ω
■ 1	1200-Ω	■ 1	4700-Ω
■ 1	1800-Ω	■ 1	5600-Ω

Miscellaneous:

■ SPST switch

NOTE: Your instructor will distribute the resistors, switch, and hookup wire for you. Some of these parts will be defective.

PROCEDURE

A. Series Circuit

A1. Using the components assigned to you by your instructor, connect the circuit shown in Figure 19–4 (p. 135). Make sure switch S is **open** and the power supply is turned **off.** The total rated resistance of the three resistors you receive from your instructor should not exceed 3000 Ω. Check the values by color code only; if the total exceeds 3000 Ω, notify your instructor. (If acceptable, this value is needed in step A4.)

A2. Draw a diagram of your circuit. Show the rated values of all resistors on the diagram. (Use the color code to determine the rated values.)

A3. Turn on the supply and **close** S. Adjust the power supply until there is 15 V across its output terminals. Show this value on your circuit diagram. Maintain the output at 15 V throughout this experiment. Check the voltage periodically and adjust if necessary.

A4. Using $V = 15$ V and the rated values of the resistors, calculate the voltage drop across each resistor, the current through each resistor, and the total resistance of the circuit. Record your answers in the appropriate "Calculated" space in Table 19–1 (p. 137).

A5. Troubleshoot the circuit by measuring each of the voltages and currents noted in Table 19–1.

A6. Determine which component is defective and briefly state your reasons for this choice. Record this information in Table 19–1. After troubleshooting the circuit, **open** S and turn **off** the supply.

B. Parallel Circuit

B1. Using the components assigned to you by your instructor, connect the circuit shown in Figure 19–5. Make sure switch S is **open** and the power supply is turned **off.** The three resistors you receive from your instructor should have a total parallel resistance of more than 500 Ω. Calculate the total resistance using the rated values (as indicated by their color code) only; if it is less than 500 Ω, notify your instructor. (If acceptable, this value is needed in step B4.)

B2. Draw a diagram of your circuit. Show the rated values of all the resistors on the diagram. (Use the color code to determine the rated values.)

B3. Turn **on** the supply and **close** S. Adjust the power supply until there is 15 V across its output terminals. Show this value on your circuit diagram. Maintain the output at 15 V throughout this experiment. Adjust if necessary.

B4. Using $V = 15$ V and the rated values of the resistors, calculate the voltage drop across each resistor, the current through each resistor, the total current, and the total resistance of the circuit. Record your answers in the appropriate "Calculated" space in Table 19–2 (p. 137).

B5. Troubleshoot the circuit by measuring each of the voltages and currents noted in Table 19–2. If only one ammeter is available, you will need to break and reconnect the circuit a number of times to disconnect and reconnect the ammeter as required. Before making any circuit changes, **open** S. Check the supply voltage after each change and after **closing** S; adjust to 15 V if necessary.

B6. Determine which component is defective and briefly state your reasons for this choice. Record this information in Table 19–2. After troubleshooting the circuit, **open** S and turn **off** the power supply.

C. Series-Parallel Circuit

C1. Using the components assigned to you by your instructor, connect the circuit shown in Figure 19–6 (p. 136). Make sure switch S is **open** and the power supply is turned **off.**

C2. Draw a diagram of your circuit. Show the rated values of all the resistors on the diagram. (Use the color code to determine the rated values.)

C3. Turn on the supply and **close** S. Adjust the power supply until there is 15 V across its output terminals. Show this value on your circuit diagram. Maintain the output at 15 V throughout this experiment. Adjust if necessary.

C4. Using V = 15 V and the rated values of the resistors, calculate the voltage drop across each resistor, the current through each resistor, the total current, and the total resistance of the circuit. Record your answers in the appropriate "Calculated" space in Table 19–3 (p. 137).

C5. Troubleshoot the circuit by measuring each of the voltages and currents noted in Table 19–3. If only one ammeter is available, you will need to break and reconnect the circuit a number of times to disconnect and reconnect the ammeter as required. Before making any circuit changes, **open** S. Check the supply voltage after each change and after **closing** S; adjust to 15 V if necessary.

C6. Determine which component is defective and briefly state your reasons for this choice. Record this information in Table 19–3. After troubleshooting the circuit, **open** S and turn **off** the power supply.

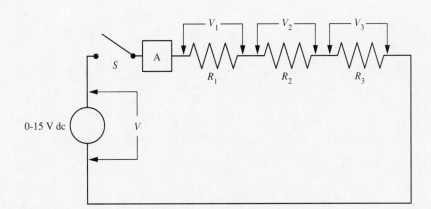

Figure 19–4. Circuit for procedure step A1. Resistors R_1, R_2, and R_3 will be assigned by the instructor.

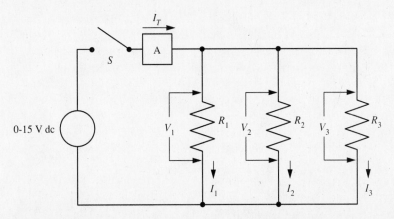

Figure 19–5. Circuit for procedure step B1. Resistors R_1, R_2, and R_3 will be assigned by the instructor.

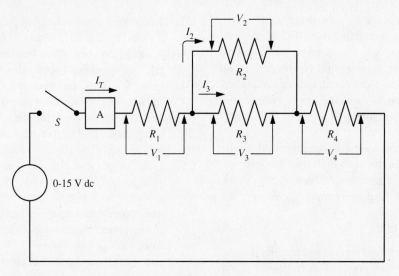

Figure 19-6. Circuit for procedure step C1. Resistors R_1, R_2, R_3 and R_4 will be assigned by the instructor.

ANSWERS TO SELF-TEST

1. (a) (i) 9.93; (ii) 21.8; (iii) 46.7; (iv) 81.5;
   (b) increased;
   (c) R_2
2. the switch
3. (a) 0.517;
   (b) increased;
   (c) (i) 0.25; (ii) 0.167; (iii) 0.1;
   (d) R_1; increased
4. (a) S; (b) open
5. (a) R_3; (b) open

TABLE 19–1. Troubleshooting a Series Circuit

	Rated Value of Resistor, Ω	Voltage, V				Current, A		
			Calculated	Measured			Calculated	Measured
R_1		V_1				I_1		
R_2		V_2				I_2		
R_3		V_3				I_3		

Applied voltage (measured): $V =$ _____   Total current (measured): $I_T =$ _____

Total calculated resistance: $R_T =$ _____   Defective component: _____

Total calculated current: $I_T =$ _____   Reason: _____

TABLE 19–2. Troubleshooting a Parallel Circuit

	Rated Value of Resistor, Ω	Voltage, V				Current, A		
			Calculated	Measured			Calculated	Measured
R_1		V_1				I_1		
R_2		V_2				I_2		
R_3		V_3				I_3		

Applied voltage (measured): $V =$ _____   Total current (measured): $I_T =$ _____

Total calculated resistance: $R_T =$ _____   Defective component: _____

Total calculated current: $I_T =$ _____   Reason: _____

TABLE 19–3. Troubleshooting a Series-Parallel Circuit

	Rated Value of Resistor, Ω	Voltage, V				Current, A		
			Calculated	Measured			Calculated	Measured
R_1		V_1				I_1		
R_2		V_2				I_2		
R_3		V_3				I_3		
R_4		V_4				I_4		

Applied voltage (measured): $V =$ _____   Total current (measured): $I_T =$ _____

Total calculated resistance: $R_T =$ _____   Defective component: _____

Total calculated current: $I_T =$ _____   Reason: _____

QUESTIONS

1. What initial assumption is made when troubleshooting a circuit?

_____

_____

2. The resistance of one resistor in a series circuit changes. It is not known whether the resistance has increased or decreased. Discuss the effect on the current in the circuit and the voltage across each resistor for each case (decreased resistance and increased resistance). Assume the source voltage remains constant.

_____

_____

_____

_____

_____

3. The resistance of one resistor in a parallel circuit changes. It is not known whether the resistance has increased or decreased. Discuss the effect on the total current in the circuit and the current in each branch (including the defective branch) for each case (decreased resistance and increased resistance). Assume the source voltage remains constant.

_____

_____

_____

_____

_____

4. Describe troubleshooting procedures to find each of the following defective parts:
   (a) A resistor (not in a circuit)   (c) Wire conductors in a circuit
   (b) SPST switch   (d) 1.5-V dry cell

_____

_____

_____

_____

_____

5. Describe the dynamic measurements you would use to find an open resistor in a series circuit.

_____

_____

_____

_____

6. Describe the dynamic measurements you would use to find an open resistor in a parallel circuit.

_____

_____

_____

_____

138 *Experiment 19*

EXPERIMENT

20

MAXIMUM POWER TRANSFER

OBJECTIVES

1. To measure power in a dc load
2. To verify experimentally that the maximum power transferred by a dc source to a load occurs when the resistance of the load equals the resistance of the source

BASIC INFORMATION

Measuring Power in a DC Load

Power is defined as the rate of doing work. The unit of power is the *watt* (W). The relationships among the power P dissipated by a resistive load R, the voltage V across R, and the current I in R are given by the following formulas:

$$P = VI$$
$$P = I^2R \qquad \textbf{(20–1)}$$
$$P = \frac{V^2}{R}$$

where P is given in watts, V in volts, I in amperes, and R in ohms.

Since the power consumed by a resistance is equal to the product of the voltage across the resistance and the current in the resistance ($P = VI$), power can be found by measuring V with a voltmeter and I with an ammeter. The two readings can then be multiplied to find power.

For example, if 12.5 V produces a current of 0.25 A in a resistance, the power consumed by the resistance is

$$P = VI$$
$$= 12.5 \times 0.25$$
$$= 3.125 \text{ W}$$

If the value of the resistance in the preceding example were known and either the voltage or current were also known, the other power formulas could be used.

If 12.5 V is measured across a 50-Ω resistance, the power consumed by the resistance is

$$P = \frac{V^2}{R}$$
$$= \frac{12.5 \times 12.5}{50} = \frac{156.25}{50}$$
$$= 3.125 \text{ W}$$

If a current of 0.25 A is measured in the 50-Ω resistance, the power consumed by the resistance is

$$P = I^2R$$
$$= 0.25 \times 0.25 \times 50$$
$$= 3.125 \text{ W}$$

NOTE: In the preceding examples, the values of I, V, and R satisfy Ohm's law. Therefore, it should not be surprising that the power calculated in each case is exactly the same. Each of the power formulas is derived by substituting Ohm's law for either V, I, or R.

Since

$$V = IR$$
$$P = VI = (IR)I = I^2R$$

and since

$$I = \frac{V}{R}$$

$$P = VI = V\left(\frac{V}{R}\right) = \frac{V^2}{R}$$

Power can also be measured directly with a *wattmeter*. The wattmeter is a four-terminal instrument. Two terminals are connected across a load (as a voltmeter would be connected), and two terminals are connected in series with the load (as an ammeter would be connected). Internally the analog wattmeter consists of two coils that interact to produce a single direct reading of power by the pointer of the meter. Although wattmeters are rarely used by electronics technicians, they are widely used as panelboard meters and portable instruments in the electric power field.

Maximum Transfer of Power

Figure 20–1 shows a circuit delivering power to a load resistance R_L from a power source V. The circuit itself contains resistance R_C, which includes the internal resistance of the power source. This simple circuit leads to a question of considerable importance in circuit design: Is there some value of circuit resistance that will allow the maximum transfer of power between the source and the load?

The power consumed by R_L is

$$P_L = I^2 R_L \qquad (20-2)$$

The value of I is determined by the voltage V (assumed to be a constant value) and the total resistance of the circuit.

$$R_T = R_C + R_L$$

where R_C is the internal resistance of the source and all other resistances in the circuit delivering power to the load.

Thus, the current in formula (20–2) is

$$I = \frac{V}{R_C + R_L}$$

and formula (20–2) becomes

$$P_L = \left(\frac{V}{R_C + R_L} \right)^2 R_L$$

$$= \frac{V^2 R_L}{(R_C + R_L)^2}$$

The direct mathematical solution of this formula that will give the value of R_C for maximum power transfer is beyond the scope of this book. However, graphical and experimental methods can lead to the same solution.

Graphically, constant values can be assigned to V and R_C and the power formula (20–2) solved for a range of values for R_L. A graph can then be plotted of P_L versus R_L. If a maximum P does occur, the graph will show the value of R_L at that point.

Assume a constant voltage of 10 V and a circuit resistance of 100 Ω. The load resistance is varied from 0 to 100 kΩ and the value of P_L is calculated at each step. The results are shown in Table 20–1. The values in this table are plotted in Figure 20–2 (p. 141).

The maximum power P_L delivered to the load R_L as seen from the table and graph is 0.25 W. This occurs when $R_L = R_C$. Further calculations with other values of R_C and V and a varying R_L will confirm the general rule that maximum power transfer occurs when $R_L = R_C$. The value of R_C can be interpreted two ways. If the resistance of the conductors connecting the constant voltage source to the load were negligible, then R_C would represent only the internal resistance of the

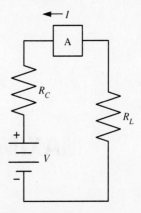

Figure 20–1. A voltage source delivers power to a load, R_L. The internal resistance of the source plus the conductors is represented by resistance R_C.

voltage source, and the rule for maximum power transfer could be stated as follows:

Maximum power transfer to a load by a constant voltage source occurs when the internal resistance of the voltage source is equal to the resistance of the load.

TABLE 20–1. Calculating Maximum Power Transfer

R_L	R_C	P_L
0	100	0
10	100	0.0826
20	100	0.139
30	100	0.178
40	100	0.204
50	100	0.222
60	100	0.234
70	100	0.242
80	100	0.247
90	100	0.249
100	100	0.250
110	100	0.249
120	100	0.248
130	100	0.246
140	100	0.243
150	100	0.240
200	100	0.222
400	100	0.160
600	100	0.122
800	100	0.0988
1,000	100	0.0826
10,000	100	0.00980
100,000	100	0.000998

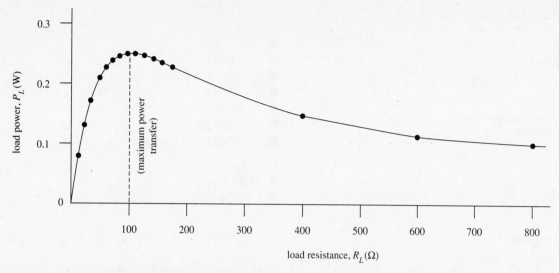

Figure 20–2. Graph of Table 20–1. Maximum power transfer occurs when the resistance of the voltage source is equal to the load resistance.

On the other hand, if the circuit delivering power to the load were more complex, then R_C would be determined by the resistance of the circuit as viewed from the output terminals of that circuit. Remember, however, that R_L cannot be measured by simply putting an ohmmeter across the output terminals of the network connected to the load, because this network has an active voltage source. Later in this manual you will learn techniques for solving such problems. At this point we will assume R_C is known or can be easily determined. With this new definition of R_C, a more general rule for maximum power transfer can be stated:

Maximum power transfer to a load from a constant voltage source occurs when the resistance of the circuit supplying the power, as viewed from the output terminals of the circuit, is equal to the load resistance.

SUMMARY

1. The power P, in watts, dissipated by a resistor R (ohms), across which there is a dc voltage V (volts), is given by the formula $P = V^2/R$.
2. The power P, in watts, dissipated by a resistor R (ohms) in which there is direct current I (amperes) is given by the formula $P = I^2R$.
3. If the dc voltage is V (volts) across a resistor and current in the resistor is I (amperes), the wattage dissipated by the resistor is given by the formula $P = V \times I$.
4. Power in a dc circuit can be measured directly with a wattmeter, or indirectly by measuring V and R, or V and I, or I and R, and substituting the measured values in the proper formula for power.
5. When power is delivered by a power source V, whose internal resistance is R_C, to a load R_L, maximum transfer of

power occurs when the load resistance equals the internal resistance of the source, that is, when $R_C = R_L$.
6. If R_C represents the resistance of a network containing a constant voltage source, maximum power transfer to a load R_L occurs when $R_C = R_L$, where R_C is the resistance of the network as viewed from its output terminals.

SELF-TEST

Check your understanding by answering the following questions:

1. The current in a 330-Ω resistor is 0.1 A. The power dissipated by the resistor is _____ W.
2. The voltage across a resistor is 12 V, and the current in the resistor is 0.05 A. The power dissipated by the resistor is _____ W.
3. The voltage across a 220-Ω resistor is 5.5 V. The power dissipated in the resistor is _____ W.
4. A power supply with an internal resistance of 25 Ω delivers power to a 50-Ω load connected across its terminals. If the voltage delivered by the supply without load is 15 V, the power dissipated by the load is _____ W.
5. A power supply with an internal resistance of 120 Ω delivers power to a resistive load. If the no-load voltage at the output of the supply is 12 V, the maximum power would be delivered to a load whose resistance is _____ Ω.
6. The power delivered by the supply to the load in question 5 is _____ W.

MATERIALS REQUIRED

Power Supply:
- Variable 0–15 V dc, regulated

Instruments:
- DMM or VOM
- 0–100-mA milliammeter

Resistors (½W, 5%):
- 2   100-Ω
- 1   330-Ω
- 1   470-Ω
- 1   1000-Ω
- 1   2200-Ω
- 1   10-kΩ potentiometer

Miscellaneous:
- SPST switch
- DPST switch

PROCEDURE

A. Measuring Power in a DC Circuit

A1. With power **off** and switch S_1 **open,** connect the circuit of Figure 20–3.

A2. Turn the power **on.** Adjust the power supply so that $V_{PS} = 10$ V. **Close** S_1. Measure V_{PS}, I_L, and the voltage V_L across the load resistor R_L. Record the values in Table 20–2 (p. 145). **Open** S_1; turn the power **off.**

A3. Disconnect R_C from the circuit and measure its resistance with an ohmmeter. Record its value in Table 20–2. Similarly, disconnect R_L from the circuit and measure its resistance with an ohmmeter. Record its value in Table 20–2.

A4. Calculate the power P_L consumed by the load resistor R_L using the measured values of V_{PS}, V_L, I_L, R_C, and R_L, of Table 20–2. Show all calculations for parts (a), (b), and (c), as follows:
  (a) Calculate P_L using V_L and I_L.
  (b) Calculate P_L using V_L and R_L.
  (c) Calculate P_L using I_L and R_L.
  Record your answers in Table 20–2.

A5. Calculate the P_T delivered by the power supply. Show all calculations. Record your answer in Table 20–2.

B. Maximum Power Transfer

B1. Measure the resistance of the 1000-Ω (rated) resistor and record its value in all the spaces in the first column of Table 20–3 (p. 145). This is the value of R_C. Use this resistor in step B2.

B2. With power **off,** and switch S_2 **open,** connect the circuit of Figure 20–4.

B3. Turn the power **on;** switch S_2 **open.** Adjust the power supply so that $V_{PS} = 10$ V. This voltage should be maintained throughout the rest of the steps in this experiment.

B4. With S_2 **open,** connect an ohmmeter across points AB of the potentiometer. Adjust the potentiometer until $R_L = 0$ Ω. Disconnect the ohmmeter. **Close** S_2. Measure the voltage across the power supply and adjust if necessary to 10 V. Measure the voltage V_L across the load R_L. Record the value in Table 20–3.

B5. Repeat step 3 for each of the values of load given in Table 20–3 (100 Ω to 10,000 Ω). Follow the sequence of operations in step 3 carefully.
  (a) With S_2 **open,** adjust the potentiometer for the load resistance in Table 20–3. Measure this value with an ohmmeter and record in Table 20–3.

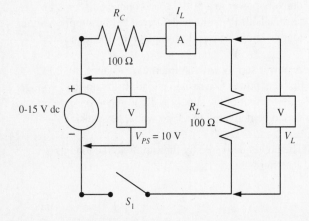

Figure 20–3. Circuit for procedure step A1.

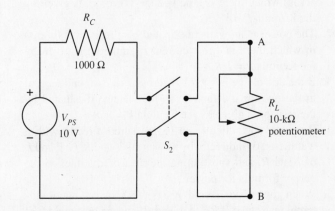

Figure 20–4. Circuit for procedure step B2.

(b) Without changing the potentiometer setting, **close** S_2.

(c) Adjust V_{PS} to 10 V if necessary.

(d) Measure V_L and record the value in Table 20–3.

B6. Using the corresponding measured values of R_C, R_L, and V_L, calculate P_L for each row of Table 20–3. Use the formula

$$P_L = \frac{V_L^2}{R_L}$$

Convert your answers to milliwatts and record the values in Table 20–3.

B7. Using the corresponding measured values of R_C, R_L, and V_{PS} (which should be 10 V in each case), calculate P_{PS}, the power delivered by the power supply for each row of Table 20–3. Use the formula

$$P_{PS} = \frac{V_{PS}^2}{R_C + R_L}$$

Convert your answers to milliwatts and record the values in Table 20–3.

C. Determining R_L for Maximum Power Transfer

C1. With S_2 **open**, connect the resistors, switch, and potentiometer, as shown in Figure 20–5. The load resistor R_L is a 10-kΩ potentiometer. Connect an ohmmeter across R_L and adjust the potentiometer so that $R_L = 0$. Do not change this setting.

C2. With S_2 **closed**, measure the resistance R_T across AB using an ohmmeter. Since this is the total resistance

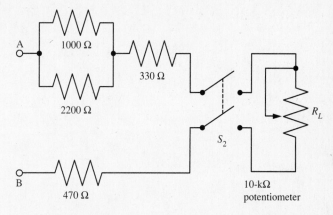

Figure 20–5. Connection of resistors, switch, and potentiometer for procedure step C1.

of the circuit feeding R_L, maximum power transfer will occur when $R_L = R_T$. Record the values of R_T and R_L in the first row of Table 20–4 (p. 145). **Open** S_2.

C3. Using an ohmmeter, adjust the resistance of the potentiometer so that $R_L = R_T$. Do not change this setting.

C4. With power **off** and S_1 **open**, connect the power supply, milliammeter, and S_1 across the circuit of Figure 20–5 as shown in Figure 20–6.

C5. Power **on**; S_1 **open**. Adjust V_{PS} to obtain a voltage between 10 and 15 V.

C6. **Close** S_1 and S_2. Measure I_T, V_{PS}, and V_L. Record the values in Table 20–4.

C7. Calculate the values of P_{PS} and P_L using the formulas shown in Table 20–4. Record your answers in the table.

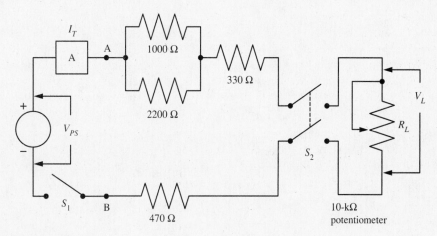

Figure 20–6. Circuit for procedure step C4.

ANSWERS TO SELF-TEST

1. 3.3	4. 2.0
2. 0.6	5. 120
3. 0.138	6. 6

TABLE 20–2. Measuring Power in a DC Circuit: Part A

Steps	V_{PS}, V	V_L, V	I_L, A	R_C, Ω	R_L, Ω
2, 3					

4	Power Formula		Power, W		
	(a) $P_L =$				
	(b) $P_L =$				
	(c) $P_L =$				
5	$P_T =$				

TABLE 20–3. Experimental Data to Determine Maximum Power in R_L: Part B

R_C, Ω	R_L, Ω	$R_C + R_L$, Ω	V_L, V	$P_L = \dfrac{V_L^2}{R_L}$, mW	$P_{PS} = \dfrac{V_{PS}^2}{R_C + R_L}$, mW
	0				
	100				
	200				
	400				
	600				
	800				
	850				
	900				
	950				
	1,000				
	1,100				
	1,200				
	1,500				
	1,700				
	2,000				
	4,000				
	6,000				
	8,000				
	10,000				

TABLE 20–4. Determining R_L for Maximum Power Transfer: Part C

R_T (measured), Ω	$R_L = R_T$, Ω	V_L, V	V_{PS}, V	I_T, A	$P_L = \dfrac{V_L^2}{R_L}$	$P_{PS} = V_{PS} \times I_T$	R_T (calculated), Ω

Maximum Power Transfer

QUESTIONS

1. Explain, in your own words, the relationship between load resistance and the transfer of power between a dc source and a load.

_____

_____

_____

_____

2. Refer to your data in Table 20–3. For what value of R_L was there maximum power transfer? Does this value confirm the relationship discussed in Question 1. Explain any discrepancies.

_____

_____

_____

_____

_____

_____

3. Refer to your data in Table 20–3. Discuss the relationship between the voltage, V_L, across the load and the resistance of the load, R_L.

_____

_____

_____

4. Refer to your data in Table 20–3. What is the relationship between the power delivered by the power supply P_{PS}, and the resistance of the load, R_L?

_____

_____

_____

_____

5. How does the power delivered to the load, P_L, vary with the resistance of the load, R_L?

_____

_____

_____

_____

6. On a separate sheet of $8\frac{1}{2} \times 11$ graph paper, plot a graph of P_L versus R_L using your data from Table 20–3. Let R_L be the horizontal (or x) axis and P_L be the vertical (or y) axis. Label the graph "P_L vs. R_L."

7. On the graph paper used in Question 6, plot a graph of P_{PS} versus R_L. Use the same axes as in Question 6. Label the graph "P_{PS} vs. R_L."

21

SOLVING CIRCUITS USING MESH CURRENTS

OBJECTIVES

1. To learn what is meant by a linear circuit
2. To verify experimentally the calculated currents found by using the mesh current method

BASIC INFORMATION

Linear Circuit Elements

Resistors are known as *linear devices* or *linear circuit elements*. If a ciruit contains only resistors or other types of resistive elements, it is known as a *linear circuit*.

A linear element is one whose voltage and current behavior obeys Ohm's law. That is, if the voltage across a device is doubled, the current through the device is doubled; if the voltage is reduced by one-third, the current decreases by one-third. Stated another way, the ratio of voltage to current is a constant. A circuit element that behaves in this fashion is the resistor.

The meaning of the word linear can be demonstrated more clearly if a graph of the voltage-current relationship is plotted. Using the circuit of Figure 21–1, the behavior of a 1000-Ω resistor was observed. A dc voltage was varied from 5 to 25 V in steps of 5 V. The current was measured at each step and recorded in Table 21–1. From these data the V and I points were plotted and the graph was drawn (Figure 21–2). The straight-line graph gives rise to the term linear as applied to the ordinary resistor. In later experiments you will use circuit elements that are not linear.

Mesh Current Method

Series-parallel circuits can be solved using Kirchhoff's voltage and current laws and Ohm's law. However, these methods are laborious and time consuming if the circuit contains more than two branches and more than one voltage source. The *mesh current method* of solving circuits uses Kirchhoff's voltage law in such a way as to eliminate much of the mathematical labor. It does this by establishing voltage equations for complete circuit loops, or meshes, and solving these equations simultaneously.

TABLE 21–1. Current/Voltage Relationship in a 1000-Ω Resistor

Voltage, V	Current, mA
0	0
5	5
10	10
15	15
20	20
25	25

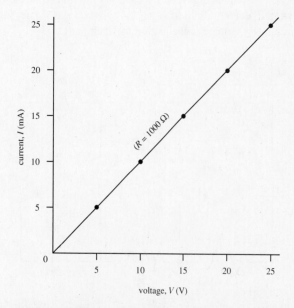

Figure 21–2. A plot of the voltage-current characteristics of the circuit of Figure 21–1.

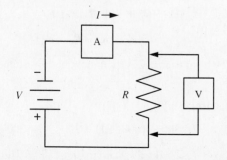

Figure 21–1. Verifying the characteristics of a linear circuit.

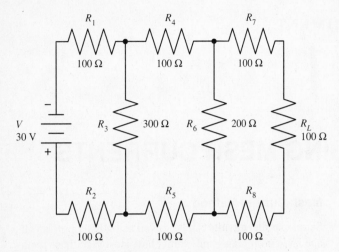

Figure 21–3. A three-branch series-parallel circuit.

Mesh Current Equations

The circuit of Figure 21–3 contains three branches and one voltage source. Suppose we wish to find the current in R_L. It can be found using the methods previously learned—that is, combining series and parallel resistors until we obtain the total resistance. We can then solve for the total current and using Kirchhoff's and Ohm's laws solve for I_L.

The mesh current method provides a direct procedure for finding current in any resistor using simultaneous equations. The first step in the procedure is to identify *closed paths* (also called *loops* or *meshes*) in the original circuit. The path need not contain voltage sources, but all voltage sources must be included when choosing the closed paths. We select the fewest paths possible that will include all resistors and voltage sources. Each closed path is assumed to have a circulating current. It is customary to assume current direction in each case as being in the clockwise direction. Using this assumed current, called a *mesh current,* we write the Kirchhoff's voltage law equation for each path.

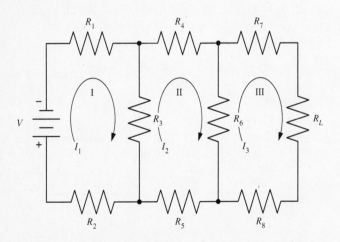

Figure 21–4. Establishing meshes and mesh currents in the circuit of Figure 21–3.

The loops often include resistors that are part of other loops. The voltage drops caused by the current in these other loops must be taken into account when writing the Kirchhoff's equation. An equation must be written for each loop chosen. This will result in a system of related equations that must then be solved simultaneously. The currents found by this procedure are the currents in the various resistors. If more than one current is found in a resistor, their algebraic sum is the actual current in that resistor.

A sample problem demonstrates this method.

Problem. Find the load current I_L in the load resistor R_L of the circuit in Figure 21–3.

Solution. Figure 21–3 is redrawn in Figure 21–4 showing the assumed mesh currents. In this case three meshes will be necessary to include all resistors and voltage sources.

The voltages around mesh I are

$$I_1R_1 + I_1R_3 + I_1R_2 - I_2R_3 = V \quad \textbf{(21–1)}$$

Because the voltage drop due to I_2 in R_3 (in mesh II) is opposite the voltage drop I_1R_3, I_2R_3 is subtracted from the other voltage drops. This equation can be simplified to

$$I_1(R_1 + R_2 + R_3) - I_2R_3 = V$$

Similarly, the voltages around meshes II and III can be written using I_2 and I_3. However, there are no voltage sources in meshes II and III.

For mesh II,

$$I_2R_3 + I_2R_4 + I_2R_6 + I_2R_5 - I_1R_3 - I_3R_6 = 0$$

Rearranging the terms and simplifying the equation, we have

$$-I_1R_3 + I_2(R_3 + R_4 + R_5 + R_6) - I_3R_6 = 0 \quad \textbf{(21–2)}$$

For mesh III,

$$I_3R_6 + I_3R_7 + I_3R_L + I_3R_8 - I_2R_6 = 0$$

Rearranging and simplifying the equation, we have

$$-I_2R_6 + I_3(R_6 + R_7 + R_8 + R_L) = 0 \quad \textbf{(21–3)}$$

The three mesh equations can now be rewritten as a system of simultaneous equations.

$$I_1(R_1 + R_2 + R_3) - I_2R_3 \qquad\qquad = V \ \textbf{(21–1)}$$
$$-I_1R_3 \quad + I_2(R_3 + R_4 + R_5 + R_6) - I_3R_6 \quad = 0 \ \textbf{(21–2)}$$
$$-I_2R_6 \quad + I_3(R_6 + R_7 + R_8 + R_L) \quad = 0 \ \textbf{(21–3)}$$

The values from Figure 21–3 can now be substituted in the equations.

$$+I_1(100 + 100 + 300) - I_2(300) \qquad\qquad = 30$$
$$-I_1(300) \quad + I_2(300 + 100 + 100 + 200) - I_3(200) = 0$$
$$-I_2(200) \quad + I_3(200 + 100 + 100 + 100) = 0$$

which becomes

$$500 I_1 - 300 I_2 \qquad\qquad = 30$$
$$-300 I_1 + 700 I_2 - 200 I_3 = 0$$
$$\qquad\qquad -200 I_2 + 500 I_3 = 0$$

The methods used to solve simultaneous equations will not be covered here. If necessary, review the methods for solving simultaneous equations in any standard algebra text.

Solution of the set of simultaneous equations (21–1), (21–2), and (21–3) results in the following currents:

$$I_1 = 0.0845 \text{ A}$$
$$I_2 = 0.0409 \text{ A}$$
$$I_3 = 0.0164 \text{ A}$$

Because I_3 is the only current through R_L, it is also I_L. Therefore, the answer to the problem is 0.0164 A, or 16.4 mA. Note also that $I_1 = 0.0845$ A represents the total current delivered to the circuit by V.

Although it was not called for in this sample problem, we can find the current in each resistor and the voltage across each resistor using the values of I_1, I_2, and I_3.

The current delivered by V is $I_1 = 0.0845$ A. Since $V = 30$ V, the total resistance of the circuit is

$$R_T = \frac{V}{I_1} = \frac{30}{0.0845} = 355 \ \Omega$$

The current in R_1 is also the current in R_2. The voltage drops across R_1 and R_2 are

$$V_{R1} = V_{R2} = IR = 0.0845 \times 100 = 8.45 \text{ V}$$

The current in mesh II, I_2, is 0.0409 A. Recall that to find the voltage across R_3, we subtracted I_2 from I_1. Therefore, the actual current across R_3 is 0.0845 – 0.0409 = 0.0436 A. The positive sign for this current indicates that the current through R_3 is in the direction indicated by mesh current I_1.

The voltage across R_3 is $IR_3 = 0.0436 \times 300 = 13.1$ V. This can be verified from the voltage drops across R_1 and R_2.

Total voltage drop = 8.45 + 8.45 = 16.9 V

Voltage across R_3 = 30.0 – 16.9 = 13.1 V

The current in R_4 and R_5 is equal to mesh current I_2, or 0.0409 A. The current in R_6 is $I_2 - I_3$ = 0.0409 – 0.0164 = 0.0245 A in the direction of I_2.

The voltage across R_4 is equal to the voltage drop across R_5.

$$V_{R4} = V_{R5} = 0.0409 \times 100 = 4.09 \text{ V}$$

The voltage drop across R_6 is $0.0245 \times 200 = 4.900$ V. Again, this can be verified from previous voltage drops. The voltage across R_3 is reduced by the voltage drops across R_4 and R_5.

$$13.1 - 2(4.09) = 13.1 - 8.18 = 4.920 \text{ V}$$

(The difference comes from rounding the voltage 13.1 up from 13.08.)

Finally, the current in R_7, R_8, and R_L is 0.0164. The voltage drops across R_7, R_8, and R_L are equal:

$$0.0164 \times 100 = 1.64 \text{ V}$$

The total voltage drop is therefore equal to the voltage drop across R_6,

$$1.64 \times 3 = 4.920 \text{ V}$$

which checks with the value of R_6 previously calculated.

SUMMARY

1. A linear circuit element is one in which the ratio of voltage to current in the element is constant.
2. Resistors are linear circuit elements.
3. Circuits containing only linear elements are called linear circuits.
4. The mesh current method is an application of Kirchhoff's voltage law.
5. Mesh current equations are solved simultaneously to find the individual mesh currents.
6. When more than one mesh current is assumed in any circuit element, the actual current in that element is the algebraic sum of the mesh currents.

SELF-TEST

Check your understanding by answering the following questions:

1. A circuit containing only resistors is known as a _____ circuit.
2. The voltage-current graph of a resistor is a _____ .
3. If the voltage in Figure 21–1 were 10 V and the original resistor were replaced by one having twice the resistance, the current would (drop/rise) _____ by _____ percent.
4. (True/False) The mesh current method should be used to solve the circuit of Figure 21–1. _____
5. (True/False) Kirchhoff's current law is the basis for writing loop current equations. _____
6. If the direction of mesh current I_2 were taken as counterclockwise instead of clockwise, the actual direction of current in R_L _____ (would, would not) change.
7. The least number of mesh currents to use in solving the circuit of Figure 21–5 is _____ .
8. Use the mesh current method to find the current delivered to the circuit by V in Figure 21–5. All resistors are 10-Ω; $V = 10$ V. $I = $ _____ A.

MATERIALS REQUIRED

Power Supply:
- Variable 0–15 V dc, regulated

Instruments:
- VOM
- DMM or EVM

Resistors:
- 7 100-Ω, ½-W, 5%

Miscellaneous:
- SPST switch

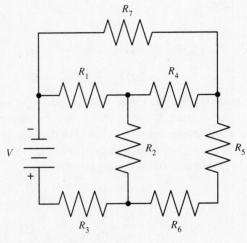

Figure 21–5. Circuit for questions 7 and 8.

PROCEDURE

1. Using an ohmmeter, measure the resistance of each resistor and record the value in Table 21–2 (p. 151).
2. With the power supply **off,** and switch S_1 **open,** connect the circuit of Figure 21–6.
3. Turn **on** the power supply and **close** S_1. Adjust the output of the supply to 10 V. Maintain this voltage throughout this experiment. Check the voltage from time to time and adjust it if necessary.
4. Measure the voltage across each resistor R_1 through R_6 and R_L. Record the values in Table 21–2.

5. Using Ohm's law and the measured value of resistance, calculate the current through each resistor. Record your answers in Table 21–2.
6. Using the rated value of the resistors and the three meshes shown in Figure 21–6, calculate mesh currents $I_1, I_2,$ and I_3. Record your answers in Table 21–2. Show all calculations on a separate sheet of paper.
7. Using your answers for $I_1, I_2,$ and I_3, calculate the current in resistors R_2 and R_4. Record your answers in Table 21–2.

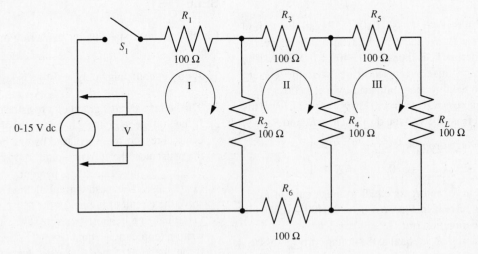

Figure 21–6. Circuit for procedure step 2.

ANSWERS TO SELF-TEST

1. linear
2. staight line
3. drop; 50
4. false

5. false
6. would not
7. 3
8. 0.458

TABLE 21–2. Verifying Mesh Current Calculations

Resistor	Resistance		Voltage Drop Measured, V	Current Calculated, A	Mesh Current Calculated, A	
	Rated Ω	Measured Ω				
R_1	100				I_1	
R_2	100					
R_3	100				I_2	
R_4	100					
R_5	100				I_3	
R_6	100				I_2	
R_L	100				I_3	

$I_{R_2} = I_1 - I_2 = $ _____ A

$I_{R_4} = I_2 - I_3 = $ _____ A

QUESTIONS

1. Explain, in your own words, the characteristics of a linear circuit.

_____

_____

_____

_____

_____

_____

_____

_____

_____

_____

_____

_____

_____

_____

_____

_____

_____

_____

_____

Solving Circuits Using Mesh Currents

2. Refer to Figure 21–6. Write the three mesh current equations but assume I_2 (the current in mesh II) is counterclockwise. Solve for the currents in R_1, R_2, R_3, R_4, R_5, R_6, and R_L using these three mesh current equations. Compare your answers with the answers and measurements in Table 21–2. Discuss any discrepancies.

3. Discuss the effect that reversing the polarity of the voltage source would have on the direction and magnitude of the current in R_1, R_2, R_3, R_4, R_5, R_6, and R_L.

_____

_____

_____

_____

_____

_____

_____

_____

_____

_____

_____

_____

_____

_____

EXPERIMENT

22

BALANCED-BRIDGE CIRCUIT

OBJECTIVES

1. To find the relationship among the resistors in a balanced-bridge circuit
2. To use the relationship among resistors in a balanced-bridge circuit to measure an unknown resistance

BASIC INFORMATION

Resistance can be measured directly using the ohms scales of a volt-ohm-milliammeter or digital multimeter. On the VOM a calibrated scale and a moving needle (pointer) indicate the value of resistance. On the DMM a digital display shows the numerical value of resistance directly.

Greater accuracy in measuring resistance can be obtained using a bridge circuit. A resistance bridge circuit, called a *Wheatstone bridge,* is shown in Figure 22–1. This type of bridge uses a highly sensitive galvanometer and a calibrated variable-resistance standard. The galvanometer is used to indicate that point A is at the exact same potential as point C. The voltages at A and C are varied by the variable-resistance

standard. If a galvanometer with a centered zero scale is used, the pointer will indicate not only when V_A and V_C are equal but also the direction of current flow when the voltages are unequal.

In Figure 22–1 the Wheatstone bridge is connected to measure the resistance of unknown resistor R_X. Resistors R_1 and R_2 are precision fixed resistors. They are called the *ratio arms* of the bridge. A precision variable resistance R_3 is called the *standard arm.* When the voltage difference between A and C is zero, no current will flow through the galvanometer, and the pointer will be centered at zero. In this condition, the bridge is said to be *balanced.* Power to the circuit is supplied by a regulated dc power supply.

The Balanced Bridge

The circuit of Figure 22–1 shows the currents in each of the arms of the Wheatstone bridge. The voltage drops in each arm can therefore be found using Ohm's law:

$$V_{AB} = I_1 R_1$$
$$V_{DA} = I_2 R_2 \qquad\qquad (22-1)$$
$$V_{CB} = I_3 R_3$$
$$V_{DC} = I_X R_X$$

For the bridge to be balanced—that is, $V_{AC} = V_{CA} = 0$—the voltage at A with respect to B, V_{AB}, must be equal to the voltage at C with respect to B, V_{CB}, or

$$V_{AB} = V_{CB}$$

Substituting the *IR* values for these voltages from formula (22–1), we have

$$I_1 R_1 = I_3 R_3$$

Dividing both sides by $I_3 R_1$ leads to the ratio

$$\frac{I_1}{I_3} = \frac{R_3}{R_1} \qquad\qquad (22-2)$$

Similarly,

$$V_{DA} = V_{DC}$$

and

$$I_2 R_2 = I_X R_X$$

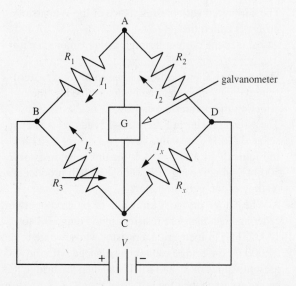

Figure 22–1. Wheatstone bridge circuit used for making accurate measurements of resistance.

from which

$$\frac{I_2}{I_X} = \frac{R_X}{R_2} \qquad (22\text{--}3)$$

At balance there is no current in line AC. From Kirchhoff's current law,

$$I_1 = I_2 \qquad (22\text{--}4)$$

$$I_3 = I_X$$

From formula (22–4), I_X is substituted for I_3 in formula (22–2).

$$\frac{I_1}{I_X} = \frac{R_3}{R_1} \qquad (22\text{--}5)$$

From formula (22–4), I_1 is substituted for I_2 in formula (22–3).

$$\frac{I_1}{I_X} = \frac{R_X}{R_2} \qquad (22\text{--}6)$$

Combining formulas (22–5) and (22–6), we obtain

$$\frac{I_1}{I_X} = \frac{R_3}{R_1} = \frac{R_X}{R_2}$$

or

$$\frac{R_3}{R_1} = \frac{R_X}{R_2} \qquad (22\text{--}7)$$

Solving for R_X yields

$$R_X = \frac{R_2 \times R_3}{R_1} = \frac{R_2}{R_1}(R_3) \qquad (22\text{--}8)$$

Formula (22–8) can be used to find the value of an unknown resistor in a bridge circuit if the values of the other three arms of the bridge are known. The formula states that the unknown resistor is equal to the ratio of R_2 to R_1 multiplied by the value of the variable resistor R_3.

Since R_2 and R_1 are fixed precision resistors, the value of R_2/R_1 can be set as required. For maximum accuracy and sensitivity in adjusting R_3, R_2 should be equal to R_1. In this case $R_2/R_1 = 1$ and formula (22–8) becomes

$$R_X = R_3$$

The limit of R_X measurements is therefore the maximum value of R_3. It can be seen that the ratio R_2/R_1 determines the maximum value of R_X that can be measured. For example, if $R_2/R_1 = 3$, then $R_X = 3R_3$, or the maximum value of R_X can be three times the maximum value of R_3. Such a ratio will also decrease the sensitivity of the measurement, because any slight adjustment of R_3 will not be easily detected by the galvanometer, thus decreasing the accuracy of the resistance value obtained from reading R_3. It is also obvious that when the ratio R_2/R_1 is less than 1, the range of resistance values that can be measured will be less than the maximum value of R_3.

SUMMARY

1. A Wheatstone bridge is used for measuring resistance values with greater accuracy than that possible with the volt-ohm-milliammeter or digital multimeter.
2. The ratio arms R_1 and R_2 in Figure 22–1 are precision resistors. Their ratio, R_2/R_1, determines the range of resistance measurements of the bridge.
3. The standard arm R_3 is a variable calibrated precision resistance by means of which the bridge is balanced when measuring an unknown resistance R_X.
4. A highly sensitive zero-center galvanometer is used as the indicator. A dc voltage source powers the bridge.
5. When the voltage across the galvanometer (V_{AC} in Figure 22–1) is zero, there is no current in the galvanometer, which then reads zero. At this point the bridge is balanced.
6. At balance in a Wheatstone bridge, the product of the resistors of opposite arms is equal; that is, in Figure 22–1,

$$R_1 R_X = R_2 R_3$$

and the value of R_X is thus

$$R_X = \frac{R_2}{R_1}(R_3) \qquad (22\text{--}8)$$

Formula (22–8) states that at balance the value of the unknown resistor R_X is equal to the resistance of the calibrated variable resistor R_3 multiplied by the ratio R_2/R_1.

SELF-TEST

Check your understanding by answering the following questions.

1. The calibrated variable-resistance of the Wheatstone bridge is called the _____ .
2. At balance the current through the galvanometer is _____ .
3. (True/False) The direction of current through a zero-centered galvanometer can be determined by the direction the pointer moves. _____
4. The galvanometer in Figure 22–2 reads zero when R_3 is set at 3750 Ω. The value of R_X is _____ Ω.
5. In Figure 22–1 $R_2/R_1 = 10$. If the unknown resistor $R_X = 5000$ Ω, what setting of R_3 (in ohms) will produce a zero reading of the galvanometer? _____ .
6. A Wheatstone bridge is known to be in good operating order and can measure resistors in the range 0.1 to 100,000 Ω. In measuring a resistor whose value is about 50,000 Ω, the bridge cannot be balanced with the present values of R_1 and R_2. To balance the bridge it is necessary to change the setting of the _____ arms.

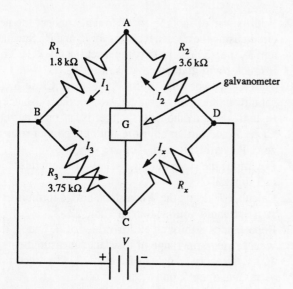

Figure 22–2. Circuit for question 4.

MATERIALS REQUIRED

Power Supply:
- Variable 0–15 V dc, regulated

Instruments:
- VOM or DMM
- Zero-center galvanometer

Resistors (½-W, 5%):
- 1   5600 Ω
- 2   5100 Ω (If available, 1% resistors are preferred)
- 1   10,000 Ω potentiometer or decade resistance box (1 Ω–10,000 Ω)

The following resistors should be available from an assortment in the shop or laboratory (at least one of each value):

560-Ω	1800-Ω	3600-Ω
620-Ω	2000-Ω	3900-Ω
750-Ω	2200-Ω	4700-Ω
1000-Ω	2400-Ω	6800-Ω
1100-Ω	3000-Ω	8200-Ω
1200-Ω	3300-Ω	10,000-Ω

- Instructor-supplied resistors for steps 8–13 (these resistors should have values greater than 10,000-Ω but less than 100 kΩ)

Miscellaneous:
- 2   SPST switches

PROCEDURE

The purpose of this experiment is to verify the bridge formulas and simulate the use of the Wheatstone bridge in measuring resistance. It is not intended to duplicate the accuracy of laboratory quality or commercial Wheatstone bridge instruments.

1. With power **off** and S_1 and S_2 **open,** connect the circuit of Figure 22–3. The potentiometer or resistance box should be set at its maximum value (10 kΩ) initially. The 5600-Ω resistor in the galvanometer circuit will reduce the sensitivity of the galvanometer during the initial stages of adjusting the standard arm resistance. The unknown resistor R_x should be chosen from the assortment noted in the Materials Required list.

2. Turn the Power **on.** Adjust V_{PS} to 6 V. **Close** S_1 (S_2 should remain **open**). The galvanometer pointer should move up scale from the zero point. (If it does not, check to see that $R_3 = 10$ kΩ.)

3. Gradually reduce the resistance of R_3. The galvanometer pointer should move down scale, approaching zero. When the pointer is at or very near zero, **close** S_2, shorting out the 5600-Ω resistor. Continue to adjust R_3 until the pointer rests on zero. Power **off,** S_1 and S_2 **open.**

4. If R_3 is a decade resistance box, read the setting of the dials and record this measured value in Table 22–1 (p. 157) for resistor 1, measured value. If R_3 is a potentiometer, remove it from the circuit without changing

its setting. Measure the resistance of the potentiometer using an ohmmeter and record the value in Table 22–1 as noted before.

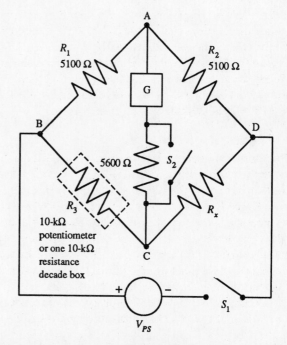

Figure 22–3. Circuit for procedure step 1.

5. Remove resistor 1, R_X, from the circuit. Record its rated resistance and tolerance values (based on the color code) in Table 22–1.
6. Using the Wheatstone bridge formula

$$R_X = \frac{R_2}{R_1}(R_3)$$

calculate the resistance of resistor X. Record your answer in Table 22–1.
7. Repeat steps 1 through 6 for five additional resistors chosen from the assortment available. For each unknown resistor, record the value of R_3, the rated value of resistance and tolerance for R_X, and the calculated value of R_X in Table 22–1.
8. Using the basic bridge circuit of Figure 22–3, with the standard arm used in steps 1 through 7, choose new resistors for the ratio arms R_1 and R_2 such that the bridge can then be used to measure resistances up to 30 kΩ. Measure R_1 and R_2 with an ohmmeter and record their values in Table 22–2.

9. With power **off** and S_1 and S_2 **open,** connect the new circuit using the chosen resistors for R_1 and R_2. Standard arm R_3 should be at its maximum setting (10 kΩ). Your instructor will give you a resistor for R_X.
10. Turn the power **on.** Adjust V_{PS} to 6 V. **Close** S_1 (S_2 should remain **open**). Gradually reduce the resistance of R_3 until the galvanometer pointer is on or near zero. **Close** S_2 and further adjust R_3 until the pointer rests on zero. Power **off;** S_1 and S_2 **open.**
11. Determine the value of R_3 as in step 4. Record this value in Table 22–2.
12. Calculate R_X using the Wheatstone bridge formula as in step 6. Record your answer in Table 22–2.
13. Repeat steps 8 through 12 for values of R_1 and R_2 that would extend the range of resistance measurements to 100 kΩ. Record all data in Table 22–2. After recording your data, **open** S_1 and S_2; power **off.**

ANSWERS TO SELF-TEST

1. standard arm
2. zero
3. true
4. 7500 Ω
5. 500 Ω
6. ratio

Experiment 22   Name _____ Date _____

TABLE 22–1. Bridge Measurement of R_x: $R_2/R_1 = 1$

Resistor Number	1	2	3	4	5	6
R_3, Ω						
Rated (color code) value, R_x, Ω						
Percent tolerance, R_x, %						
Calculated value, R_x, Ω						

TABLE 22–2. Bridge Multipliers

Steps	Maximum Measurable Resistance, kΩ	Measured Value R_1, Ω	Measured Value R_2, Ω	Ratio $\frac{R_2}{R_1}$	R_3, Ω	R_x Rated Value, Ω	R_x Rated Tolerance, %	R_x Calculated Value, Ω
8, 9, 10, 11, 12	30							
13	100							

QUESTIONS

1. Explain, in your own words, the relationship among the resistors in a balanced-bridge circuit.

_____

_____

_____

_____

_____

_____

_____

_____

2. Give four factors that determine the accuracy of resistance measurements made with a balanced-bridge circuit.

_____

_____

_____

_____

_____

_____

_____

_____

_____

Balanced-Bridge Circuit

3. Why is it important to use a very sensitive galvanometer in a balanced-bridge circuit?

_____

_____

_____

_____

_____

_____

_____

_____

4. Refer to Table 22–1. Were the resistance measurements within the tolerance ratings of the resistors? Explain any discrepancies.

_____

_____

_____

_____

_____

_____

_____

_____

5. Is the 6-V power supply critical in this experiment? Would a higher or lower voltage have affected your results? What would happen if the supply voltage were zero? Discuss fully.

_____

_____

_____

_____

_____

_____

_____

23

SUPERPOSITION THEOREM

OBJECTIVES

To verify experimentally the superposition theorem

BASIC INFORMATION

We have applied Ohm's and Kirchhoff's laws and mesh methods to study simple resistive circuits—that is, circuits containing series, parallel, or series-parallel combinations of resistors. However, there are more complex circuits in electricity. Solving such circuits may become difficult using the methods we have learned thus far. For such complex circuits, more-powerful methods of solution are helpful.

Superposition Theorem

The superposition theorem states that

In a linear circuit containing more than one voltage source, the current in any one element of the circuit is the algebraic sum of the currents produced by each voltage source acting alone. Further, the voltage across any element is the algebraic sum of the voltages produced by each voltage source acting alone.

To apply this theorem to the solution of a problem, we must understand what is meant by "each source acting alone." Suppose a network, say Figure 23–1, has two voltage sources

V_1 and V_2 and we wish to find the effect on the circuit of each source acting alone. To determine the effect of V_1 we must replace V_2 by its internal resistance and solve the modified circuit. If any of the voltage sources are considered ideal (i.e., having no internal resistance) or if the internal resistance is very low compared with other circuit elements, the voltage source can be replaced with a short circuit. In Figure 23–2 (p. 160) we have replaced V_2 with a short circuit. The circuit of Figure 23–2 is a series-parallel circuit with one voltage source V_1. We can solve for the currents in each of the resistors R_1 through R_5 and the current supplied by V_1 using methods learned in previous experiments. We can also find the voltage across each resistor of the circuit. To determine the effect of V_2, we replace V_1 with its internal resistance, again a short circuit, as in Figure 23–3 and solve the circuit of Figure 23–3 (p. 160). Again, we will find the currents in R_1 through R_5 and the current supplied by V_2. Similarly, we can find the voltage across each resistor. The final step is to add algebraically the two currents in each of the resistors to find the total current in the resistor. The voltage across each resistor will also be the algebraic sum of two voltages. The current supplied by each voltage source will be the algebraic sum of the currents in the short circuits that replaced the voltage sources and the current supplied by the voltage source itself.

A sample problem demonstrates this procedure.

Problem. Using the values shown in Figure 23–1, find the current in each resistor, the voltage across each resistor, and the current supplied by each voltage source.

Solution. The first step in the solution is to replace V_2 with a short circuit and solve the new circuit shown in Figure 23–2.

Using the mesh method, we write

$$I_I(R_1 + R_2 + R_3) - I_{II}(R_3) = V_1$$

$$-I_I(R_3) + I_{II}(R_3 + R_4 + R_5) = 0$$

$$I_I(200) - I_{II}(100) = 20$$

$$-I_I(100) + I_{II}(200) = 0$$

Solving the simultaneous equations for I_I and I_{II}, we obtain

$$I_I = 0.133 \text{ A} = 133 \text{ mA}$$

$$I_{II} = 0.0667 \text{ A} = 66.7 \text{ mA}$$

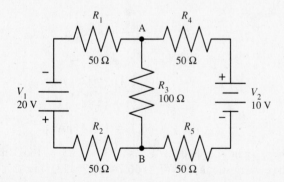

Figure 23–1. Resistor circuit with two voltage sources.

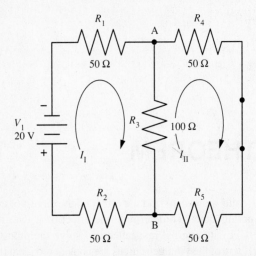

Figure 23-2. The first step when using superposition is to replace one of the voltage sources with its internal resistance. In this circuit V_2 is an ideal source (no internal resistance), so that V_2 is replaced by a short circuit.

The currents due to V_1 alone are

$$I(R_1) = 133 \text{ mA}$$

$$I(R_2) = 133 \text{ mA}$$

$$I(R_3) = 133 - 66.7 = 66.3 \text{ mA}$$

$$I(R_4) = 66.7 \text{ mA}$$

$$I(R_5) = 66.7 \text{ mA}$$

The current supplied by V_1 is 133 mA, and the current through the short circuit that replaced V_2 is 66.7 mA.

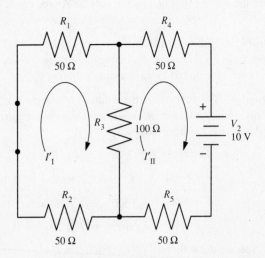

Figure 23-3. After the circuit of Figure 23-2 is solved, V_1 is replaced by a short circuit and the new circuit is solved.

The voltages across each resistor can be found using Ohm's law:

$$V(R_1) = R_1 \times I_1 = 50 \times 0.133 = 6.65 \text{ V}$$

$$V(R_2) = R_2 \times I_1 = 100 \times 0.0663 = 6.63 \text{ V}$$

$$V(R_3) = R_3 \times I_T = 100 \times 0.0663 = 6.63 \text{ V}$$

$$V(R_4) = R_4 \times I_{II} = 50 \times 0.0667 = 3.34 \text{ V}$$

$$V(R_5) = R_5 \times I_{II} = 50 \times 0.0667 = 3.34 \text{ V}$$

Next, we replace V_1 with a short circuit (as in Figure 23-3) and solve the circuit with V_2. Using the mesh method to solve for I_1' and I_{II}' results in the following:

$$I_1' = 0.0333 \text{ A} = 33.3 \text{ mA}$$

$$I_{II}' = 0.0667 \text{ A} = 66.7 \text{ mA}$$

The current through each resistor in this case will be

$$I'(R_1) = 33.3 \text{ mA}$$

$$I'(R_2) = 33.3 \text{ mA}$$

$$I'(R_3) = 33.3 - 66.7 = -33.4 \text{ mA}$$

$$I'(R_4) = 66.7 \text{ mA}$$

$$I'(R_5) = 66.7 \text{ mA}$$

The current supplied by V_2 is 66.7 mA, and the current through the short circuit that replaced V_1 is 33.3 mA.

Using Ohm's law, we can find the voltage across each resistor:

$$V'(R_1) = 1.67 \text{ V}$$

$$V'(R_2) = 1.67 \text{ V}$$

$$V'(R_3) = -1.67 \text{ V}$$

$$V'(R_4) = 3.34 \text{ V}$$

$$V'(R_5) = 3.34 \text{ V}$$

By combining each of the currents, as stated by the superposition theorem, we can find the actual current due to both voltage sources:

$$I_1 = 133 + 33.3 = 166.3 \text{ mA}$$

$$I_2 = 133 + 33.3 = 166.3 \text{ mA}$$

$$I_3 = 66.3 + (-33.4) = 32.9 \text{ mA}$$

$$I_4 = 66.7 + 66.7 = 133.4 \text{ mA}$$

$$I_5 = 66.7 + 66.7 = 133.4 \text{ mA}$$

The current supplied by V_1 is $133 + 33.3 = 166.3$ mA. The current supplied by V_2 is $66.7 + 66.7 = 133.4$ mA.

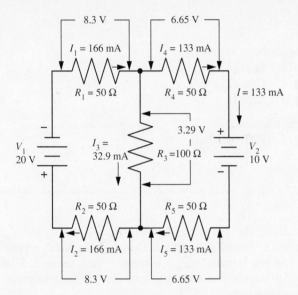

Figure 23–4. The voltages and currents of the circuit in the sample problem.

The voltages across each resistor can now be found using Ohm's law.

$$V_1 = 0.166 \times 50 = 8.3 \text{ V}$$

$$V_2 = 0.166 \times 50 = 8.3 \text{ V}$$

$$V_3 = 0.0329 \times 100 = 3.29 \text{ V}$$

$$V_4 = 0.133 \times 50 = 6.65 \text{ V}$$

$$V_5 = 0.133 \times 50 = 6.65 \text{ V}$$

The currents and voltages are shown in Figure 23–4. You should verify these values using Kirchhoff's voltage and current laws.

NOTE: Because answers were rounded to three significant figures, verification of current and voltage values may not agree in the third significant figure.

SUMMARY

1. The superposition theorem is most useful when applied to linear circuits that have two or more voltage sources.
2. To find the current in or voltage across an element in a linear circuit, all the voltage sources except one are removed and replaced by their internal resistances (short circuits in the case of regulated or ideal voltage sources), and the effect of the one remaining source on the circuit is determined. This process is repeated for each source. The actual currents and voltages produced by all sources is the algebraic sum of the individual currents and voltages.

SELF-TEST

Check your understanding by answering the following questions:

1. (True/False) The superposition theorem may be applied to the circuit of Figure 23–5. _____
2. In Figure 23–2 the graph of voltage across R_4 versus current in R_4 is _____ .
3. The 10-V source in Figure 23–1 causes_____ (more/less) current in R_3 than would be the case if V_1 were the only voltage source and V_2 were replaced by a short circuit.
4. In applying the superposition theorem to the solution of the circuit in Figure 23–1, V_2 is replaced by a _____ .
5. In the circuit of Figure 23–1 the voltage across R_5 is _____ V.
6. The current through R_1 in Figure 23–6 is_____ A.

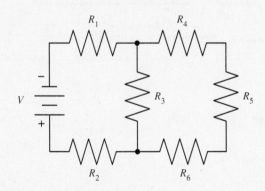

Figure 23–5. Circuit for question 1.

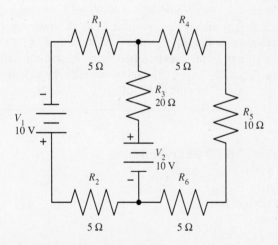

Figure 23–6. Circuit for question 6.

MATERIALS REQUIRED

Power Supply:
■ 2 variable 0–15 V dc, regulated

Instruments:
■ DMM or VOM
■ 0–100-mA milliammeter

Resistors (½-W, 5%):
■ 1   820-Ω
■ 1   1200-Ω
■ 1   2200-Ω

Miscellaneous
■ 2 SPDT switches

PROCEDURE

CAUTION: This experiment requires current readings in three different parts of a circuit. If only one ammeter is available, turn off power in both supplies before connecting and disconnecting the meter.

1. With power **off** in both power supplies and both switches S_1 and S_2 in the B position, connect the circuit of Figure 23–7. Note the polarity of the supplies carefully.

2. Turn power supply 1 **on**. Adjust its voltage so that V_{PS1} = 15 V. Set switch S_1 to position A and S_2 to position B. This will apply power to R_1, R_2, and R_3. Measure I_1, I_2, I_3, voltage V_1 across R_1, voltage V_2 across R_2, and volt-age V_3 across R_3. Record the values in Table 23–1 (p. 163). It is important to note the direction of current (by using + and – signs) as well as its value. Also note the direction of the voltage drops V_1, V_2, and V_3. (The voltage sign should always be opposite that of the current sign. Thus if V_1 is –, I_1 will be +.) Turn V_{PS1} **off.**

3. Set S_1 to position B. Adjust power supply 2 so that V_{PS2} = 10 V. Set S_2 to position A. This will apply power to R_1, R_2, and R_3 from V_{PS2}. Measure I_1, I_2, I_3, voltage V_1 across R_1, voltage V_2 across R_2, and voltage V_3 across R_3. Record the values in Table 23–2 (p. 163). Note carefully the polarity of each measurement, as in step 2.

4. With V_{PS1} = 15 V and V_{PS2} = 10 V, set S_1 to position A (S_2 should already be in position B). Both power supplies are now feeding R_1, R_2, and R_3. Measure I_1, I_2, I_3, V_1, V_2, and V_3 as in the previous two steps. Record the values in Table 23–3 (p. 163), noting carefully the polarity of each value. Turn the power **off.**

5. Using the measured values of R_1, R_2, and R_3 and V_{PS1} = 15 V and V_{PS2} = 10 V, calculate I_1, I_2, and I_3 supplied by two sources using the superposition theorem. Show all calculations and diagrams. Record your calculated values in Table 23–3.

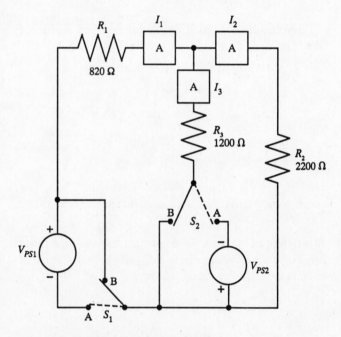

Figure 23–7. Circuit for procedure step 1.

ANSWERS TO SELF-TEST

1. false
2. a straight line
3. less
4. short circuit
5. 6.65
6. 0.75

TABLE 23–1.   Effect of V_{PS1} Alone

Current, mA	Voltage, V
I_1:	V_1:
I_2:	V_2:
I_3:	V_3:

TABLE 23–2.   Effect of V_{PS2} Alone

Current, mA	Voltage, V
I_1:	V_1:
I_2:	V_2:
I_3:	V_3:

TABLE 23–3.   Effect of V_{PS1} and V_{PS2} Acting Together

Measured Values		Calculated Values					
		V_{PS1} Only		V_{PS2} Only		V_{PS1} and V_{PS2} Together	
Current, mA	Voltage, V	Current, mA	Voltage, V	Current, mA	Voltage, V	Current, mA	Voltage, V
I_1	V_1	I_1	V_1	I_1	V_1	I_1	V_1
I_2	V_2	I_2	V_2	I_2	V_2	I_2	V_2
I_3	V_3	I_3	V_3	I_3	V_3	I_3	V_3

QUESTIONS

1. Explain, in your own words, how the superposition theorem is used to find the currents in a circuit supplied by more than one voltage source.

_____

_____

_____

_____

_____

_____

_____

_____

_____

_____

_____

_____

_____

_____

_____

_____

_____

     Superposition Theorem   **163**

2. Refer your data in Tables 23–1, 23–2, 23–3. Do the results of your experiment confirm the superposition theorem? Cite specific data from the tables.

_____

_____

_____

_____

_____

_____

_____

_____

_____

3. Why is it important to include the polarity sign when recording the value of current in step 2?

_____

_____

_____

_____

_____

_____

_____

_____

_____

_____

4. If the polarity of both power supplies in Figure 23–7 were reversed, how would the current in R_2 be affected?

_____

_____

_____

_____

_____

_____

_____

_____

_____

_____

_____

24

THEVENIN'S THEOREM

OBJECTIVES

1. To determine the Thevenin equivalent voltage (V_{TH}) and resistance (R_{TH}) of a dc circuit with a single voltage source
2. To verify experimentally the values of V_{TH} and R_{TH} in solving a series-parallel circuit

BASIC INFORMATION

Thevenin's theorem is another mathematical tool that is very helpful in solving complex linear circuit problems. The theorem makes it possible to determine the voltage or current in any portion of a circuit. The technique employed involves reducing the complex circuit to a simple equivalent circuit.

Thevenin's Theorem

Thevenin's theorem states that any linear two-terminal network may be replaced by a simple equivalent circuit consisting of a Thevenin voltage source V_{TH} in series with an internal resistance R_{TH} causing current to flow through the load. Thus, the Thevenin equivalent for the circuit of Figure 24–1(a) (p. 166) feeding the load R_L is the circuit of Figure 24–1(d). If we knew how to calculate the values V_{TH} and R_{TH}, the process of finding the current I_L in R_L would be a simple application of Ohm's law.

The rules for determining V_{TH} and R_{TH} are as follows:

1. The voltage V_{TH} is the voltage "seen" across the load terminals in the original network, with the load resistance removed (open-circuit voltage); that is, it is the voltage we would measure if we placed a voltmeter across AB in Figure 24–1(a) with the load resistance removed.
2. The resistance R_{TH} is the resistance seen from the terminals of the open load, looking into the original network when the voltage sources in the circuit are short-circuited and replaced by their internal resistance.

The development of the Thevenin equivalent circuit of Figure 24–1(a) is as follows:

1. [Figure 24–1(b)] The load resistance R_L has been removed, and the voltage across AB has been calculated. In this case the voltage drop across R_3 is one-half the source

voltage V, because R_3 is one-half the total resistance in the series circuit consisting of V, R_1 (assumed to be the internal resistance of the voltage source V), R_2, and R_3. The Thevenin equivalent voltage $V_{TH} = 6$ V.

2. [Figure 24–1(c)] The voltage source V has been shorted and only its internal resistance remains in the circuit. The equivalent resistance of the parallel circuit across AB is now calculated.

$$R_{TH} = \frac{(R_1 + R_2) \times R_3}{R_1 + R_2 + R_3} = \frac{(5 + 195) \times 200}{5 + 195 + 200}$$

$$= \frac{40,000}{400} = 100 \ \Omega$$

3. [Figure 24–1(d)] The Thevenin equivalent voltage and resistance are connected in series with the load resistance R_L to form a simple series circuit.

The load current I_L can now be found by using Ohm's law.

$$I_L = \frac{V_{TH}}{R_L + R_{TH}} = \frac{6}{50 + 100} = \frac{6}{150}$$

$$I_L = 0.040 \ A$$

It might seem that the Thevenin method unnecessarily adds work to solving a circuit and that Ohm's and Kirchhoff's laws could solve the problem more quickly and easily. Of course, the sample problem was intentionally made simple to illustrate the method more clearly, but even a simple circuit can prove the worth of the method. Suppose it was necessary to find I_L for a range of 10 values of R_L while the rest of the circuit remained unchanged. It would be quite laborious to apply Ohm's and Kirchhoff's laws 10 times to calculate each value of I_L. With just a single calculation of the Thevenin equivalent circuit, the current I_L for any value of R_L can be calculated quickly through a single application of Ohm's law.

Solving An Unbalanced-Bridge Circuit by Thevenin's Theorem

Figure 24–2(a) (p. 167) is an unbalanced-bridge circuit. It is required to find the current I in R_5. Thevenin's theorem lends itself readily to a solution of this problem.

For this exercise, consider R_5 the load. The problem then is to transform the circuit into its Thevenin equivalent supplying current to R_5, as in Figure 24–2(b).

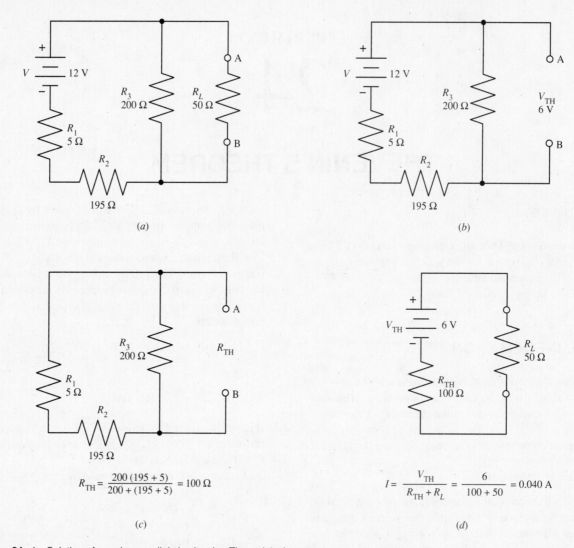

$$R_{TH} = \frac{200\,(195 + 5)}{200 + (195 + 5)} = 100\ \Omega$$

(c)

$$I = \frac{V_{TH}}{R_{TH} + R_L} = \frac{6}{100 + 50} = 0.040\ A$$

(d)

Figure 24–1. Solution of a series-parallel circuit using Thevenin's theorem.

The Thevenin voltage V_{TH} is found by removing R_5 from the circuit and solving for V_{BC} in Figure 24–2(c). The difference in voltage between BD and CD will be V_{BC}. Voltages V_{BD} and V_{CD} can be found directly using resistance ratios.

$$V_{BD} = \frac{R_4}{(R_1 + R_4)} \times V$$

$$= \frac{160}{200} \times 60 = 48\ V$$

$$V_{CD} = \frac{R_3}{(R_2 + R_3)} \times V$$

$$= \frac{120}{180} \times 60 = 40\ V$$

$$V_{BD} - V_{CD} = V_{BC} = 48 - 40 = 8\ V = V_{TH}$$

The Thevenin resistance R_{TH} is found by shorting the voltage source and replacing it with its internal resistance. In this case it is assumed V is an ideal voltage source, and its internal resistance is therefore zero. Thus, AD is, in effect, shorted. The resistance across BC (the Thevenin equivalent resistance) can be visualized more easily if the circuit of Figure 24–2(c) is redrawn with AD shorted. Figure 24–2(d) shows BC more clearly and thus makes it easier to find R_{BC}.

Resistor R_1 in parallel with R_4 becomes

$$\frac{40 \times 160}{40 + 160} = \frac{6400}{200} = 32\ \Omega$$

Resistor R_2 in parallel with R_3 becomes

$$\frac{60 \times 120}{60 + 120} = \frac{7200}{180} = 40\ \Omega$$

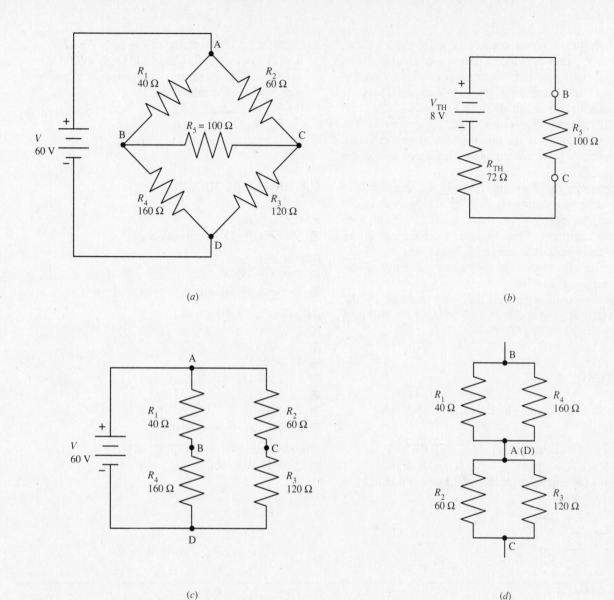

Figure 24–2. Solving an unbalanced bridge circuit using Thevenin's theorem.

Thus,

$$R_{BC} = 32 + 40 = 72 \ \Omega = R_{TH}$$

Substituting these values in the Thevenin equivalent circuit [Figure 24–2(b)] and solving for I, we find

$$I = \frac{V_{TH}}{R_{TH} + R_5} = \frac{8}{172} = 0.0465 \ A$$

Experimental Verification of Thevenin's Theorem

It is possible to determine the values of V_{TH} and R_{TH} for a load R_L in a specific network by measurement. Then we can experimentally set the output of a voltage-regulated power supply to V_{TH} by connecting a resistor whose value is R_{TH} in series with V_{TH} and with R_L. We can measure I in this equivalent circuit. If the measured value of I_L in R_L in the original network is the same as the I measured in the Thevenin equivalent, we have one verification of Thevenin's theorem. For a more complete verification, this process would have to be repeated many times with random circuits.

SUMMARY

1. Thevenin's theorem states that any linear two-terminal network may be replaced by a simple equivalent circuit that acts like the original circuit across the load connected to the two terminals.

2. The equivalent circuit consists of a Thevenin voltage source (V_{TH}) in series with the Thevenin internal resistance (R_{TH}) in series with the two terminals of the load. Thus, for the complex circuit of Figure 24-2(a), the Thevenin equivalent at the terminals BC is Figure 24-2(b).

3. To determine the Thevenin voltage V_{TH}, open the load at the two terminals in the original network and calculate the voltage at these two terminals. This *open-load* voltage is V_{TH}.

4. To determine the Thevenin resistance R_{TH}, keep the load open at the two terminals in the original network and short the original power source and replace it with its own internal resistance. Then calculate the resistance at the open-load terminals, looking back into the original circuit.

5. Thevenin's theorem can be used with a circuit having one or more power sources.

6. Once the complex network has been replaced by the Thevenin equivalent circuit, the current in the load may be found by applying Ohm's law.

SELF-TEST

Check your understanding by answering the following questions:

1. Consider the circuit of Figure 24-1(a). V = 24 V, R_1 = 30 Ω, R_2 = 270 Ω, R_3 = 500 Ω, and R_L = 560 Ω. Assume the internal resistance of the supply is zero. Find each value.
   (a) V_{TH} = _____ V
   (b) R_{TH} = _____ Ω
   (c) I_L = _____ A

2. Consider the circuit of Figure 24-2(a). V = 12 V, R_1 = 200 Ω, R_2 = 500 Ω, R_3 = 300 Ω, R_4 = 600 Ω, and R_5 = 100 Ω. Assume a voltage-regulated power supply. Find each value.
   (a) V_{TH} = _____ V
   (b) R_{TH} = _____ Ω
   (c) I_5 = _____ A (the current in R_5)

MATERIALS REQUIRED

Power Supply:
■ Variable, 0–15 V dc, regulated

Instruments:
■ DMM or VOM
■ 0–5-mA milliammeter

Resistors (½-W, 5%):
■ 1   330-Ω
■ 1   390-Ω
■ 1   470-Ω
■ 1   1000-Ω
■ 1   1200-Ω
■ 2   3300-Ω
■ 1   5-kΩ, 2-W potentiometer

Miscellaneous:
■ 2 SPST switches

PROCEDURE

1. Using an ohmmeter, measure the resistance of each of the 7 resistors supplied. Record the values in Table 24-1 (p. 171).

2. With power **off** and both S_1 and S_2 **open**, connect the circuit of Figure 24-3 with R_L = 330 Ω. Turn the power **on**; **close** S_1. Adjust V_{PS} to 15 V. **Close** S_2 and measure I_L, the current through the load resistor R_L. Record this value in Table 24-2 (p. 171) in the 330-Ω row under "Original Circuit." **Open** S_2. S_1 should remain **closed.**

3. With S_1 **closed** and S_2 **open**, measure the voltage across BC (Figure 24-3). This is voltage V_{TH}; record the value in Table 24-2 in the 330-Ω row under the "V_{TH} Measured" column. **Open** S_1; turn **off** the power supply.

4. Remove the power supply from the circuit by disconnecting it across AD. Short AD by connecting a wire across the two points.

5. With S_2 still open, connect an ohmmeter across BC to measure the resistance across points B and C. This is R_{TH}. Record the value in Table 24-2 in the 330-Ω row under "R_{TH} Measured."

6. Adjust the power supply so that V_{PS} = V_{TH}. Connect the ohmmeter across the potentiometer, and adjust the resistance so that the resistance across the potentiometer is R_{TH}.

7. Disconnect the 330-Ω load resistor, S_2, and the milliammeter from the circuit of Figure 24-3 and connect them as shown in Figure 24-4. With S_2 **open,** and power **on,** check to see that V_{PS} = V_{TH}.

8. **Close** S_2. Measure I_L and record the value in Table 24-2 in the 330-Ω row under "Thevenin Equivalent Circuit, Measured." **Open** S_2; turn the power **off.**

9. Using the measured values of V_{PS}, R_1, R_2, R_3, and R_4 (Table 24-1), calculate V_{TH} for the circuit of Figure 24-3. Record your answer in Table 24-2 in the 330-Ω row under "V_{TH}, Calculated."

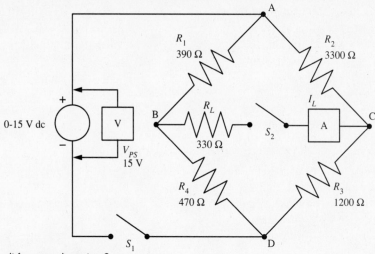

Figure 24-3. Circuit for procedure step 2.

10. Calculate R_{TH} in Figure 24-3 using the measured values for R_1, R_2, R_3, and R_4. (Voltage-regulated power supplies normally have negligible internal resistance.) Record your answer in Table 24-2 in the 330-Ω row under "R_{TH}, Calculated."

11. Using the calculated values of V_{TH} and R_{TH} from steps 9 and 10 as recorded in Table 24-2, calculate I_L. Record your answer in Table 24-2 in the 330-Ω row under "I_L, Calculated."

12. Substitute a 1000-Ω resistor for R_L in the circuit of Figure 24-3. Turn the power **on** and adjust V_{PS} to 15 V; **close** S_1 and S_2. Measure I_L and record the value in Table 24-2 in the 1000-Ω row under "I_L, Measured, Original Circuit." **Open** S_2.

13. Remove the 1000-Ω load resistor R_L and connect the 3300-Ω load resistor. Adjust V_{PS} to 15 V, if necessary. **Close** S_2. Measure I_L and record the value in the 3300-Ω row under "I_L Measured, Original Circuit." **Open** S_1 and S_2; turn the power **off**.

14. Connect the Thevenin equivalent circuit as shown in Figure 24-4 using the 1000-Ω load resistor in place of the 330-Ω resistor. V_{TH} and R_{TH} should be the measured values recorded in Table 24-2 in the 330-Ω row.

15. Turn the power **on**; adjust V_{PS} to V_{TH}. **Close** S_2 and measure I_L. Record the value in Table 24-2 in the 1000-Ω row under "I_L, Measured, Thevenin Equivalent Circuit." **Open** S_2.

16. Remove the 1000-Ω resistor and connect the 3300-Ω load resistor. **Close** S_2 and measure I_L. Record the value in Table 24-2 in the "3300-Ω" row, under "I_L, Measured, Thevenin Equivalent Circuit." **Open** S_2; power **off**.

17. Calculate I_L for $R_L = 1000$-Ω and $R_L = 3300$ Ω for the circuit of Figure 24-3, using the measured values for R_1, R_2, R_3, R_4, and R_L. Record your answers in Table 24-2.

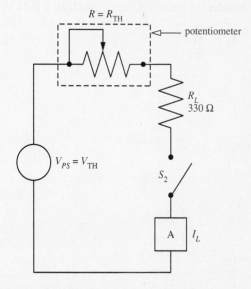

Figure 24-4. Thevenin's equivalent circuit for procedure step 7. The values of V_{TH} and R_{TH} are determined in steps 3 and 5.

ANSWERS TO SELF-TEST

1. (a) 15; (b) 188; (c) 0.02
2. (a) 4.5; (b) 338; (c) 0.01

Experiment 24 Name ______________________ Date ____________

TABLE 24–1. Measured Resistor Values

Resistor	*Rated Value, Ω*	*Measured Value, Ω*
R_1	390	
R_2	3300	
R_3	1200	
R_4	470	
R_L	330	
R_L	1000	
R_L	3300	

TABLE 24–2. Measurements to Verify Thevenin's Theorem

	V_{TH}, V		R_{TH}, Ω		I_L, mA		
					Measured		
R_L, Ω	*Measured*	*Calculated*	*Measured*	*Calculated*	*Original Circuit*	*Thevenin Equivalent Circuit*	*Calculated*
330							
1000	X	X	X	X			
3300	X	X	X	X			

QUESTIONS

1. Explain, in your own words, how Thevenin's theorem is used to convert any linear two-terminal network into a simple equivalent circuit consisting of a resistance in series with a voltage source.

2. Refer to your data in Table 24–2. How do the values of I_L measured in the original circuit (Figure 24–3) compare with those measured in the Thevenin equivalent circuit (Figure 24–4)? Should the comparable measurements be the same? Explain why.

_____

_____

_____

_____

_____

_____

_____

_____

3. Refer to Table 24–2. Compare calculated and measured values of R_{TH}. Are the results as you would have expected? Explain. Make the same comparison with the two values of V_{TH}.

_____

_____

_____

_____

_____

_____

_____

_____

4. Explain an advantage of using Thevenin's theorem when finding load currents in a dc circuit.

_____

_____

_____

_____

_____

_____

_____

25

NORTON'S THEOREM

OBJECTIVES

1. To determine the values of Norton's constant-current source I_N and Norton's current-source resistance R_N in a dc circuit containing one or two voltage sources
2. To verify experimentally the values I_N and R_N in the solution of complex dc networks containing two voltage sources

BASIC INFORMATION

Norton's Theorem

Thevenin's theorem simplifies the analysis of complex networks by reducing the original circuit to a simple equivalent circuit containing a constant-voltage source V_{TH} in series with an internal resistance R_{TH}. Norton's theorem utilizes a similar technique of simplification. The Norton source, however, delivers a constant current.

Norton's theorem states that any two-terminal linear network may be replaced by a simple equivalent circuit consisting of a constant-current source I_N in parallel with an internal resistance R_N. Figure 25–1(a) shows an actual network terminated by a load resistance R_L. Figure 25–1(b) shows the Norton equivalent circuit. The Norton current I_N is distributed between resistance R_N and the load R_L.

With reference to Figure 25–1(a), the rules for determining the constants in the Norton equivalent circuit are as follows:

1. The constant current I_N is the current that would flow in AB if the load resistance between A and B were replaced by a short circuit.
2. The Norton resistance R_N is the resistance seen from terminals AB with the load removed and the voltage sources shorted and replaced with their internal resistance. Thus, R_N is defined in exactly the same way as the Thevenin resistance R_{TH}.

Applications

Consider the circuit of Figure 25–2(a) (p. 174). We wish to find the current I_L in R_L using Norton's theorem. (Of course, this circuit can also be solved using Ohm's and Kirchhoff's laws as well as mesh and Thevenin's methods. Such solutions are recommended as student exercises.)

The development of the Norton equivalent circuit for Figure 25–2(a) may be followed from Figure 25–2(b), (c), and (d).

1. [Figure 25–2(b)] Load resistor R_L is short-circuited, thus shorting R_3. The current produced by V is I_N.

$$I_N = \frac{V}{R_1 + R_2} = \frac{20}{5 + 195} = \frac{20}{200}$$

$$I_N = 0.1 \text{ A}$$

2. [Figure 25–2(c)] The voltage source V is shorted and replaced by its internal resistance. With R_L removed, the resistance across AB is calculated. The resistance is R_N.

$$R_N = \frac{(R_1 + R_2) \times (R_3)}{R_1 + R_2 + R_3} = \frac{(5 + 195) \times (200)}{5 + 195 + 200}$$

$$R_N = \frac{40,000}{400} = 100 \ \Omega$$

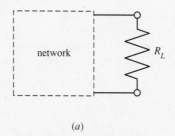

(a)

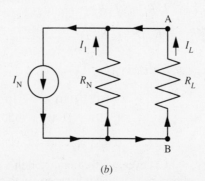

(b)

Figure 25–1. Norton's equivalent circuit consists of a constant current source, I_N, and a shunt resistance, R_N.

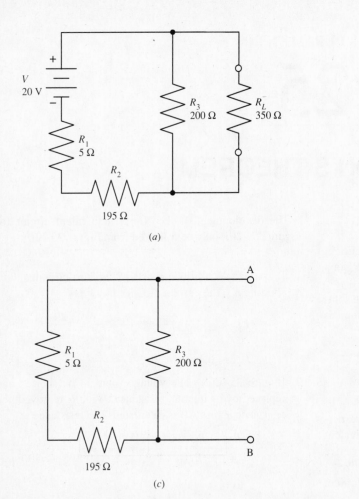

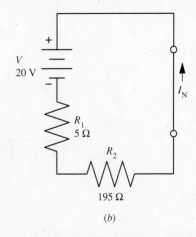

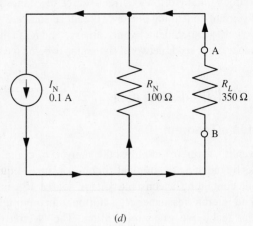

Figure 25–2. Applying Norton's theorem to the solution of a dc network.

3. [Figure 25–2(d)] The original circuit is replaced by the Norton constant-current source $I_N = 0.1$ A in parallel with the Norton resistance $R_N = 100$ Ω. The load resistance R_L is connected across the Norton equivalent circuit.

From Ohm's law, the value of I_L can now be calculated.

$$V_{RN} = V_L = I_L R_L = I_{RN} R_N$$

$$I_{RN} = I_N - I_L$$

$$I_L R_L = (I_N - I_L) R_N$$

$$I_L R_L = I_N R_N - I_N R_N$$

$$I_L (R_L + R_N) = I_N R_N$$

$$I_L = \frac{I_N R_N}{R_L + R_N}$$

$$= \frac{(0.1)(100)}{(350) + (100)} = \frac{10}{450}$$

$$I_L = 0.022 \text{ A}$$

As in the case of Thevenin's theorem, Norton's theorem is useful in applications in which it is necessary to calculate the load current as the load resistance varies over a wide range of values.

Solution of a DC Network With Two Voltage Sources

The solution of complex dc networks containing two or more voltage sources is possible using any of the methods discussed in this and previous experiments.

Norton's theorem is used to solve the following problem.

Problem. Develop a formula for finding the load current in the circuit of Figure 25–3(a) over a range of different load resistors. Using this formula find I_L for $R_L = 100$ Ω, 500 Ω, 1000 Ω. Assume that V_1 and V_2 are constant-voltage sources.

Solution. The first step is to find the Norton constant-current source I_N. This is done by replacing R_L with a short across FG and solving for the current through FG. Using the mesh currents I_1 and I_N [Figure 25–3(b)], we have

$$I_1(R_1 + R_2) - I_N R_2 = -V_1 - V_2$$

$$-I_1 R_2 + I_N R_2 = V_2$$

$$320 I_1 - 220 I_N = -30$$

$$-220 I_1 + 220 I_N = 20$$

Solving for I_N, we obtain

$$I_N = 0.009 \text{ A}$$

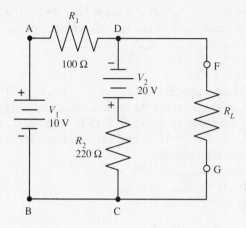

(a)

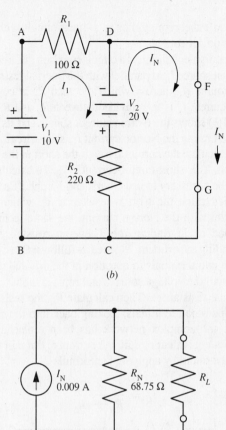

(b)

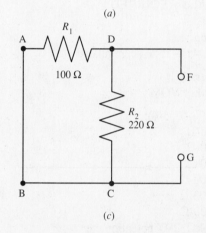

(c)

(d)

Figure 25–3. Applying Norton's theorem to a circuit with two voltage sources.

While I_N is negative, we are concerned with its value, not its direction.

The Norton resistance R_N is the resistance measured across FG in Figure 25–3(c). This is the same as the Thevenin resistance R_{TH}. It is found by shorting all voltage sources and replacing them with their internal resistance. In this problem we assume that V_1 and V_2 are ideal voltage sources (that is, constant voltage sources), so that their internal resistance is zero. The resistance across FG in this case is R_1 in parallel with R_2

$$R_N = \frac{100 \times 220}{100 + 220} = \frac{22,000}{320}$$

$$R_N = 68.75 \ \Omega$$

To find I_L we can use the formula from the previous example:

$$I_L = \frac{I_N R_N}{R_N + R_L}$$

$$I_L = \frac{0.009 \times 68.75}{68.75 + R_L} = \frac{0.619}{68.75 + R_L}$$

We can now calculate the values of I_L for each of the values of R_L.

For $R_L = 100 \ \Omega$,

$$I_L = \frac{0.619}{68.75 + 100} = \frac{0.619}{168.75} = 0.004 \ \text{A}$$

For $R_L = 500 \ \Omega$,

$$I_L = \frac{0.619}{68.75 + 500} = \frac{0.619}{568.75} = 0.001 \ \text{A}$$

For $R_L = 1000 \ \Omega$,

$$I_L = \frac{0.619}{68.75 + 1000} = \frac{0.619}{1068.75} = 0.0006 \ \text{A}$$

SUMMARY

1. Norton's theorem offers another method of solving complex linear circuits. Norton's theorem allows a complex two-terminal network to be replaced by a simple equivalent circuit that acts like the original circuit at the load connected to the two terminals.

2. Norton's theorem applies to a linear circuit with one or more power sources.
3. The equivalent circuit is a circuit consisting of a constant-current source I_N in parallel with the internal resistance of the source R_N, across which the load R_L is connected. The current I_N is then divided between R_N and R_L. Figure 25-1(b) shows the Norton current source and its load.
4. To determine the Norton current I_N, short-circuit the load and calculate the current through the short in the original circuit. This short-circuit current is I_N. To calculate I_N, it may be necessary to use Ohm's and Kirchhoff's laws.
5. To determine the Norton resistance R_N, which acts in parallel with the Norton current, the same technique is applied as in finding the Thevenin resistance in the preceding experiment. Proceed as follows: Open the load at the two terminals in question in the original network. Short all the voltage sources and replace them with their internal resistances. Then calculate R_N, the resistance at the open-load terminals, looking back into the circuit.
6. Once the complex network has been replaced by the Norton equivalent circuit, the current I_L through the load may be found by applying the formula

$$I_L = \frac{I_N \times R_N}{R_N + R_L}$$

SELF-TEST

Check your understanding by answering the following questions:
1. Consider the circuit of Figure 25-2(a). $V = 12$ V, $R_1 = 1\ \Omega$, $R_2 = 39\ \Omega$, $R_3 = 60\ \Omega$, and $R_L = 27\ \Omega$. Assume the internal resistance of the voltage source V is zero. Find each value in the equivalent Norton circuit.
   (a) $I_N =$ _____ A
   (b) $R_N =$ _____ Ω
   (c) $I_L =$ _____ A
2. Consider the circuit of Figure 25-3(a). $V_1 = 30$ V, $V_2 = 30$ V. Assume the internal resistance of these voltage sources is zero. $R_1 = 45\ \Omega$, $R_2 = 150\ \Omega$, and R_L (the load) $= 47\ \Omega$. Find each value in the Norton equivalent circuit.
   (a) $I_N =$ _____ A
   (b) $R_N =$ _____ Ω
   (c) $I_L =$ _____ A

MATERIALS REQUIRED

Power Supply:
■ 2 variable 0–15 V dc, regulated

Instruments:
■ 2 DMM or VOM

Resistors (½-W, 5%):
■ 1   390-Ω
■ 1   560-Ω
■ 1   680-Ω
■ 1   1200-Ω
■ 1   1800-Ω
■ 1   2700-Ω
■ 1   10-kΩ, 2-W potentiometer

Miscellaneous:
■ 2 SPST switches
■ 3 SPDT switches

PROCEDURE

A. Determining I_N and R_N

A1. With power **off** in both supplies; S_4 and S_5 **open**; and switches S_1, S_2, and S_3 in position Ⓐ, connect the circuit of Figure 25-4.

A2. Power **on** V_{PS1} and V_{PS2}. Adjust the supply voltages so that $V_{PS1} = 12$ V and $V_{PS2} = 6$ V. (Note carefully the correct polarity of the connections.) Maintain these voltages throughout the experiment. **Close** S_4 and S_5. Measure I_L through R_L and record the results in Table 25-1 (p. 179) in the 1200-Ω row under "I_L, Measured, Original Circuit."

A3. Replace R_L in turn with a 390-Ω, 560-Ω, and 1800-Ω resistor. In each case, measure I_L and record the values in the "I_L, Measured, Original Circuit" column.

A4. Move S_3 to position Ⓑ. This, in effect, replaces R_L with a short. The current measured by the meter is the short-circuit current of the Norton equivalent generator I_N.

Record the value in Table 25-1 in the 1200-Ω row under "I_N, Measured."

A5. Turn the power **off**. Move S_1, S_2, and S_3 to position Ⓑ and open S_5. This, in effect, replaces the voltage sources with short circuits and opens the load circuit between D and E. (Regulated power supplies are considered to have negligible resistance.) S_4 is still **closed**.

A6. With an ohmmeter, measure the resistance across CF. This is the resistance across the Norton equivalent generator R_N. Record this value in Table 25-1 in the 1200-Ω row under "R_N, Measured."

A7. From the circuit of Figure 25-4, calculate the value of the Norton current I_N and record it in Table 25-1 in the 1200-Ω row under "I_N, Calculated."

A8. From the circuit of Figure 25-4, calculate the value of the Norton shunt resistance R_N and record it in Table 25-1 in the 1200-Ω row under "R_N, Calculated."

A9. Using the values of I_N and R_N from steps A7 and A8, calculate the load current I_L for the 1200-Ω, 390-Ω, 560-Ω, and 1800-Ω load resistors in Figure 25–4. Record these values in Table 25–1 under "I_L, Calculated."

B. Using the Norton Equivalent Circuit

B1. With power **off** and S_1 **open,** connect the circuit of Figure 25–5. Meter A1 will measure the Norton current I_N, whereas meter A2 will measure the load current I_L. The potentiometer will serve as R_N. Use an ohmmeter while adjusting the potentiometer until its resistance measures that of R_N found in step A6.

B2. Adjust the power supply to its lowest output voltage. Turn power **on** and **close** S_1. Slowly increase the output of the power supply until the *current* measured by ammeter A1 is equal to the value of I_N you found in step A4 and recorded in Table 25–1.

B3. With meter A1 measuring I_N, record the load current I_L measured by meter A2 in Table 25–1 in the 1200-Ω row under "I_L, Measured, Norton Equivalent Circuit." S_1 **open;** power **off.**

B4. For each of the other load resistors in Table 25–1, connect the Norton equivalent circuit (Figure 25–5) and measure I_L for each value of R_L. Record your values in Table 25–1 under "I_L, Measured, Norton Equivalent Circuit." **Open** S_1; power **off.**

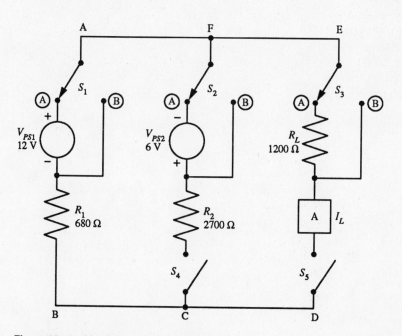

Figure 25–4. Circuit for procedure step A1.

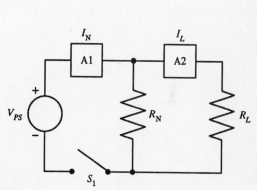

Figure 25–5. Circuit for procedure step B1.

ANSWERS TO SELF-TEST

1. (a) 0.3; (b) 24; (c) 0.14
2. (a) 0.47; (b) 34.6; (c) 0.199

TABLE 25–1.　Measurements to Verify Norton's Theorem

R_L, Ω	I_N, mA Measured	I_N, mA Calculated	R_N, Ω Measured	R_N, Ω Calculated	I_L, mA Measured Original Circuit	I_L, mA Measured Norton Equivalent Circuit	I_L, mA Calculated
1200							
390							
560							
1800							

QUESTIONS

1. Explain, in your own words, how Norton's theorem is used to convert any two-terminal linear network into a simple circuit consisting of a constant current source in parallel with a resistance.

_____

_____

_____

_____

_____

_____

_____

_____

_____

_____

2. Refer to your data in Table 25–1. How do the values of I_L measured in the original circuit (Figure 25–4) compare with those measured in the Norton equivalent circuit (Figure 25–5)? Should the comparable measurements be the same? Explain why.

_____

_____

_____

_____

_____

_____

_____

_____

　　　　　　Norton's Theorem　　**179**

3. Refer to Table 25–1. Compare measured and calculated values for I_N. Are the results as you would have expected? Explain. Make the same comparison with the two values of R_N.

_____

_____

_____

_____

_____

_____

_____

_____

_____

_____

_____

_____

_____

_____

_____

_____

4. Explain an advantage of using Norton's theorem when finding load currents in a dc circuit.

_____

_____

_____

_____

_____

_____

_____

_____

_____

_____

_____

_____

_____

_____

26

MILLMAN'S THEOREM

OBJECTIVES

To verify Millman's theorem experimentally

BASIC INFORMATION

It is often necessary to find the voltage across two points to which a number of parallel branches and sources are connected. In previous experiments various techniques involving Kirchhoff's laws and superposition were used. Millman's theorem provides another method for solving such circuits. In many cases it provides a more direct and quicker method than either Kirchhoff's laws or superposition.

Millman's Theorem

If a circuit can be redrawn or viewed as having two common lines (for example, a "hot" line and a ground), Millman's theorem can be used to find the voltage between the two lines. If the circuit contains only one voltage source, the usual methods used to solve parallel circuits are best used. However, if a number of the parallel branches contain voltage sources, the usual methods become cumbersome and time consuming. Millman's theorem provides a shortcut method in those cases.

Figure 26–1 is a circuit containing two sources. If the voltage from X to ground must be found, any of the techniques covered in earlier experiments can be used. However, by redrawing the circuit as in Figure 26–2, Millman's theorem can be readily applied.

Millman's theorem is in the form of a formula:

$$V_{XG} = \frac{\dfrac{V_1}{R_1} + \dfrac{V_2}{R_2} + \dfrac{V_3}{R_3}}{\dfrac{1}{R_1} + \dfrac{1}{R_2} + \dfrac{1}{R_3}} \qquad (26\text{--}1)$$

where

V_{XG} is the voltage across the common lines
V_1 is the total voltage sources in the first branch
R_1 is the total resistance in the first branch
V_2 is the total voltage sources in the second branch
R_2 is the total resistance in the second branch
V_3 is the total voltage sources in the third branch
R_3 is the total resistance in the third branch

If any branch does not contain a voltage source, the voltage is equal to zero. Formula (26–1) can be extended to any number of branches by simply adding the V_n/R_n term to the numerator and the $1/R_n$ term to the denominator.

To illustrate the application of Millman's theorem, the circuit of Figure 26–2 will be solved using formula (26–1).

$$V_{XG} = \frac{\dfrac{10}{10} + \dfrac{0}{20} - \dfrac{10}{20}}{\dfrac{1}{10} + \dfrac{1}{20} + \dfrac{1}{20}}$$

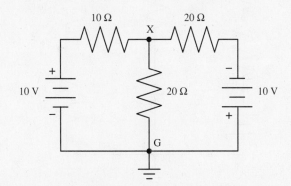

Figure 26–1. Millman's theorem can be used to solve circuits with more than one voltage source.

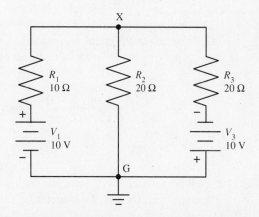

Figure 26–2. The circuit of Figure 26–1 redrawn to show its applicability to Millman's theorem.

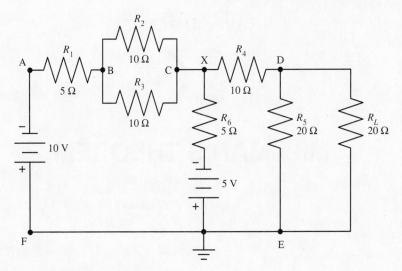

Figure 26–3. A series-parallel circuit with two voltage sources.

The polarity of V_3 is negative, since it would make point X negative with respect to ground. (Of course, if ground was considered positive, V_3 would be positive, and V_1 would be written as negative.)

Simplifying, we have

$$V_{XG} = \frac{1 + 0 - 0.5}{0.1 + 0.05 + 0.05} = \frac{0.5}{0.2}$$

$$V_{XG} = 2.5 \text{ V}$$

To use the Millman formula, the branches must all be in parallel. Thus, a series-parallel circuit cannot be solved directly by the Millman formula. Sometimes a series-parallel circuit lends itself to simplification, so that a two-step calculation can be used. A second example illustrates this process.

Figure 26–3 is a series-parallel circuit with two voltage sources. The current through the load R_L will be found using the Millman formula. Because the circuit as shown cannot be represented by a pure parallel circuit, it is necessary to combine some components. By combining R_2 and R_3 into a single resistance and R_5 and R_L into a single resistance, the circuit can be solved by Millman's theorem.

Since R_2 and R_3 are equal, their equivalent resistance $R_{2,3}$ is 10/2, or 5 Ω. Similarly, combining R_5 and R_L results in 20/2, or 10 Ω. Now the parallel circuit is complete, as shown in Figure 26–4. Applying the Millman formula, we obtain

$$V_{XG} = \frac{\dfrac{10}{10} + \dfrac{5}{5} + \dfrac{0}{20}}{\dfrac{1}{10} + \dfrac{1}{5} + \dfrac{1}{20}} = \frac{1 + 1 + 0}{0.1 + 0.2 + 0.05} = \frac{2}{0.35}$$

$$V_{XG} = 5.71 \text{ V}$$

This is the voltage across XE in the original circuit of Figure 26–3. The current in this part of the circuit is

$$I = \frac{5.71}{20} = 0.286 \text{ A}$$

At point D the current divides in half, one-half to R_5, the other half through R_L:

$$I_{RL} = \frac{0.286}{2} = 0.143 \text{ A}$$

SUMMARY

1. The Millman theorem formula is

$$V_{AB} = \frac{\dfrac{V_1}{R_1} + \dfrac{V_2}{R_2} + \dfrac{V_3}{R_3} + \cdots + \dfrac{V_n}{R_n}}{\dfrac{1}{R_1} + \dfrac{1}{R_2} + \dfrac{1}{R_3} + \cdots + \dfrac{1}{R_n}}$$

where the Vs are the total voltage sources in each branch and the Rs are the total resistance in each branch.

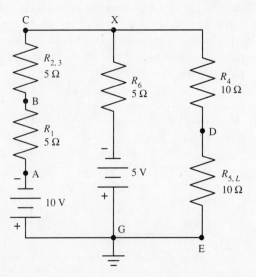

Figure 26–4. Figure 26–3 redrawn to show its three parallel branches

2. Millman's formula can be used only in pure parallel circuits.

3. Millman's formula is most useful in parallel circuits in which more than one branch has a voltage source.

4. If polarity is assigned to the two parallel lines of the circuit, any voltage sources connected to the lines with reverse polarity must be represented in the Millman formula with a minus sign.

5. Millman's formula can be used as a shortcut in solving circuit problems requiring more than one step.

SELF-TEST

Check your understanding by answering the following questions.

1. The Millman theorem can be used to solve circuit problems that can also be solved using _____ laws and _____.

2. Millman's theorem can be used to solve only pure _____ circuit problems.

3. The answer to Millman's formula is given in _____ units.

4. (Refer to Figure 26–2.) If $R_1 = R_2 = R_3 = 10 \ \Omega$, $V_1 = 5$ V, and $V_3 = 10$ V, find the current in R_2 using Millman's formula. $I_2 =$ _____ A. The direction of current is _____ (downward from X to G/upward from G to X).

MATERIALS REQUIRED

Power Supplies:
- 2 variable 0–15 V dc, regulated

Instruments:
- DMM or EVM
- VOM
- 0–100-mA dc ammeter (The VOM or DMM can be substituted for the ammeter if they have the necessary dc current range.)

Resistors ($\frac{1}{2}$-W, 5%):
- 4 68-Ω
- 1 100-Ω

Miscellaneous:
- 2 SPST switches

PROCEDURE

1. Using an ohmmeter, measure the resistance of the five resistors in this experiment. Label the resistors R_1 through R_4 and R_L, and connect them in the circuits that follow, according to their labels. Record the measured resistance values in Table 26–1 (p. 185).

2. With power supply V_1 **off** and S_1 **open**, connect the circuit of Figure 26–5. Use the resistors labeled R_1 and R_2. This is circuit 1.

3. Turn **on** the power supply and **close** S_1. Increase the output voltage of V_1 to 15 V. Measure the current in the circuit and record the value in Table 26–2 (p. 185). **Open** S_1 and turn **off** the power supply.

4. With power supply V_2 **off** and S_2 **open,** connect the circuit of Figure 26–6. Use resistor labeled R_3 in this circuit. This is circuit 2.

5. Turn **on** the power supply and **close** S_2. Increase the output voltage of V_2 to 10 V. Measure the current in the circuit and record the value in Table 26–2. **Open** S_2 and turn **off** the power supply.

6. Connect the circuit of Figure 26–7 (p. 184). Use the resistors labeled R_4 and R_L. Record the reading of the ammeter in Table 26–2. This is circuit 3.

7. Reconnect the circuits of Figures 26–5, 6, and 7 without the switches and power supplies, as shown in Figure 26–8 (p. 184). In reconnecting the circuits be sure to observe the locations of each of the labeled resistors. This is circuit 4.

8. Using an ohmmeter, measure the resistance of the circuit across CG. Record this value in Table 26–2.

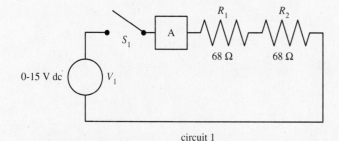

circuit 1

Figure 26–5. Circuit 1 for procedure step 2.

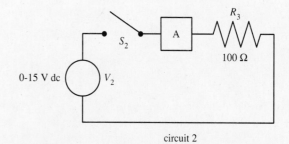

circuit 2

Figure 26–6. Circuit 2 for procedure step 4.

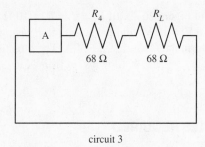

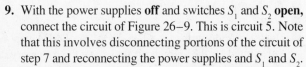

circuit 3

Figure 26–7. Circuit 3 for procedure step 6.

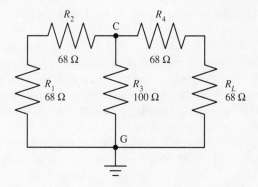

Figure 26–8. Circuit 4 for procedure step 7.

9. With the power supplies **off** and switches S_1 and S_2 **open,** connect the circuit of Figure 26–9. This is circuit 5. Note that this involves disconnecting portions of the circuit of step 7 and reconnecting the power supplies and S_1 and S_2.

10. Turn **on** V_1 and adjust its output voltage to 15 V.

11. Turn **on** V_2 and adjust its output voltage to 10 V.

12. **Close** S_2, then **close** S_1. If necessary, readjust V_1 to 15 V and V_2 to 10 V.

13. Measure the voltage across CG and record the value in Table 26–2. After you have completed the necessary measurement, **open** S_1 and S_2 and turn **off** the power supplies.

14. Using the total measured current of circuits 1, 2, and 3 and the total resistance of circuit 4, calculate the voltage across CG.

$$V_{CG} = \text{total measured current} \times \text{total resistance}$$

Record your answer in Table 26–2.

15. Using the rated value of the resistors, calculate the voltage across CG using Millman's theorem formula. Record your answer in Table 26–2.

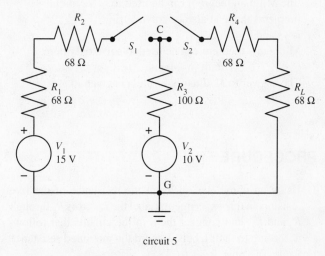

circuit 5

Figure 26–9. Circuit 5 for procedure step 9.

ANSWERS TO SELF-TEST

1. Kirchhoff's; superposition
2. parallel
3. voltage
4. −0.167; downward from X to G

TABLE 26-1.   Resistance Measurements

Resistor	R_1	R_2	R_3	R_4	R_L
Rated value, Ω	68	68	100	68	68
Measured value, Ω					

TABLE 26-2.   Verification of Millman's Theorem

Circuit	Voltage, V	Measured Current, A	Measured Total Resistance, Ω	Measured Voltage across CG, V	Calculated V_{CG} using Measured Current and Resistance	Calculated V_{CG} using Millman's Theorem and Rated Values
1	15					
2	10					
3	0					
4	0					
5						

QUESTIONS

1. Explain, in your own words, how Millman's theorem is used to solve a dc circuit. Note any limitations and restrictions on its application.

_____

_____

_____

_____

_____

_____

_____

_____

_____

_____

_____

_____

_____

_____

_____

_____

_____

_____

_____

_____

_____

2. Compare the three values of V_{CG} in Table 26–2 for circuit 5 (Figure 26–9). Should they be equal? Explain why.

_____

_____

_____

_____

_____

_____

_____

_____

_____

_____

_____

_____

_____

_____

_____

_____

_____

3. Explain an advantage of using Millman's theorem to solve a dc circuit.

_____

_____

_____

_____

_____

_____

_____

_____

_____

_____

_____

_____

_____

27

MAGNETIC FIELD ASSOCIATED WITH CURRENT IN A WIRE

OBJECTIVES

1. To verify experimentally that a magnetic field exists around a wire carrying current
2. To find experimentally the direction of the magnetic field around a current-carrying wire
3. To determine experimentally the pattern of magnetic lines of force about a coil

BASIC INFORMATION

Current Producing a Magnetic Field

The physicist Hans Christian Oersted observed that a compass needle was deflected when placed in the vicinity of a wire carrying an electric current. Moreover, he observed that the direction in which the compass needle pointed depended on its position relative to the wire and on the direction of the current.

These observations are depicted in Figure 27–1(a) (p. 188). A current-carrying wire is shown coming out of the plane of the page. Its cross section is represented by a circle W. When a compass is placed in position 1, its needle points in the direction shown. As the compass is moved from position 1 to 2, 3, and 4, the needle also changes direction, as shown.

If the wire is not carrying current, as in Figure 27 - 1(b), the compass will point to north whether it is in position 1, 2, 3, or 4.

If the direction of current in the wire is reversed from that in Figure 27–1(a), then the compass will point in a direction opposite to the one in Figure 27–1(a), as shown in Figure 27–1(c).

These observations led to the following conclusions:

1. A magnetic field exists around a wire carrying current.
2. The direction of the magnetic field depends on the direction of current in the wire.
3. The magnetic field appears to be circular around the wire.

Later research confirmed that the magnetic field lies in a plane perpendicular to the current-carrying wire, that the magnetic field is circular around the wire, and that the

magnetic field is strongest close to the wire. Research also showed that a magnetic field surrounds any moving electric charge whether it is being carried by a wire or not. For example, the electron beam in a cathode-ray tube produces the same type of circular magnetic field as a current-carrying wire.

Direction of Magnetic Field

Observations such as those illustrated in Figures 27–1(a) and (c) lead to a rule for predicting the direction of the magnetic field around a current-carrying wire. The rule, which was developed when current was considered to flow from + to – (conventional current flow), is referred to in many textbooks as the *right-hand rule*. Because this text is based on current flow from – to + (electron-flow current), the rule is modified to use the left hand. The *left-hand rule* states:

If the current-carrying wire is grasped by the fingers of the left hand with the thumb extended and pointing in the direction of electron-flow current, the fingers encircle the wire in the direction of the magnetic lines of force (Figure 27–2) (p. 188).

The right-hand rule is identical except that the right hand is used and the thumb points in the direction of conventional current flow. No matter which rule is used, the direction of the magnetic lines of force will be the same.

Magnetic Field Produced by a Coil

When two magnetic fields are brought near each other, they interact and the lines of force are distorted. Further, the lines of force of two magnetic fields *aid each other* when their lines of force are in the *same direction,* and *oppose (cancel)* each other when their lines of force are in *opposite directions*.

If a current-carrying wire is wrapped in the form of a coil, as in Figure 27–3 (p. 188), the principle of magnetic fields aiding and opposing will apply. The circular lines of force around the conductors will combine, forming the pattern of lines of force around the coil shown in Figure 27–3. The lines of force entering and leaving the coil in effect form a south and a north pole. Again, a left-hand rule can be used to

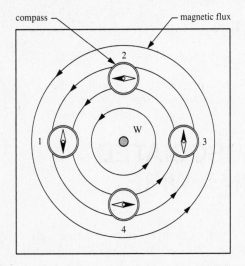

(a)

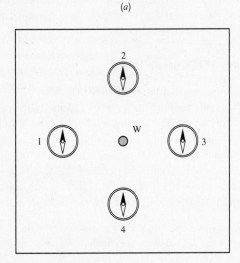

(b)

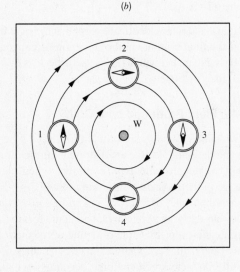

(c)

Figure 27–1. The magnetic field surrounding a current-carrying conductor.

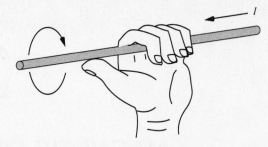

Figure 27–2. Left-hand rule used to determine the direction of magnetic lines of force around a current-carrying conductor.

determine the direction of the lines of force and therefore the south- and north-pole ends of the coil. The left-hand rule for coils is as follows:

With the thumb extended, grasp the coil with the fingers of the left hand, in the direction of electron-flow current. The extended thumb will point in the direction of the magnetic lines of force. Therefore, the end of the coil to which the thumb is pointing may be considered a north pole and the other end of the coil, the south pole (Figure 27–4).

The wires forming the coil may be wrapped around any type of material, but generally the form around which the wire is wrapped is an insulating, nonmagnetic circular, square, or rectangular tube. The tube itself may be hollow, in which case the coil is said to have an *air core*. In many applications a soft-iron bar is placed in the tube. Because the soft iron is a magnetic material, it provides a good path for the lines of force in the core of the coil, thus increasing flux density and forming a strong magnet. However, the soft iron will develop its magnetic strength only when the coil is energized. Soft iron does not retain its magnetic strength when it is not in a magnetic field, as is the case when the coil is deenergized. The arrangement of coil and soft-iron core is therefore used to provide a temporary magnet, or an *electromagnet*.

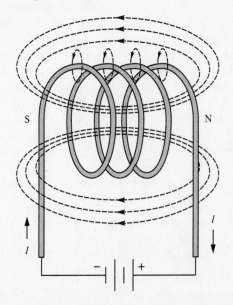

Figure 27–3. Magnetic field around a current-carrying coil.

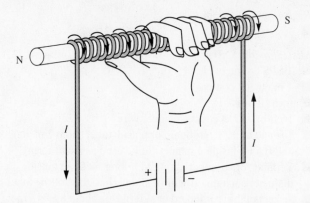

Figure 27–4. Left-hand rule for determining the poles formed by a current-carrying coil.

The strength of the magnet formed by a coil depends on the number of turns of wire in the coil and the amount of current in the coil. The strength of the magnet is given in terms of magnetomotive force, mmf, which is analogous to voltage in an electric circuit. The formula for magnetomotive force (mmf) is

$$mmf = IN \qquad (27-1)$$

where mmf is magnetomotive force in ampere-turns

    I is the current in the coil in amperes

    N is the number of turns in the coil

In the magnetic circuit, magnetomotive force is dependent upon the lines of force produced by the coil ø and the reluctance of the material through which the lines of force must pass $\mathcal{R}$. Thus, mmf can be expressed with the following formula:

$$mmf = \text{ø}\,\mathcal{R} \qquad (27-2)$$

If we think of lines of force ø as analogous to current, and reluctance $\mathcal{R}$ analogous to resistance, formula (27–2) can be considered the "Ohm's law" of magnetic circuits.

Coils with soft-iron cores are widely used in industry for such devices as relays, motor-control contactors, alarms, and similar devices. Coils with movable soft-iron cores are used as operating devices to open and close doors, move material, actuate other devices, and the like. Iron-core coils are frequently referred to as *solenoids*.

SUMMARY

1. A magnetic field is developed by a moving electric charge.
2. Circular magnetic lines of force appear around a wire carrying current. The magnetic field is at right angles to the wire and surrounds it. The magnetic field extends the entire length of the wire.
3. The direction of the lines of force depends on the direction of current. If the wire is grasped in the left hand with the thumb pointing in the direction of electron-flow current,

the fingers point in the direction of the circular magnetic lines of force (Figure 27–2).
4. If the wire is wound in the form of a coil and current is flowing in the coil, the coil will produce a magnetic field (Figure 27–3).
5. The polarity of the poles formed by a coil may be determined by grasping the coil with the left hand, with the fingers pointing in the direction of current in the windings. The extended thumb then points to the north pole of the magnet (Figure 27–4).
6. The poles of a coil may be reversed by (a) reversing the current or (b) reversing the direction the wire is wound.
7. Coils are frequently wound on soft-iron cores. When current is turned on, the iron core acts like a temporary bar magnet, but it loses its magnetism when current is turned off.
8. The strength of the magnetic field of a coil may be increased by increasing the current in the coil or the number of turns of wire or both.
9. A magnetic circuit may be compared to an electric circuit, where (a) mmf (magnetomotive force) is the equivalent of V (electromotive force); (b) ø, the resulting lines of magnetic flux, is the equivalent of I, and (c) $\mathcal{R}$, the opposition to ø, is the equivalent of R.
10. In a magnetic circuit

$$mmf = \text{ø}\,\mathcal{R}$$

and

$$mmf = IN$$

where mmf is magnetomotive force in ampere-turns

    ø is magnetic flux

    $\mathcal{R}$ is magnetic reluctance

    I is current in amperes

    N is the number of turns of wire

11. The strength of an electromagnet is directly proportional to I, the current in the coil; that is, mmf increases as I increases.

SELF-TEST

Check your understanding by answering the following questions:

1. (True/False) An electric current creates a magnetic field.
   _____
2. The magnetic field at any one point about a wire carrying current lies in a plane _____ to the wire at that point.
3. If a wire were perpendicular to this page and electron-flow current were in a direction toward you, the magnetic field around the wire would be in the _____ (clockwise/counterclockwise) direction.

4. (True/False) Soft iron is more permeable than air. Therefore, a coil with a soft-iron core has a more intense field in the core than does a coil with an air core.

_____

5. The left-hand rule can be used to determine the location of the _____ formed by current in a coil.

6. The strength of an electromagnet can be increased by increasing (a) _____ or (b) _____ .

MATERIALS REQUIRED

Power Supply:
- Variable 0–15 V dc 1 A, regulated

Instruments:
- DMM, VOM, or ammeter with 1 A scale

Resistors (½-W, 5%):
- 1   15-Ω, 25-W

Miscellaneous:
- 1 SPST switch
- 1 magnetic compass
- Iron filings
- 16 ft of no. 18 copper magnet wire
- 1 round hollow cardboard or plastic tube, about 2 in. long and 0.5 in. inside diameter (to be used as a coil form)
- 1 round soft-iron bar about 2 in. long, with an outside diameter such that it will fit snugly into the hollow tube
- 1 sheet of thin cardboard or stiff plastic about 8½ × 11 in.
- Electrical tape

PROCEDURE

Solenoid with Air Core

1. Construct a solenoid coil by wrapping the no. 18 wire tightly around the hollow tube. By wrapping the turns tightly against one another, you will probably need two or three layers of wire to produce a 100-turn coil. Leave about 8 in. of wire free before starting to wrap the wire around the hollow tube. After wrapping the 100 turns around the tube, bring out at least an 8-in. length of wire. Wrap electrical tape around the coil to prevent it from unraveling (Figure 27–5). Draw an arrow on the outside of the tape to indicate the direction of the winding. Label the ends of the coil as "start" and "end" of winding.

2. With power **off** and S_1 **open,** connect the coil constructed in step 1 in the circuit of Figure 27–6. The coil is shown in simplified form to indicate the direction of the winding. Magnet wire is usually insulated with a thin coat of plastic

or varnish. Make sure you remove this insulation before connecting the ends of the coil to the rest of the circuit.

3. Rest the length of the coil on a table and place a magnetic compass 2 in. from the end of the coil as shown in Figure 27–7. Make sure the compass pointer is oriented so that the west-east line is centered on the coil as shown. Note that the start of the winding is connected to the positive side of the power supply. Make sure there are no magnets or other magnetic materials in the immediate vicinity of your setup. Moving the coil around the compass should have no effect on the compass pointer.

4. Adjust the power supply control to minimum output voltage. Turn power **on.**

5. Increase the output voltage until the ammeter reads 0.75 A. Note the new position of the compass pointer. Record in Table 27–1 (p. 193) the direction indicated by the north-seeking end of the pointer (indicate the direction as N, S, E, or W).

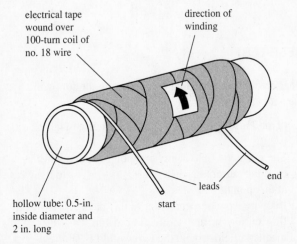

Figure 27–5. Construction of solenoid coil for procedure step 1.

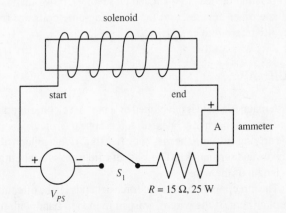

Figure 27–6. Circuit with solenoid for procedure step 2.

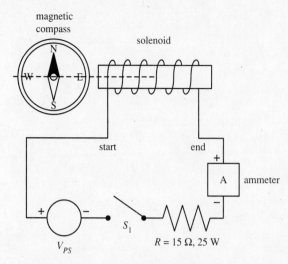

magnetic compass

solenoid

start   end

+

A ammeter

−

+ −

S_1

V_{PS}

$R = 15\ \Omega,\ 25\ W$

Figure 27–7. Position of compass and solenoid for procedure step 3.

6. Move the coil to within 0.5 in. of the compass. Record the position of the compass pointer in Table 27–1. Do not leave the coil in this position for more than a few seconds. Go immediately to the next step.

7. With the coil still 0.5 in. from the compass, decrease the voltage until the ammeter reads 0.10 A. Record the position of the compass pointer in Table 27–1.

8. **Open** S_1. The current in the coil is now zero. Record the position of the pointer in Table 27–1.

Solenoid with Soft-Iron Core

9. Insert a 2-in. soft-iron core inside the solenoid coil. Arrange the coil with iron core and magnetic compass as in step 3 (see Figure 27–7).

10. With the power supply adjusted to its lowest voltage, **close** S_1. Increase the supply voltage slowly until the ammeter reads 0.75 A. Observe the compass pointer and record its direction (N, S, E, or W) in Table 27–1.

11. Move the solenoid to within 0.5 in. of the compass as in step 6. Record the position of the compass pointer in Table 27–1. Do not hold this position for more than a few seconds. Proceed immediately to step 12.

12. Decrease the voltage of the supply until the ammeter reads 0.10 A. Record the position of the compass pointer in Table 27–1.

13. **Open** S_1. The current in the coil is now zero. Record the position of the pointer in Table 27–1.

14. Reverse the direction of current in the solenoid by reversing the connections to the start and end leads of the coil. **Close** S_1 and increase the voltage until the ammeter reads 0.75 A. Record the position of the pointer in Table 27–1. Proceed quickly to step 15.

15. **Open** S_1; power off. Record the position of the pointer in Table 27–1.

Magnetic Field of a Solenoid

16. With the solenoid in the same position as in step 15 but without the compass, place a thin cardboard or plastic sheet directly over the coil without touching it. Sprinkle a very thin layer of iron filings evenly on the cardboard.

17. Turn **on** the power and **close** S_1. Adjust the voltage until the ammeter measures 0.75 A. Notice the movement of the iron filings when S_1 is closed. Gently tap the top of the cardboard until the iron filings align themselves in a steady pattern. Sketch the pattern on a separate piece of paper. **Open** S_1; power **off**.

ANSWERS TO SELF-TEST

1. true
2. perpendicular
3. clockwise
4. true
5. poles
6. (a) number of turns;
   (b) amount of current

TABLE 27–1. Magnetic Field About a Solenoid

Step	Current in Solenoid, A	Solenoid Distance from E, in.	Current Polarity		Direction of Pointer
			Start of Winding	End of Winding	
5	0.75	2	+	−	
6	0.75	0.5	+	−	
7	0.10	0.5	+	−	
8	0	0.5	+	−	
10	0.75	2	+	−	
11	0.75	0.5	+	−	
12	0.10	0.5	+	−	
13	0	0.5	+	−	
14	0.75	0.5	−	+	
15	0	0.5	−	+	

QUESTIONS

1. Explain, in your own words, the relationship between current in a wire and a magnetic field. Cite specific experimental data that confirm these relationships.

2. Is it possible to determine the north and south poles at the ends of the solenoid in Figure 27–7 even before S_1 is closed? If not, what additional information must be known? If it is possible, what is the pole of the solenoid end closest to the compass?

3. Did step 14 in the procedure affect your answer to Question 2? If so, how?

4. Explain the left-hand rule for determining the poles of a solenoid. What would be the effect of using the right hand for determining the poles?

_____

_____

_____

_____

_____

_____

_____

_____

_____

_____

_____

_____

_____

_____

5. Explain how each of the following affects the strength of an electromagnet:
   (a) The amount of current in the coil
   (b) The number of turns of wire in the coil
   (c) The type of material used in the core

_____

_____

_____

_____

_____

_____

_____

_____

_____

_____

_____

_____

_____

INDUCING VOLTAGE IN A COIL

OBJECTIVES

1. To verify experimentally that a voltage is induced in a coil when the lines of force of a magnet cut across its windings
2. To verify experimentally that the polarity of the induced voltage depends on the direction in which magnetic lines of force cut the coil windings

BASIC INFORMATION

Electromagnetic Induction

If a zero-center galvanometer or a low-reading digital ammeter is connected to the ends of a straight piece of copper conductor and the conductor is placed a few inches away from the poles of a horseshoe magnet, as in Figure 28–1(*a*) (p. 196), there will be no reading on the meter. If the conductor is then moved down quickly between the poles of the magnet, as in Figure 28–1(*b*), the meter will show a reading and then return to zero when the conductor stops moving. If the conductor is then moved up through the magnetic field of the horseshoe magnet, the meter will again have a reading but this time with the opposite polarity.

The same reaction of the meter will occur if the conductor is stationary and the horseshoe magnet is moved across the conductor.

Finally, if the conductor is allowed to remain stationary in the center of the stationary magnetic field, there will be no reading at all on the meter.

From the preceding discussion, we can conclude that when a magnetic field is cut by a conductor, a voltage is induced in the conductor, and if the conductor is part of a complete circuit, a current will flow in the conductor. This process is true whether the conductor moves across the magnetic field or the magnetic field is moved across a conductor.

The point to remember is that the conductor must *cut across* the magnetic field. If the conductor is held parallel to the direction of the field and moves through it in that fashion, the lines of force will not be cut and no voltage will be induced.

Polarity of Induced Voltage

The fact that the polarity of the meter reading changes depending on the direction of movement of the conductor is evidence that the polarity of the voltage induced in the conductor depends on the *direction* of cutting of the lines of force. The polarity of the induced voltage can be established by *Lenz' law.*

Lenz' law states that the polarity of the induced voltage must be such that the direction of the resulting current in the conductor produces a magnetic field around the conductor that will oppose the *motion* of the inducing field. An example illustrates the meaning of Lenz' law.

The ends of the air-core coil in Figure 28–2 (p. 196) are connected to a sensitive ammeter capable of indicating current direction. The north pole of a bar magnet is pushed quickly into the center of the coil from the left, inducing a voltage in the winding with the polarity shown. The resulting current sets up a magnetic field in the coil, so that a north pole appears on the left end of the coil. The north pole thus produced opposes the north pole of the bar magnet being pushed into the left side of the coil. Therefore, the original motion of the bar magnet produced a magnetic field that now opposes that motion. This satisfies Lenz' law. Experimentally, the reading of the ammeter will verify the polarity of the induced voltage and the direction of current. Using the left-hand rule for coils will establish the poles at the ends of the coil.

If the bar magnet is now pulled to the left out of the core of the coil, the polarity of the voltage induced in the coil will be reversed, and the current will flow in the opposite direction. This will reverse the poles at the ends of the coil. A south pole will appear at the left end of the coil. As the bar magnet is pulled from the core, the south pole of the coil will attract the north pole of the magnet back into the core, thus opposing the pulling motion of the bar magnet out of the core. Again this condition satisfies Lenz' law.

Counter Electromotive Force

To this point we have discussed that a voltage is induced in a conductor when there is relative motion between the conductor and the magnetic field of a permanent magnet—

that is, when the conductor cuts the lines of force of the magnet. In fact, the effect of induction in a conductor is achieved when there is relative motion between the conductor and the field of *any* magnet, including an electromagnet.

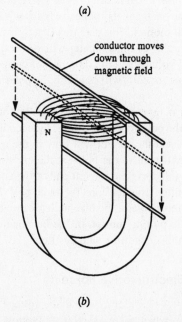

(a)

(b)

Figure 28–1. The meter indicates current in the circuit when the conductor is moved quickly through the magnetic field.

In a previous experiment it was demonstrated that a magnetic field exists about a current-carrying coil. This coil is in fact an electromagnet and may be substituted for the bar magnet in the preceding example.

The term *relative motion* is not limited merely to the physical movement of a magnet near a conductor or of a conductor in a magnetic field. Relative motion may exist without any physical movement whatsoever. Consider a coil through which an increasing or decreasing (not a steady-state) current is flowing. In the case of an increasing current, the magnetic field about the coil is increasing or expanding; that is, it is a *moving field*. When the current in the coil is decreasing, the magnetic field about the coil is decreasing or collapsing. Again, it is a moving magnetic field. A conductor held stationary in this moving magnetic field is in effect being *cut* by the magnetic lines of force. Therefore, a voltage will be induced in the conductor. Expanding and collapsing magnetic fields will cause voltages of opposite polarity to be induced in the conductor.

The expanding or collapsing (moving) magnetic lines of force about the coil with an increasing or decreasing current *cut the windings of the coil itself.* Accordingly, a voltage is induced in the coil, which by Lenz' law will oppose the inducing force. That is, the voltage induced in the coil will have a polarity opposite that of the voltage that initially caused current to flow in the coil. The voltage induced in the coil may therefore be called a *counter electromotive force*, or simply a *counter emf.*

Magnitude of Induced Voltage

The value of the voltage induced in a coil depends directly on (1) the *number* of turns N of wire in the coil and (2) the *rate* at which the lines of force are cut by the windings of the coil. Thus, the more turns of wire, the higher will be the induced voltage across the ends of the coil. Similarly, the faster the magnetic flux is cut by the windings, the higher will be the voltage produced across the coil.

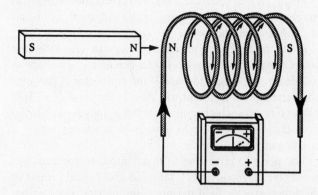

Figure 28–2. The bar magnet pushed into the center of the coil in the direction shown generates a current that produces magnetic poles; these poles tend to repel the motion of the bar magnet. This verifies Lenz' law.

SUMMARY

1. A voltage is induced in a conductor when the conductor cuts the lines of force in a magnetic field. Cutting the lines of force can be achieved by moving either the conductor or the magnetic field.
2. The polarity of the voltage induced in the conductor is determined by the direction in which the lines of force are cut. Thus, if a positive voltage is induced, say, by a conductor cutting the lines of force in a *downward* direction, a negative voltage will be induced by an *upward* cutting of these same lines of force.
3. The polarity of the induced voltage can be predicted by Lenz' law, which states: The polarity of the voltage induced in a conductor must be such that the *magnetic field set up by the resulting current* in the conductor will oppose the motion of the magnetic field that originally produced it.
4. The value of the voltage induced in a coil when its windings cut magnetic lines of force depends directly on the number of turns in the winding and on the rate at which the magnetic flux lines are cut by the windings.

SELF-TEST

Check your understanding by answering the following questions:

1. A moving magnetic field will _____ a _____ in a conductor if the magnetic lines of force _____ the conductor.

2. The polarity of voltage induced in a conductor will depend on the _____ of cutting of the lines of force by the conductor.
3. A conductor moving parallel to the lines of force in a magnetic field _____ (will/will not) have a voltage induced in it.
4. (True/False) The more turns there are in a coil whose windings cut the lines of force in a magnetic field, the greater will be the voltage induced in it, all other things being equal. _____
5. _____ law can be used to predict the polarity of voltage induced in a coil if the polarity of the field inducing the voltage is known.

MATERIALS REQUIRED

Instruments:
■ Zero-center galvanometer (a zero-center sensitive microammeter can be used)

Miscellaneous:
■ Solenoid coil consisting of 100 turns of no. 18 magnet wire wound on a hollow cardboard or plastic cylindrical form 3 in. long and 1 in. inside diameter.
■ Bar magnet, about 4 in. long and narrow enough to slide easily inside the coil form.

PROCEDURE

1. Connect the coil and galvanometer as shown in Figure 28–3. Note the direction of the winding around the core form. Place the bar magnet lengthwise as shown with its north pole facing the coil and about 2 in. away from the end of the coil.
2. With the magnet and coil stationary, observe the reading of the galvanometer. Record this value and the polarity of the reading in Table 28–1 (p. 199).
3. Turn the coil on its end and rest it on the lab table. The start of the winding should be at the bottom of the coil. While observing the galvanometer, insert the north pole of the bar magnet down quickly into the core of the coil. Record the highest reading of the galvanometer and the polatity of the reading in Table 28–1.
4. With the magnet resting in the core of the coil, observe the galvanometer and record the maximum reading and polarity in Table 28–1.
5. While observing the galvanometer, withdraw the magnet quickly from the core of the coil. Record the highest

reading and its polarity in Table 28–1.
6. Reverse the bar magnet. While observing the galvanometer, insert the south pole down quickly into the core of the coil. Record the highest reading of the galvanometer and its polarity.

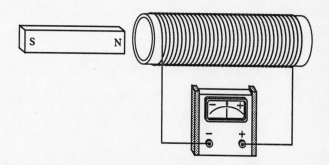

Figure 28–3. The coil and galvanometer are connected in series for procedure step 1.

7. With the magnet stationary within the core of the coil, record the reading of the galvanometer and its polarity in Table 28–1.

8. While observing the galvanometer, withdraw the magnet quickly from the core. Record the highest reading and polarity in Table 28–1.

9. Repeat step 6, but insert the magnet much more quickly than before. Record the highest reading and polarity in Table 28–1.

10. Repeat step 6, but insert the magnet much more slowly than before. Record the highest reading and polarity in Table 28–1.

11. Stand the bar magnet vertically on end with the north pole up. With the galvanometer still connected to the coil, draw the coil quickly down over the magnet. The start of the coil winding should be at the upper end of the coil. Be careful not to move or shake the galvanometer. Record the highest reading and polarity in Table 28–1.

12. While observing the galvanometer, quickly lift the coil over and away from the magnet. Record the highest reading and polarity in Table 28–1.

ANSWERS TO SELF-TEST

1. induce; voltage; cut
2. direction
3. will not
4. true
5. Lenz'

Name _____ Date _____

TABLE 28–1. Inducing Voltage in a Coil

Step	Condition	Voltage, Polarity	Current, Highest Reading, μA
2	Magnet stationary		
3	North pole of magnet inserted into solenoid		
4	Magnet stationary within the solenoid core		
5	Magnet withdrawn from solenoid		
6	South pole of magnet inserted into solenoid		
7	Magnet stationary within the solenoid core		
8	Magnet withdrawn from solenoid		
9	Same as step 6 but more rapidly		
10	Same as step 6 but more slowly		
11	Solenoid plunged down over magnet		
12	Solenoid pulled up, away from magnet		

QUESTIONS

1. Explain fully, in your own words, the result of lines of force cutting across the conductors in a winding. Describe specific conditions and results of the process.

_____

_____

_____

_____

_____

_____

_____

_____

2. Explain Lenz' law in your own words.

_____

_____

_____

_____

_____

_____

_____

_____

3. Does the result of your experiment verify Lenz' law? If not, what steps would need to be performed in order to verify the law? If your experimental results did verify the law, discuss these results.

_____

_____

_____

_____

_____

_____

_____

_____

4. Refer to your data in Table 28–1. What results verify that it is relative motion between the magnetic field and the windings of the solenoid that produces an induced voltage?

_____

_____

_____

_____

_____

_____

_____

_____

_____

5. Faraday's law states that the magnitude of the voltage induced in a winding depends directly on the number of turns of wire in the winding and the rate of speed at which the magnetic lines of force cut the winding. Do your results verify this law? If so, discuss your results. If not, what additonal steps are needed to verify the law?

_____

_____

_____

_____

_____

_____

_____

_____

_____

_____

200 *Experiment 28*

APPLICATIONS OF THE DC RELAY

OBJECTIVES

1. To study the characteristics and operation of a dc relay
2. To explore the uses of dc relays in electric circuits

BASIC INFORMATION

Applications of Magnetism

One of the earliest applications of magnetism was the magnetic compass. It can be said that the compass ushered in the age of modern marine navigation and exploration, for mariners were then no longer dependent on sighting the stars. With the compass they could find their way at any time of day or night and in any weather.

The next dramatic development in the application of magnetism was in the generation of electricity. The generator was designed to take advantage of the emf created when a conductor cuts magnetic lines of force. In the early dc generators used by the power companies, the armature of the generator contained many windings of copper wire rotating in the magnetic field produced by field windings and acted much like an electromagnet. In the modern generator used by utilities that arrangement is reversed: The armature windings are embedded in the stationary frame of the generator, whereas the field windings that produce the magnetic lines of force rotate. In both cases, conductors are being cut by magnetic flux, thus producing a voltage.

The electric motor also works on the principle of the interaction between the magnetic fields associated with its armature and field windings. There are electric motors in every home: the motor that circulates the furnace heat, the air-conditioner motor, the washing machine and dryer motors, the electric clock motor, and the like.

Electronics, too, depends in many cases on magnetic fields. The cathode-ray tube, for example, requires magnetic fields of various types to operate effectively. Circuits of all types use inductors and transformers, two devices whose operation is a function of electromagnetism.

The Relay

The *relay,* another electromagnetic device, performs countless tasks in industry and in the home. Relays are used for switching, for indicating, for transmitting, and for protecting circuits. Although modern solid-state electronic devices have replaced electromagnetic mechanical relays for many applications, it is not likely that the thousands of other applications of these relays could be easily transferred economically or physically to solid-state devices in the foreseeable future.

Relays are electromagnetically operated, remotely controlled switches, with one or more sets of contacts. When energized, the relay operates, opening or closing its contacts or opening some contacts and closing others. Contacts that are open when the relay is not energized are called *normally open* (NO) contacts. Contacts that are closed when the relay is not energized are called *normally closed* (NC) contacts. Figure 29–1 shows three types of schematic symbols used for open and closed contacts.

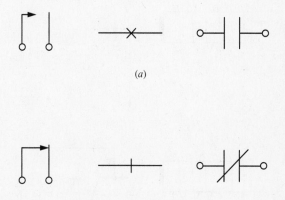

Figure 29–1. Relay symbols. The arrowhead indicates the stationary contact. (*a*) Normally open (NO) contacts; when the relay is energized, they *make* contact. (*b*) Normally closed (NC) contacts; when the relay is energized they *break* contact.

There are certain terms usually associated with relays that the technician should understand. A relay is said to "pick up" when it is energized. The *pickup value* is the smallest value of actuating current required to close a normally open contact or open a normally closed contact. When a relay is deenergized, it is said to "drop out." Relay contacts are held in their normal position either by springs or by some gravity-actuated mechanism.

A typical relay consists of an electromagnet, a small movable plate called an armature, and a set of contacts attached to the armature. When the electromagnet is energized, the armature is attracted to the electromagnet, which causes the contacts to open or close depending on whether their deenergized state is closed or open.

Figure 29–2(a) shows the major elements of a relay. The spring is used to return the armature to its normal deenergized position. The tension-adjusting screw can provide fine-tuning for the magnetic force, and thus the coil current, that is necessary to move the armature when the coil is energized.

When the coil in Figure 29–2(a) is in its normal deenergized state, the circuit between terminals 3 and 4 resembles an open switch. When the electromagnet is energized, the armature is attracted to the coil, and the movable contact connected to terminal 3 closes on the stationary contact (the arrowhead) connected to terminal 4. The circuit between 3 and 4 now resembles a closed switch. The entire action is similar to that of a single-pole, single-throw (SPST) switch.

In Figure 29–2(b) two mechanically connected, movable contacts are attached to the armature. The figure shows the deenergized state of the relay. In this condition the contact between 4 and 5 is normally closed and the contact between 3 and 4 is normally open. When the coil is energized the contacts move to the position shown in Figure 29–2(c). In this condition the circuit between 3 and 4 is closed and the circuit between 4 and 5 is open. The action of the relay in Figure 29–2(b) and (c) is that of a single-pole, double-throw switch.

Remote-Control Operations

One of the major advantages of relays is their ability to be operated from remote locations at relatively low voltages. The electromagnet coil of the relay can be energized with low current and voltage from a source some distance away from the relay itself. The circuit connected to the contacts, on the other hand, can be rated for much higher voltages and current. The circuit of Figure 29–3 uses a 12-V dc source to control a 277-V lighting circuit. The relay control switch is located on the wall, whereas the 277-V circuit and relay are located in the ceiling of the room.

Relays can be used to open circuits as well as close them. In Figure 29–4 (p. 203) a relay coil is connected to a heat sensor. The main contact of the relay is in series with the line feeding an electric heater. When the sensor detects an overheating condition, it produces enough current to energize the relay coil. This opens the NC contacts of the relay, thereby opening the circuit and turning off the heater.

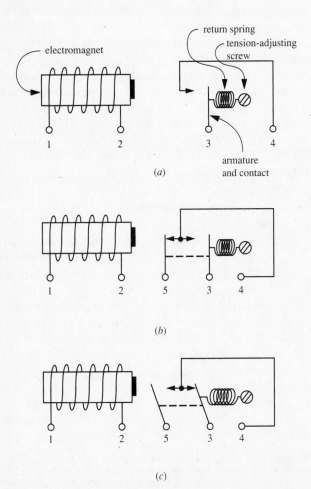

Figure 29–2. Pictorial representation of a relay. (a) SPST relay. (b) SPDT relay. (c) SPDT relay in energized state.

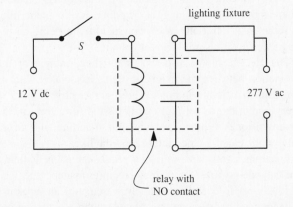

Figure 29–3. Control of high-voltage ac lighting circuit with a low-voltage dc relay.

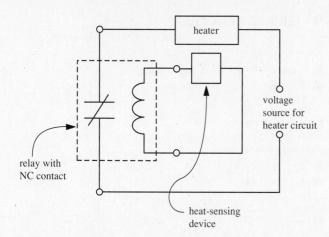

Figure 29–4. Relay with an NC contact used to prevent an electric heater from overheating.

Relay Specifications

Relay manufacturers supply a specification sheet with each of their relays. This "spec" sheet contains relay ratings, designates whether the relay is dc or ac, and specifies the location and the ratings of the contacts. For example, a relay may be specified as a dc SPDT relay with a relay coil resistance of 400 Ω. The contacts of the relay may be rated at 5 A.

The relay coil and contact terminals can usually be located by inspection. If a relay is enclosed in a sealed unit, this may not be possible. An ohmmeter may then be used to identify the terminals. The ohmmeter is used to measure the resistance between any two terminals. There are only a limited number of two-terminal combinations possible. The

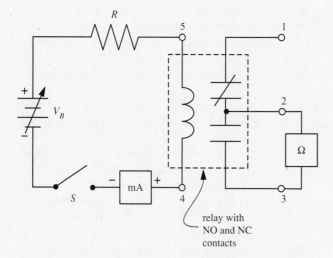

Figure 29–5. Circuit for determining the pickup value of a dc relay.

resistance of the coil is measured between the two terminals that are connected to the relay coil. Relay coils will vary from very low to very high values of resistance. Normally closed contacts will measure zero resistance. Normally open contacts will read infinite resistance. For example, in the relay in Figure 29–5, the coil resistance can be measured with an ohmmeter between terminals 4 and 5. There will be zero resistance between terminal 2 and 1 and infinite resistance between terminals 2 and 3. The resistance between terminals 1 and 3 will also be infinite. If you cannot determine by inspection whether contact 2 or 1 is the movable contact, the relay should be tripped. A resistance check will then show, as in the case of Figure 29–5, that there is now zero resistance between contacts 2 and 3 and infinite resistance between contacts 2 and 1 and contacts 3 and 1. It is therefore evident that contact 2 is the movable one.

CAUTION: In making resistance checks of relay contacts, be certain the power (to the load) is **off;** otherwise you may damage the ohmmeter.

SUMMARY

1. Relays are electromagnetic switches used as protective devices, as indicating devices, and as transmitting devices.
2. Protective relays protect good components from the effects of circuit components that have failed.
3. Transmission relays are used in communication systems.
4. Indicating relays may be used to identify a component that has failed, or they may be used with attention-getting devices such as bells or buzzers.
5. Relays may be used as SPST switches, or they may have complex switching arrangements.
6. The switch contacts of relays may be normally open (NO) or normally closed (NC). The contacts are held in their normal positions by springs or by some gravity-actuated mechanism.
7. Relays are either ac- or dc-operated and generally consist of an electromagnet, a movable armature, and contacts.
8. Terminals are provided on a relay for connections to the winding of the electromagnet and for the relay switch contacts.
9. An advantage of a relay over an ordinary switch is that a low-power source may be used to turn a relay on and off. As a result of this action, heavy-duty relay contacts open and close the circuit for a high-power load.
10. A relay system may be designed so that its load circuit is open when the relay is energized, called an open-circuit system. Or a relay may be designed as a closed-circuit system, in which the load circuit is open when the relay is energized.

SELF-TEST

Check your understanding by answering the following questions:

1. A dc relay is one in which _____ current in the relay coil actuates the relay mechanism.
2. The movable arm of a relay is called the _____ .
3. A relay that "picks up" at 10 mA is one that is turned _____ (on/off) when 10 mA flows in the _____ .
4. Some important electrical specifications of a relay are the
   (a) _____ and _____ of the coil;
   (b) pickup _____ of the coil;
   (c) current-handling capacity of the _____ .
5. In Figure 29–2 (b), switch contacts _____ and _____ complete the load circuit when the relay is energized.
6. (True/False) A switch is more efficient than a relay in turning a remotely located high-power load on and off.

_____

MATERIALS REQUIRED

Power Supplies:
■ Variable 0–15 V dc, regulated
■ 120-V, 60-Hz line voltage

Instrument:
■ VOM or DMM

Resistors (½-W, 5%):
■ 1   560-Ω
■ 1   1800-Ω

Relay:
■ 1 dc, SPDT, 12-V coil voltage, 300–400-Ω field, 120-V ac, 0.5-A contacts

Miscellaneous:
■ Test lamp set consisting of 60-W incandescent lamp, on-off switch, 1-A fuse, line cord, and polarized plug

PROCEDURE

1. You will be given a SPDT relay for this experiment. Inspect and test the relay to determine the following (the terminals have been numbered for reference purposes):
   (a) Relay (field) coil terminals
   (b) NO contact terminals
   (c) NC contact terminals
   (d) Nameplate data (if any)
   (e) Relay coil resistance, measured
   Record all information in Table 29–1 (p. 207).

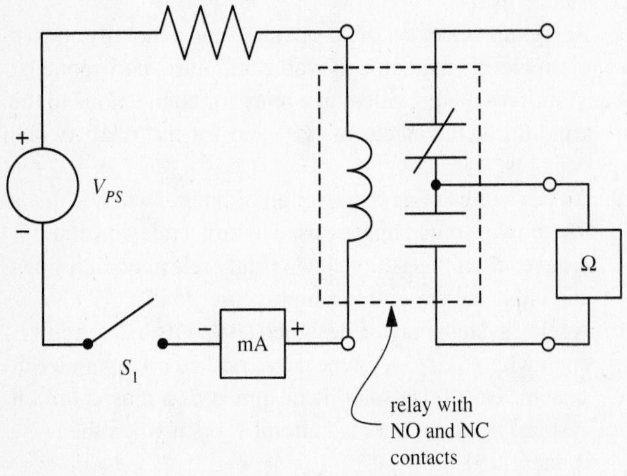

Figure 29–6. Circuit with relay for procedure step 2.

2. With power **off** and S_1 **open**, connect the circuit of Figure 29–6. Set the ohmmeter range for a low-resistance reading. Adjust the power supply to its lowest voltage.
3. Turn power **on** and **close** S_1. While observing the ohmmeter and listening and watching for any changes in the relay, slowly increase the power supply voltage. The ammeter will read the increasing current in the relay, or field, coil. At some point the relay coil will be energized enough for the relay to pick up. You might hear a click as the armature moves and the relay contacts close. The NO contacts across the ohmmeter will close and the ohmmeter reading will immediately drop to zero. At the point when this occurs, the ammeter will be measuring the pickup current. Record this value in Table 29–1. Continue to increase the voltage until the rated voltage is applied to the relay coil.
4. Slowly reduce the voltage across the relay coil while observing the ohmmeter and listening to and watching the relay. At some point you might hear a click from the relay as the armature returns to its original deenergized condition and the NC contacts close again. The NO contacts will open and the ohmmeter will measure infinite resistance. At this point the relay has dropped out (often referred to as being "tripped"). The ammeter will be measuring the dropout current. Record this value in Table 29–1.
5. Verify your original values of pickup and dropout currents by repeating steps 3 and 4. If care was taken to note the exact points at which the ohmmeter readings changed, the

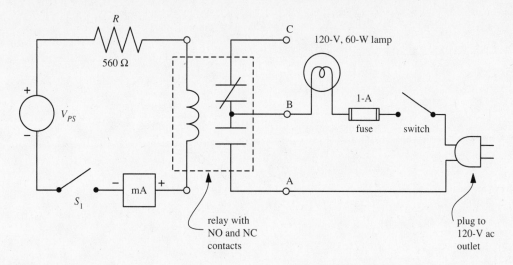

Figure 29–7. Lamp circuit connected to the relay circuit of Figure 29–6 for procedure step 6.

pickup and dropout current values should not vary significantly from your original readings. **Open** S_1; turn the power **off.**

6. With power **off,** S_1 **open,** and the lamp circuit unplugged, with its switch **off,** connect the circuit of Figure 29–7. The lamp circuit is connected across the NO relay terminals A and B, as shown on the diagram. (Your relay may not be marked with these letters.)

7. CAUTION: This part of the experiment involves potentially lethal voltage. Use extreme care when operating this circuit. Do not touch any parts of the circuitry with both hands at the same time.

Adjust V_{PS} to the rated voltage of the relay coils. Turn **on** the power supply, but leave S_1 **open.** Plug the 60-W ac lamp circuit into a 120-V outlet and close the lamp circuit switch. Record the state of the lamp in Table 29–2 (p. 207) by checking the "on" or "off" box.

8. **Close** S_1. Observe the state of the lamp and record whether it is on or off in Table 29–2.

9. **Open** the lamp circuit switch and unplug the 60-W lamp circuit from the 120-V ac outlet. Reconnect the circuit across the NC relay terminals B and C. (Your relay may not be marked with these letters.) Plug the 60-W lamp circuit into a 120-V ac outlet. **Close** the lamp circuit switch. Record the state of the lamp in Table 29–2.

10. **Open** S_1. Record the state of the 60-W lamp in Table 29–2. **Open** the lamp circuit switch and unplug the circuit from the 120-V outlet. Turn **off** V_{PS}.

ANSWERS TO SELF-TEST

1. direct
2. armature
3. on; relay coil
4. (a) resistance, voltage (b) current (c) contacts
5. 3, 4
6. false

Experiment 29 Name ______________________ Date ____________

TABLE 29–1. Relay Characteristics

Function	*Terminal Connection*	*Relay Characteristics*	
Relay coil		Relay coil resistance, Ω	
Normally open contacts		Pickup current, mA	
Normally closed contacts		Dropout current, mA	
Nameplate data:			

TABLE 29–2. Operation of SPDT Relay

Step	S_1	*60-W Lamp Circuit Connection to Relay*	*State of 60-W Lamp*	
			On	*Off*
7	Open	Across A-B		
8	Closed	Across A-B		
9	Closed	Across B-C		
10	Open	Across B-C		

QUESTIONS

1. Expl:in, your own words, what the pickup value of a relay is.

2. Is it possible to change the pickup value of the relay you used in this experiment? If so how can this be done? If not, why?

_____

_____

_____

_____

_____

_____

_____

_____

_____

_____

3. When a relay's contacts return to their deenergized condition, the relay is said to be reset, or tripped. What is another term for the reset current of your relay? How was this value obtained experimentally?

_____

_____

_____

_____

_____

_____

_____

_____

_____

4. Could the relay you used in this experiment be used for on-off control of a 1200-W heater? Explain.

_____

_____

_____

_____

_____

_____

_____

_____

30

OSCILLOSCOPE OPERATION—TRIGGERED SCOPE

OBJECTIVES

1. To identify the operating controls of a triggered oscilloscope
2. To set up the oscilloscope and adjust the controls properly to observe an ac voltage waveform

BASIC INFORMATION

The cathode-ray oscilloscope (CRO), or "scope," as it is familiarly known, is one of the most versatile instruments in electronics. The technician must therefore be able to operate this instrument and understand how and where it is used.

Oscilloscopes can be classified as *triggered* or *nontriggered*. Triggered oscilloscopes are the more sophisticated of the two, can do more, and are widely used in industry. They are used in applications requiring the study of low- and high-frequency waveforms, precise measurement of time, and timing relationships.

The nontriggered scope is historically important because it was the first oscilloscope developed. It was used to view ac waveforms at test points in a circuit where knowledge of the existence or nonexistence of a waveform was usually all that was required. Precise time measurements could not be made with this elementary instrument. Nontriggered scopes were popular with TV service technicians. However, they have almost all been replaced with triggered oscilloscopes.

What an Oscilloscope Does

An oscilloscope displays the instantaneous amplitude of an ac voltage waveform versus time on the screen of a cathoderay tube (CRT). Inside the cathode-ray tube are an electron gun assembly, vertical and horizontal deflection plates, and a phosphorescent screen.

The electron gun emits a high-velocity, low-inertia beam of electrons that strikes the chemical coating on the inside face of the CRT, causing it to emit light. The intensity of the light given off at the screen of the CRT is determined by the voltages in the electron gun assembly. The intensity (called *brightness*) can be varied by a control located on the oscilloscope panel.

The motion of the beam over the CRT screen is controlled by deflection voltages generated in the oscilloscope circuits outside the CRT and the deflection plates inside the CRT to which the deflection voltages are applied.

Figure 30–1 (p. 210) is an elementary block diagram of an oscilloscope. The CRT provides the screen on which waveforms of electrical signals are viewed. These signal waveforms are applied to the vertical input on the oscilloscope and are processed by vertical amplifiers in the circuitry of the oscilloscope external to the CRT. Because the oscilloscope must handle a wide range of signal-voltage amplitudes, a vertical attenuator—a variable voltage divider—sets up the proper signal level for viewing. The signal voltage applied to the vertical deflection plates causes the electron beam of the CRT to be deflected vertically. The resulting up-and-down movement of the beam on the screen, called the *trace*, is significant in that *the extent of vertical deflection is directly proportional to the amplitude of the signal voltage applied to the vertical, or V, input.*

To make it possible for the oscilloscope to graph a time-varying voltage, a linear time-base deflection voltage is applied to the horizontal deflection plates. This voltage is also developed, in the oscilloscope circuits external to the CRT, by a time-base or sweep generator. It is this sweep generator that is either triggered or nontriggered.

Dual-Trace Oscilloscopes

Triggered oscilloscopes with *two* traces are in common use. The two traces are developed on the screen of the scope by means of electronic switching. Dual-trace oscilloscopes make it possible to observe simultaneously two time-related waveforms at different points in an electronic circuit.

Operating Controls of a Triggered Oscilloscope

The type, location, and function of the front panel controls of an oscilloscope differ from manufacturer to manufacturer and from model to model. The descriptions that follow apply to the broadest range of general-use scope models.

Intensity. This control sets the level of brightness or intensity of the light trace on the CRT. Rotation in a clockwise (CW) direction increases the brightness. Too high an

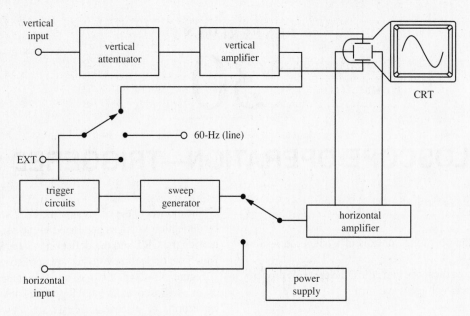

Figure 30–1. Elementary block diagram of an oscilloscope.

intensity can damage the phosphorescent coating on the inside of the CRT screen.

Focus. This control is adjusted in conjunction with the intensity control to give the sharpest trace on the screen. There is interaction between these two controls, so adjustment of one may require readjustment of the other.

Astigmatism. This is another beam-focusing control found on some oscilloscopes that operates in conjunction with the focus control for the sharpest trace. The astigmatism control is sometimes a screwdriver adjustment rather than a manual control.

Horizontal and Vertical Positioning or Centering. These are trace-positioning controls. They are adjusted so that the trace is positioned or centered both vertically and horizontally on the screen. In front of the CRT screen is a faceplate called the *graticule,* on which is etched a grid of horizontal and vertical lines. Calibration markings are sometimes placed on the center vertical and horizontal lines on this faceplate.

Volts/Div. This control attenuates the vertical input signal waveform that is to be viewed on the screen. This is frequently a click-stop control that provides step adjustment of vertical sensitivity. A separate Volts/Div. control is available for each channel of a dual-trace scope. Some scopes mark this control Volts/cm.

Variable. In some scopes this is a concentric control in the center of the Volts/Div. control. In other scopes this is a separately located control. In either case, the functions are similar. The variable control works with the Volts/Div. control to provide a more sensitive control of the vertical height of the waveform on the screen. The variable control also has

a calibrated position (CAL) either at the extreme counterclockwise or clockwise position. In the CAL position the Volts/Div. control is calibrated at some set value—for example, 5 mV/div., 10 mV/div., or 2 V/div. This allows the scope to be used for peak-to-peak voltage measurements of the vertical input signal. Dual-trace scopes have a separate variable control for each channel.

Time/Div. This is usually two concentric controls that affect the timing of the horizontal sweep or time-base generator. The outer control is a click-stop switch that provides step selection of the sweep rate. The center control provides a more sensitive adjustment of the sweep rate on a continuous basis. In its extreme clockwise position, usually marked CAL, the sweep rate is calibrated. Each step of the outer control is therefore equal to an exact time unit per scale division. Thus, the time it takes the trace to move horizontally across one division of the screen graticule is known. Dual-trace scopes generally have one Time/Div. control. Some scopes mark this control Time/cm.

Vert. Pos. This control is used to adjust the vertical position of the trace. Each channel of a dual-trace scope generally has its own Vert. Pos. control.

X-Y Switch. When this switch is engaged, one channel of the dual-trace scope becomes the horizontal, or *X,* input, while the other channel becomes the vertical, or *Y,* input. In this condition the trigger source is disabled.

Triggering Controls. The typical dual-trace scope has a number of controls associated with the selection of the triggering source, the method by which it is coupled, the level at which the sweep is triggered, and the selection of the slope at which triggering takes place:

1. *Level Control.* This is a rotary control which determines the point on the triggering waveform where the sweep is triggered. When no triggering signal is present, no trace will appear on the screen. Associated with the level control is an Auto switch, which is often an integral part of the level rotary control. In the Auto position the rotary control is disengaged and automatic triggering takes place. In this case a sweep is always generated and therefore a trace will appear on the screen even in the absence of a triggering signal. When a triggering signal is present, the normal triggering process takes over.

2. *Coupling.* This control is used to select the manner in which the triggering is coupled to the signal. The types of coupling and the way they are labeled vary from one manufacturer and model to another. For example, dc coupling usually indicates the use of capacitive coupling that blocks dc; line coupling indicates the 50- or 60-Hz line voltage is the trigger. If the oscilloscope was designed for television testing, the coupling control might be marked for triggering by the horizontal or vertical sync pulses.

3. *Source.* The trigger signal may be external or internal. As already noted, the line voltage may also be used as the triggering signal.

4. *Slope.* This control determines whether triggering of the sweep occurs at the positive or negative portion of the triggering signal. The switch itself is usually labeled positive or negative, or simply + or –.

Probes. The input signal to the oscilloscope is connected to the vertical signal input jack using various types of connectors and a shielded coaxial cable. An insulated probe is used to connect the scope to the circuit being investigated. The probes themselves may contain circuits that affect the waveform viewed on the screen of the scope. The probe may produce a direct reading or may reduce the reading by some set ratio. Because the probe itself may introduce some distortions in the waveform being observed, an adjusting screw may be provided on the probe.

SUMMARY

1. A triggered oscilloscope can be used to measure dc, as well as low- and high-frequency ac voltages, waveforms, and time.
2. A nontriggered scope is normally used to observe low-frequency waveforms but cannot measure time directly.
3. An oscilloscope displays the amplitude of an ac waveform versus time.
4. A cathode-ray tube is the screen of an oscilloscope.
5. The purpose of the electron gun in a CRT is to emit and deflect an electron beam that strikes the phosphorescent coating on the inside of the screen and causes it to give off light.

6. The signal voltages applied to the vertical deflection plates of a CRT cause the beam to be deflected up and down.
7. The horizontal deflection plates receive the linear sweep voltage, which generates the time base.
8. The intensity control is used to set the brightness of the trace.
9. The focus control is used to narrow the beam into the sharpest trace. There may be an auxiliary astigmatism control for focusing.
10. The horizontal and vertical centering or positioning controls are used to position the trace on the CRT screen.
11. The etched faceplate in front of the CRT face that appears as vertical and horizontal graph lines is called the graticule. Linear calibration markers (height and width) are frequently etched on the graticule.
12. The Volts/Div. control is calibrated to measure the amplitude of signal waveforms along the vertical axis.
13. The Time/Div. control is calibrated to measure time along the horizontal axis.
14. The triggering controls determine the manner in which a trigger pulse is initiated to start the sweep generator.
15. The trigger can be run automatically, the mode that is frequently used.
16. The trigger circuit can be actuated by a signal from internal oscilloscope circuits.
17. The trigger circuit can also be actuated by an external signal voltage.
18. The trigger circuit can also be actuated by a power line–derived voltage from internal oscilloscope circuits.

SELF-TEST

Check your understanding by answering the following questions:

1. (True/False) Oscilloscopes are used for the observation of ac waveforms. _____
2. (True/False) Precise time measurements can be made with both triggered and nontriggered scopes. _____
3. The waveforms seen on the screen of a CRT show the _____ versus _____ .
4. The _____ deflection plates of a CRT are the signal plates; the _____ deflection plates are for the time-base voltage.
5. Of the controls on an oscilloscope, those that affect the height of the signal are called _____ .
6. Those controls that affect the sharpness of the trace are called _____ and _____ .
7. The etched faceplate in front of the face of the CRT is called the _____ .
8. A frequently used triggering mode of a triggered oscilloscope is _____ .

9. The controls that affect the up-and-down movement of the trace are called _____ .
10. The height of the waveform displayed on the oscilloscope screen is directly proportional to the _____ of the waveform.

MATERIALS REQUIRED

Power Supply:
■ Power source for oscilloscope

Instruments:
■ Oscilloscope with triggered sweep, calibrated time base, calibrated vertical amplifier, internal voltage calibrator, and direct probe. The scope can be a single- or dual-trace oscilloscope.

Miscellaneous:
■ Operating or instruction manual for the oscilloscope
■ Assorted leads for connecting to the oscilloscope terminals and jacks

NOTE: Before performing this experiment, read the operating manual completely. The manual should be conveniently available to you throughout the experiment.

PROCEDURE

1. Power to the oscilloscope should be **off.** Examine the front panel of the oscilloscope carefully, noting the types and functions of each control and jack. Examine the back of the scope for additional jacks, switches, and controls.
2. Tabulate the information on the scope in Table 30–1 (p. 215). Some of the information will be found in the operating manual. Other information can be found by inspection.
3. Turn the oscilloscope **on.** Set the time-base switch to the EXT or X-Y position. If you are using a dual-trace scope, set the control for channel 1 (or A) operation. After a brief warmup period, a bright spot should appear on the screen. If there is no spot, adjust the intensity control clockwise.

   CAUTION: Do not set the intensity too high and do not leave the spot on too long; the screen coating can be permanently damaged.

   If the spot still does not appear, it may be off screen. Use the vertical and horizontal positioning controls until the spot becomes visible on the screen.
4. If you have not already done so, adjust the applicable controls to produce a sharp, clear, centered spot on the screen. This will involve the vertical and horizontal positioning, focus, astigmatism (use only if an adjustable control is on the front panel), and intensity controls. Note that the focus and intensity controls (and the astigmatism control, if adjustable) interact with one another. Therefore, it is necessary to adjust both (or all three) to obtain the sharpest, thinnest point of light.
5. Set the time-base switch to automatic operation (this may involve simply disengaging the X-Y switch). Adjust the time-base (Time/Div.) control to 1 ms/div and the Trigger-level control to Auto and positive slope. A line (trace) should appear on the screen. Adjust the intensity control to produce a visible but not too intense trace. Use the horizontal positioning control to move the left end of the line to the leftmost graduation of the graticule.
6. Connect the vertical input of the oscilloscope to the output of the voltage calibrator on the scope. Set the output of the calibrator to approximately 2 V. Set the Time/Div. control to calibrated (CAL). The calibrator waveform should be displayed on the screen. Adjust the Volts/Div. control so that the waveform is about three divisions peak to peak.

   NOTE: Some oscilloscopes do not permit varying the calibrator voltage. If your scope has a set calibrator voltage, use that voltage in this step and skip steps 8 and 9.

7. Set the Time/Div. control to calibrated (CAL) and vary the adjustable Time/Div. control until three cycles of the calibrator waveform appear on the screen.
8. (See step 6.) Without changing the Volts/Div., setting, decrease the calibrator output voltage to approximately 0.5 V. Observe the effect of this change on the displayed waveform.
9. (See step 6.) Do not change the Volts/Div. setting. Increase the calibrator output to approximately 5 V (or the highest voltage between 2 and 5 V). Again observe the effect of the change in voltage on the waveform.
10. Reset the calibrator to 2 V (or the single setting of your oscilloscope calibrator as in step 6). The waveform should be approximately 3 divisions peak-to-peak and display three cycles.
11. Increase the Time/Div. control one setting. (For example, if the control was set to 0.5 ms/div., set it to the next faster time, 0.1 or 0.2 ms/div.). Observe the effect of this change on the waveform.

12. Decrease the Time/Div. control one setting past its position in step 10. (For example, if the control was set to 0.5 ms/div., set it to the next slower time, 1 or 2 ms/div.) Observe the effect of the change on the waveform.

13. Turn off the oscilloscope. Have someone change all the control and switch settings. Repeat steps 3 through 12 with as little reference to the operating manual as possible.

ANSWERS TO SELF-TEST

1. true
2. false
3. amplitude; time
4. vertical; horizontal
5. Volts/Div. (or Volts/cm)
6. focus; astigmatism
7. graticule
8. automatic
9. vertical positioning, or centering
10. amplitude (voltage)

Experiment 30   Name _____ Date _____

TABLE 30–1. Oscilloscope Features, Functions, and Controls

Instrument No. _____ Manufacturer _____ Model _____

Features (Check all that apply)

Single trace _____	Calibrated time base _____
Dual trace _____	Calibrated vertical control _____
Triggered sweep _____	Two-channel +/– capability _____
Auto sweep _____	× 10 probe _____
XY operation _____	Direct probe _____
Sweep magnification _____	Other features (describe) _____
Internal voltage calibrator _____	_____

List of Controls and Their Functions

Control	Function

QUESTIONS

1. In procedure steps 8 and 9, the calibrator voltage was first decreased and then increased from the settings of steps 6 and 7. If your oscilloscope enabled you to change the calibrator voltage, what was the effect on the voltage waveform displayed?

_____

_____

_____

_____

2. After procedure step 10, the Time/Div. control was increased (step 11) and then decreased (step 12). What effect did those changes to the Time/Div. control have on the number of cycles displayed?

_____

_____

_____

_____

3. Refer to Question 2. Was the frequency of the calibration waveform changed due to steps 11 and 12? If so, describe the change. If not, was anything changed?

_____

_____

_____

_____

4. Is there any relationship between the number of cycles of the waveform displayed and the setting of the Time/Div. control? Discuss fully.

_____

_____

_____

_____

_____

_____

_____

5. Refer to the data in Table 30–1. What controls affect the following, and how?
   (a) The height of the displayed waveform
   (b) The brightness of the waveform
   (c) The sharpness of the waveform
   (d) The position of the waveform on the screen
   (e) Turning on the sweep generator

_____

_____

_____

_____

_____

_____

_____

_____

_____

_____

EXPERIMENT

31

SIGNAL-GENERATOR OPERATION

OBJECTIVES

1. To identify the controls and features of an audio-frequency (AF) generator
2. To identify the controls and features of a function generator
3. To operate the signal generators and observe their output waveforms on an oscilloscope

BASIC INFORMATION

Signal generators provide ac voltages over selected frequency ranges. They are also capable of producing various types of waveforms. Although no single generator can provide the entire range of frequencies and waveforms, they can produce frequencies lower than 1 Hz to many thousands of megahertz. Waveforms can be sine, sawtooth, or square. The voltage output of signal generators is usually low, because signal generators are normally used for testing and experimenting rather than as power supplies. A particular type of signal generator, called a *function*, or *waveform*, *generator*, produces sine, sawtooth, and square waves over a wide range of frequencies but at a relatively low voltage.

Signal generators are often identified by the frequency range they are designed to produce. Thus an AF signal generator is designed to provide frequencies in the audio (so-called sound) range, from several hertz to 20 kHz. The signal itself is a simple sine wave. The output voltage may be several volts to as much as 20 or more volts. Because signal generators are primarily test instruments, the accuracy and stability of their output is critical. The two signal generators covered in this experiment are the AF sine-wave generator and the function generator.

AF Signal Generator

Although AF signal generators are designed to operate in the audio-frequency range (up to about 20 kHz), most commercial generators have ranges in the hundreds of kilohertz. Because the signal generator is often used to provide a test signal for troubleshooting radio receivers, its frequency range extends to that of the standard FM and AM broadcast bands. Signal generators of this type are also equipped to generate modulating signals. To prevent extraneous signals in the air from affecting the signal-generator output, a shielded cable is used between the signal generator and the device to which it is connected.

In this experiment we focus on the waveform and frequency of the signal generator as well as its output voltage. We will observe all three on an oscilloscope while operating the controls of the generator.

The position and function of generator controls vary from manufacturer to manufacturer, but in general, the following are found on all generators.

1. *On-Off Switch.* Signal generators are bench-type instruments that require an ac source to operate (generally 120 V, 60 Hz). The on-off switch applies power to the circuits of the generator. Some models of portable signal generators are battery operated.
2. *Output control.* This control is used to adjust the voltage output of the generator. The output terminal of the generator is generally made to accept a shielded cable. Most outputs are unmetered; to determine the exact output level, an external voltmeter or oscilloscope is required.
3. *Range control.* In most signal generators, frequencies are grouped in a number of ranges spanning the entire frequency spectrum of the generator. For example, one manufacturer produces a generator with a frequency output of 10 Hz to 1 MHz in five decade bands: 10 Hz–100 Hz, 100 Hz–1 kHz, 1 kHz–10 kHz, 10 kHz–100 kHz, 100 kHz–1 MHz. Range controls can be click-stop rotary dials or pushbutton selector switches.
4. *Frequency control.* This is normally a continuously variable rotary dial that fine-tunes the exact frequency within the range set by the range control.

Some AF signal generators also contain modulation switches (AM, FM, or PM), and modulation level controls.

Function Generator

The purpose of the function generator is to produce signals for test and troubleshooting purposes. The signals serve as stimuli for tracing the behavior of parts of circuits or entire systems. In addition, the generated signal may in turn trigger other signals whose behavior is being studied or observed on an oscilloscope. The controls on a function generator are similar to those on an AF signal generator.

1. *On-Off Switch.* The power source for most function generators is a 120-V, 60-Hz supply, although some portable models are battery operated. The on-off switch controls power to the circuits of the generator.
2. *Waveform control.* Function generators are capable of producing two or more types of waveforms. The most common types of waveforms are sine, sawtooth, and square waves. These controls are sometimes labeled function controls.
3. *Output-level control.* This controls the voltage output, or amplitude, of the waveform.
4. *Range switch.* As in the case of the AF signal generator, the total frequency range of the generator is divided into frequency bands selected by a switch or series of switches.
5. *Frequency control.* This serves to select the exact frequency within the frequency band set by the range switch. It, in effect, fine-tunes the frequency of the output waveform.

Some function generators contain a dc offset switch (or dial) that adds a positive or negative dc component to any of the waveforms produced by the generator.

SUMMARY

1. A signal generator supplies very accurate and stable ac voltages of variable frequencies and output (voltage) levels.
2. An AF signal generator provides ac voltages in the audio (sound) range: 10 Hz to 20 kHz.
3. Commercial AF generators, however, often have ranges extending to 100 kHz and beyond.
4. The output of the typical AF generator is a sine wave, although some commercial generators also produce square waves.
5. The output (voltage) of the AF generator is generally unmetered. An external voltmeter or oscilloscope is used to set the output level.
6. The following are the typical controls and functions of an AF signal generator:

   ■ On-off power switch
   ■ Frequency range control switch or switches or rotary click-stop dial to select frequency range
   ■ Frequency dial to set the exact frequency within the range set by the range control
   ■ Output level control used to set the output voltage of the generator

7. AF signal generators provide the test signals used to test and troubleshoot radio circuits. These generators sometimes contain modulation functions for checking AM and FM circuits.
8. Function (or waveform) generators produce signals of variable frequencies in different waveforms.

9. The typical function generator can produce sine, sawtooth, and square waves.
10. The following are the typical controls and functions of a function generator:

    ■ On-off power switch
    ■ Waveform-selector switch
    ■ Switch or switches or rotary dial to select the frequency range
    ■ Frequency dial to set the exact frequency within the range set by the range switch
    ■ DC offset control to add a positive or negative dc voltage to the ac output voltage
    ■ Output level control to set the voltage level of the output

11. AF signal generators and function generators differ in features and design from one manufacturer to another. It is important to have convenient access to the operating manual of the equipment being used.

SELF-TEST

Check your understanding by answering the following questions.

1. Audio-frequency signal generators are often used to provide a signal when testing or troubleshooting _____ circuits.
2. The audio frequencies are in the range of _____ Hz to _____ kHz.
3. (True/False) The typical AF signal generator uses a built-in voltmeter to set the output voltage level. _____
4. A _____ control is used to set the band of frequencies that the AF generator can produce.
5. Fine-tuning of the exact output frequency of the generator is set by a _____ switch.
6. Another name for a function generator is a _____ generator.
7. A dc _____ control adds a negative or a positive voltage to the ac output of the function generator.
8. The most common waveforms provided by a function generator are the _____ , _____ , and _____ waveforms.

MATERIALS REQUIRED

Instruments:
■ AF signal generator (sine-wave output)
■ Function generator (with sine-, sawtooth-, and square-wave output)
■ Triggered oscilloscope (single- or dual-trace) with direct probe
■ Operating manuals for the above instruments

PROCEDURE

Before turning on the power to any of the instruments used in this experiment, study the manufacturer's operating manual supplied with the instrument. Note especially the location and function of each switch, dial, terminal, and control on the face of the instrument. Review the procedure that follows and compare the names of the controls referred to in the procedure with the actual equipment you will be using. Your instructor will review this information with you before you turn on power.

A. AF Sine-Wave Signal-Generator Operation

A1. Connect the output of the signal generator to the vertical input of the oscilloscope. (If a dual-trace scope is used, make all connections to channel 1.) Set the output (voltage) level control to the middle of its full range, and the frequency range switch or control to the 10–100 kHz range. Turn **on** the signal generator.

A2. Turn **on** the scope. Set the scope for automatic triggering. If you are using a dual-trace scope, make sure it is set to display only the channel 1 signal. Adjust the signal generator to deliver a 100-kHz sine wave to the scope. Adjust the Volts/Div. control of the scope until the wave on the screen is 6 divisions peak-to-peak and centered vertically. This may also require an adjustment of the output level control of the signal generator. Adjust the Time/Div. control of the scope until two complete sine waves are displayed on the screen (further adjusting of the Volts/Div. and output level controls may be necessary).

A3. Slowly reduce the output level of signal generator while observing the effect this has on the height of the wave displayed on the screen. Now slowly increase the output level beyond the setting of step A2 and observe the effect this has on the height of the wave displayed on the screen. Reset the output level control to restore the wave to its original 6-division peak-to-peak height.

A4. Change the frequency range switch and frequency control on the signal generator to produce a 1000-Hz signal. Readjust the Time/Div. control to display two cycles across the entire width of the screen and centered vertically on the screen. Adjust the Volts/Div. control so that the sine waves are about 6 divisions peak-to-peak. In Table 31–1 (p. 221), record the setting of the Time/Div. control and the number of divisions spanned by the sine waves.

A5. Increase the frequency output of the generator to 2000-Hz and observe the effect on the display. Do not adjust or change the setting of the Time/Div. control. In Table 31–1, record the number of cycles displayed and the number of divisions spanned by the cycles.

A6. Decrease the frequency output of the generator to 500 Hz. In Table 31–1, record the number of cycles displayed and the number of divisions spanned by the cycles. After recording the values, turn **off** the scope and the signal generator.

B. Function-Generator Operation

B1. Connect the output of the function generator to the vertical input of the scope. Set the function generator to the sine-wave output and 1000 Hz; set the output level in the middle of its full range. Turn **on** the function generator.

B2. Turn **on** the scope and adjust the Time/Div. control to produce one cycle across the width of the screen. Adjust the Volts/Div. control to produce a display about 6 divisions peak-to-peak. Sketch the waveform displayed on the screen in Table 31–2 (p. 221).

B3. Switch the frequency to 2000 Hz. Do not change or adjust the Time/Div. control. Center the display vertically and horizontally. Readjust the height to 6 divisions peak-to-peak if necessary. Sketch the waveform displayed on the screen in Table 31–2.

B4. Switch back to 1000-Hz output (do not readjust or change the Time/Div. control). Set the function switch to produce a sawtooth-wave output. If necessary, adjust the Volts/Div. control to produce a display 6 divisions peak-to-peak. Sketch the sawtooth-wave displayed on the screen in Table 31–2.

B5. Increase the sawtooth-wave frequency to 2000 Hz and repeat step B3.

B6. Switch back to 1000 Hz and set the function switch to produce a square-wave output. Do not change or readjust the Time/Div. control. Center the waveform on the screen. Sketch the waveform displayed on the screen in Table 31–2.

B7. Repeat step B3 for the square-wave output. After you have completed this step, turn **off** the scope and function generator.

ANSWERS TO SELF-TEST

1. radio
2. 10; 20
3. false
4. range

5. rotary dial
6. waveform
7. offset
8. sine; sawtooth; square

TABLE 31–1. AF Signal-Generator Operation

Frequency Setting of Signal Generator, Hz	Number of Cycles Displayed on Screen	Time/Div. Setting of Scope, Time Units/Div.
1000	2	
2000		
500		

TABLE 31–2. Function-Generator Operation

	Sine Wave	Sawtooth Wave	Square Wave
1000 Hz			
2000 Hz			

QUESTIONS

1. Discuss, in your own words, the relationship between the number of cycles of the waveform displayed on the screen and the frequency setting of the AF signal generator (with a constant Time/Div. control setting). Refer to your data for this experiment.

_____

_____

_____

_____

_____

_____

_____

_____

_____

_____

Signal Generator Operation **221**

2. In procedure step A3, the output level of the signal generator was first decreased and then increased beyond the level of step A2. Discuss the effect of the decrease and increase on the number of cycles and amplitude of the waveform displayed.

_____

_____

_____

_____

_____

3. State whether or not the AF signal generator you used in Part A had each of the following controls or functions. If it did, explain, in your own words, the type and purpose of the control or function.
   (a) On-off switch
   (b) Frequency control
   (c) Output-level control
   (d) Range control

_____

_____

_____

_____

_____

_____

_____

_____

4. State whether or not the function generator you used in Part B had each of the following controls or functions. If it did, explain, in your own words, the type and purpose of the control or function.
   (a) On-off switch
   (b) Range control
   (c) Frequency control
   (d) Output-level control
   (e) Waveform-selector switch

_____

_____

_____

_____

_____

_____

_____

_____

_____

EXPERIMENT
32

OSCILLOSCOPE VOLTAGE MEASUREMENTS

OBJECTIVES

1. To check the calibration accuracy of an oscilloscope for use as an ac voltmeter
2. To make peak-to-peak ac voltage measurements with a scope
3. To make dc voltage measurements with a scope

BASIC INFORMATION

Measuring Voltages with the Oscilloscope

An oscilloscope is used to view and measure ac waveforms. The height of the voltage waveform displayed on the oscilloscope screen is directly proportional to the peak-to-peak amplitude of the voltage. Thus, for the same settings of Volts/Div., or vertical-gain, controls, a 100-V signal will have twice the height of a 50-V signal. A 25-V signal will have one-fourth the height of a 100-V signal. Moreover, if we assume that for this same fixed setting of the vertical controls every division of vertical deflection corresponds to 25 V of input, then a 25-V signal will have 1 div. of height, a 50-V signal will have 2 div. of height, a 75-V signal will have 3 div. of height, and so on. Two and one-half divisions of vertical deflection (height) will correspond to 62.5 V, and so forth. This characteristic of the signal circuits makes it possible to use the oscilloscope for measuring ac voltages.

The same method is also used to measure dc. The ac-dc switch is placed in the dc position, and the probe is connected to a point of dc voltage in the circuit. The ground lead of the scope is connected to the ground of the circuit. The number of divisions the trace rises above or falls below the zero base setting is a measure of the + or – dc voltage.

NOTE: In this experiment we shall refer to the number of divisions of height of the voltage waveform, because these markings are found on the graticules of most oscilloscopes.

In measuring voltages with the oscilloscope, the procedure is to apply the signal voltage to the calibrated vertical input of the oscilloscope and to measure on the etched faceplate or graticule the number of divisions of height of the waveform. This number is then multiplied by the calibration

factor of the oscilloscope, giving the peak-to-peak voltage of the waveform. For example, if the oscilloscope is set for 20 V/div., 3 div. of deflection (height) measure a 60-V signal (that is, 3 div. × 20 V/div. = 60 V). Similarly, if the scope is calibrated for 10 V/div., 6 div. of height also measure a 60-V signal (Figure 32–1).

Note that there are no voltage scales on the graticule—there are simply linear markings whose value is determined by the calibration factor to which the Volts/Div. vertical-gain controls of the oscilloscope are calibrated. Thus, the Volts/Div., or vertical-gain, controls correspond to the range controls of a voltmeter; the graticule markings correspond to the scale of an analog voltmeter.

Checking Calibration Accuracy

The triggered oscilloscope has a calibrated vertical amplifier. There are two input controls that affect, by attenuation, the height of the waveform displayed on the screen. One control is continuously variable and is usually marked VAR. A switch-type Volts/Div. control changes the signal attenuation by precalibrated steps. *The VAR control must be turned to its calibrated position, which may be completely clockwise or counterclockwise. Then, the marking to which the switch-type Volts/Div. control is set is the calibration factor of the signal-measuring vertical amplifier of the oscilloscope.* Thus,

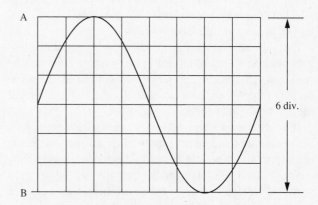

Figure 32–1. The maximum height of the sine wave (the distance between points A and B) is 6 divisions. If the scope is set at 10 V/div., then the sine wave indicates a voltage of 60 V peak-to-peak.

with VAR control set to its calibrated position and the Volts/Div. control set to 0.5 V, every 1 div. of height corresponds to 0.5 V.

NOTE: These rules apply to a scope with a direct probe. If a probe with an attenuation ratio is used, the reading must be multiplied by the attenuation factor of the probe.

The Volts/Div. control is frequently a decade-type attenuator; that is, it switches the voltage-range calibration by a factor of 10. For this type of attenuator the calibration marker settings might read 0.01 V, 0.1 V, 1.0 V, 10 V, and so on.

Some triggered oscilloscopes contain a warning light that goes on when the VAR control is not set to its calibrated position—that is, when the scope is uncalibrated. Measurements cannot be accurately made when this light is on. Before measurement, the VAR control must be turned to its calibrated position. The warning light then goes out, and the oscilloscope is again calibrated.

Triggered oscilloscopes have a self-contained voltage calibrator for checking the calibration of the Volts/Div. selector. The output of this calibrator may be a single voltage, as for example 1 V peak-to-peak, or the scope may have a range of calibration voltages selected by a switch. For this switch-type calibrator, the measured outputs usually correspond to the positions of the Volts/Div. control. Thus, for the Volts/Div. vertical-gain selector just described, the calibrator would deliver 0.01 V, 0.1 V, 1.0 V, and so on, depending on the switch setting of the calibrator.

If the scope calibrator delivers just a single voltage, say 1 V, the calibration of the vertical amplifiers may be checked on several, but not all, settings. Thus, the 1-V/div. vertical selector setting of the scope may be checked, as may the 0.5-V/div. setting. But a setting of, say, 20 V/div. cannot be checked with any degree of accuracy. We must either assume that the other attenuator settings are correct—that is, that there has been no change in the attenuator voltage divider—or use an external calibrator with variable outputs. Accurate external voltage calibrators are available, although technicians rarely require them for the triggered oscilloscopes used to troubleshoot circuits.

External calibrators provide line-derived adjustable signal voltages. The peak-to-peak output-voltage level may be read on a self-contained meter or from calibrated dials. The output of the voltage calibrator is fed to the vertical input of the oscilloscope to check its calibration.

If an external voltage calibrator is not available, a sine-wave generator and an accurately calibrated peak-to-peak-reading ac voltmeter can be used. The generator frequency control is set at 1000 Hz, and the output control is set for a specified voltage as read by the ac voltmeter used here as a standard. The measured signal acts as the calibration source and is fed to the vertical input leads of the scope.

To check the calibration accuracy of the Volts/Div. selector, the technician would apply the oscilloscope probe to the calibrator output jack and note the height of the waveform displayed. Thus, if the calibrator were set for 1.0 V output, the calibrator signal should be deflected 1 div. in height when the Volts/Div. selector is set to 1 V/div.

SUMMARY

1. An oscilloscope can be used for ac voltage measurements.
2. Triggered oscilloscopes contain calibrated signal amplifiers. These are used directly in the measurement of ac voltage waveforms and dc voltages.
3. Before a triggered oscilloscope is used to measure voltage, its calibration factors, Volts/Div. setting, should be checked by applying the measured calibration voltages available on the oscilloscope to the input terminals of the scope. If they check, the scope is ready to make measurements.
4. An accurate, known voltage source, whether provided by the oscilloscope calibrator or by an external calibrator, may be used to check the accuracy of calibration of the vertical amplifiers.
5. Oscilloscope signal amplifiers are normally calibrated to make peak-to-peak measurements.
6. Instruments called voltage calibrators may be used as external calibration sources. They deliver a measured signal voltage for calibration purposes.

SELF-TEST

Check your understanding by answering the following questions:

1. A dc voltage _____ (can/cannot) be measured with an oscilloscope.
2. For peak-to-peak waveform measurements, the oscilloscope _____ amplifiers must be _____ .
3. Triggered oscilloscopes _____ (have/do not have) calibrated vertical amplifiers.
4. The Volts/div. selector setting on a triggered scope determines the _____ factor by which the signal _____ of the waveform must be multiplied to get the voltage of the waveform.
5. The Volts/Div. vertical-gain selector is calibrated and set at 2 V/div. At this setting the height of a waveform is 3.5 div. The voltage of this waveform is _____ V.
6. If the same signal voltage as in question 5 is applied to the calibrated vertical input of the oscilloscope but the vertical-gain selector is set at 1 V/div., the height of the waveform is _____ div.
7. (True/False) Calibration voltages are generally available on triggered oscilloscopes. _____

8. If an oscilloscope does not have a self-contained source for checking voltage calibration, an _____ may be used.

MATERIALS REQUIRED

Power Supplies:
- Isolation transformer
- Variable-voltage autotransformer (Variac or equivalent)
- Variable 0–15 V dc, regulated

Instruments:
- Triggered oscilloscope with direct probe (single- or dual-trace scope may be used)
- EVM or DMM

Resistors (½-W, 5%):
- 1   5100-Ω
- 1   10,000-Ω
- 1   15,000-Ω

Miscellaneous:
- SPST switch
- Polarized line cord, with on-off switch, and fuse.

PROCEDURE

Because the types and labeling of controls vary from one oscilloscope manufacturer to another, the procedures that follow use general terminology. Check the operating instructions for your scope for specific information on comparable controls.

A. Checking Calibration Accuracy

A1. Locate the calibration voltage jacks or terminals on your scope. There may be more than one calibration voltage available on your scope.

A2. Turn **on** the scope. Connect the direct probe of the scope to the calibration voltage terminal. If there is more than one calibration voltage, use the lowest voltage first. If the scope is dual-trace, perform the tests in this procedure with the Channel 1 input only. The Volts/Div. variable control must be set on Calibrated.

A3. The calibration voltage is usually a square wave. Set the Volts/Div. control to the lowest voltage that will produce a measurable square wave on the screen. Record the number of peak-to-peak divisions of the square wave displayed. Record your measurements in Table 32–1. Set the Volts/Div. control, in turn, to each value that will produce a measurable square wave display. Record all values in Table 32–1 (p. 227).

A4. If there is more than one calibration voltage, repeat step A3 for each of the voltages and each Volts/Div. setting that produces a measurable square wave display. When all measurements have been completed, disconnect the probe and turn **off** the scope.

A5. If the scope you are using does not appear to be properly calibrated, based on your readings, notify your instructor. If the scope is reasonably accurate, proceed to Part B.

B. Measuring AC Voltages with an Oscilloscope

B1. With the line cord unplugged, its switch **off**, and switch S_1 **open**, connect the circuit of Figure 32–2 (p. 226). The autotransformer should be set at its lowest output voltage.

B2. Turn the line cord switch **on**; **close** S_1. Increase the output voltage of the autotransformer to produce a voltage of 18 V p-p across AD (voltmeter V measures effective values).

B3. Turn **on** the scope. Connect the probe of the scope to point A and the ground clip to point D. Adjust the scope's controls to produce a single sine-wave display. Use the Volts/Div. control to adjust the height of the wave to almost fill the full height of the screen. Record the peak-to-peak divisions and the Volts/Div. setting of voltage V_{AD} in Table 32–2 (p. 227).

B4. Repeat the procedure of step B3 with the probe connected across BD. Record your measurements in Table 32–2 for V_{BD}.

B5. Repeat step B3 with the probe connected across CD. Record your measurements in Table 32–2 for V_{CD}. Turn **off** and disconnect the scope. **Open** S_1 and turn the line cord **off**. Unplug the line cord.

B6. Convert the height measurements from steps B3, B4, and B5 to peak-to-peak voltages using the Volts/Div. settings in each case. Record your answers in Table 32–2.

B7. Using circuit formulas, calculate the peak-to-peak voltages V_{AD}, V_{BD}, and V_{CD}, assuming a supply voltage of 18 V p-p. Record your answers in Table 32–2.

C. Measuring DC Voltages with an Oscilloscope

C1. With the power supply **off**, connect the three series resistors of part B across the dc power supply, as in Figure 32–3 (p. 226).

C2. Turn **on** the power supply and adjust the voltage to 15 V dc across AD (V_{AD}).

C3. Turn **on** the oscilloscope. Adjust the controls to produce a horizontal trace centered vertically on the screen (on some graticules this is the *x*-axis). This trace will be the reference voltage (0).

C4. Connect the oscilloscope across AD. Measure the number of divisions the trace moves from the reference line. Record this value and the Volts/Div. setting in Table 32–3 (p. 227).

C5. Repeat step C4 for points BD (V_{BD}). Record the values in Table 32–3.

C6. Repeat step C4 for points CD (V_{CD}). Record the values in Table 32–3.

C7. Measure voltages V_{AD}, V_{BD}, and V_{CD} with a voltmeter and record the values in Table 32–3.

C8. Using the Volts/Div. setting and height measurements in Table 32–3 for V_{AD}, V_{BD}, and V_{CD} from steps C4, C5, and C6, convert the values to voltages. Record your answers in Table 32–3.

C9. Calculate voltage V_{AD}, V_{BD}, and V_{CD} using circuit formulas and a supply voltage of 15 V. Record your answers in Table 32–3.

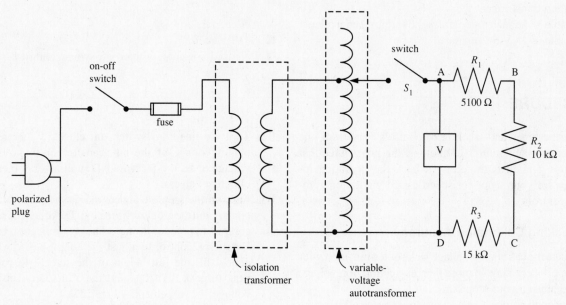

Figure 32–2. Circuit for procedure step B1.

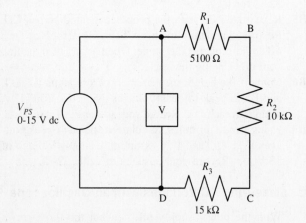

Figure 32–3. Circuit for procedure step C1.

ANSWERS TO SELF-TEST

1. can
2. vertical or signal; calibrated
3. have
4. calibration; height

5. 7
6. 7
7. true
8. external calibrator

TABLE 32–1. Checking Calibration Accuracy

Calibration Voltage, $V_{p\text{-}p}$	Peak-to-Peak Height of Square Wave, div.	Volts/Div. Setting	Calculated Voltage, $V_{p\text{-}p}$

TABLE 32–2. AC Voltage Measurement

Test Points	Voltage	Volts/Div. Setting	Peak-to-Peak Height of Wave, div.	Measured Voltage (Oscilloscope), V	Calculated Voltage (Circuit Formula), V
A to D	V_{AD}				
B to D	V_{BD}				
C to D	V_{CD}				

TABLE 32–3. DC Voltage Measurement

Test Points	Voltage	Volts/Div. Setting	Height from Reference, div.	Measured Voltage (Oscilloscope), V	Measured Voltage (Voltmeter), V	Calculated Voltage (Circuit Formulas), V
A to D	V_{AD}					
B to D	V_{BD}					
C to D	V_{CD}					

QUESTIONS

1. Explain, in your own words, the general procedure for measuring voltage using an oscilloscope.

_____

_____

_____

_____

_____

_____

_____

Oscilloscope Voltage Measurements

2. Explain the importance of procedure steps A3 and A4 in making accurate voltage measurements with an oscilloscope.

_____

_____

_____

_____

_____

3. Refer to your data in Table 32–2. Compare the voltage measurements made with the oscillosope and your calculated values. Discuss any differences.

_____

_____

_____

_____

_____

4. Refer to the data in Table 32–3. Compare the voltage measurements made with the oscilloscope and the voltmeter with your calculated values. Discuss any differences.

_____

_____

_____

_____

_____

5. Discuss the advantages of using an oscilloscope to make ac voltage measurements.

_____

_____

_____

_____

6. Discuss the disadvantages, if any, of using an oscilloscope to measure ac voltage.

_____

_____

_____

_____

_____

33

PEAK, RMS, AND AVERAGE VALUES OF AC

OBJECTIVES

1. To learn the relationships among peak, rms, and average values of alternating voltage and current
2. To verify by experiment the relationship between peak and rms values of an ac voltage

BASIC INFORMATION

Generating an AC Voltage

When a conductor cuts lines of magnetic force, a voltage is induced in the conductor. This is the principle of the electric generator. We will apply this principle to the generation of an ac voltage.

Consider a single loop of wire so arranged on a shaft that it can be rotated through the magnetic field that exists between the north (N) and south (S) poles of a magnet, as in Figure 33–1. As the loop of wire turns, it cuts across the magnetic lines of force. This induces a voltage in each side of the loop. The direction (polarity) of the induced voltage is determined by the direction of the magnetic field and the direction of motion of the conductor.

In Figure 33–1 the loop is shown rotating in a counterclockwise direction. The left side of the loop is traveling down through the magnetic field, whereas the right side of the loop is traveling up through the field. The polarity of the voltage induced in one side is therefore opposite to that induced in the other side. But the two sides of the loop are actually in series, so that the two induced voltages are additive.

The voltage induced in the loop is proportional to the rate at which the lines of force are cut by the loop. Figure 33–2 (p. 230) is an end view of the loop at various positions as it rotates between the north and south poles. Positions 1 and 5, 2 and 6, 3 and 7, 4 and 8 represent opposite sides of the same loop.

At position 1 the side of the loop is momentarily moving parallel to the lines of force. No lines are being cut, therefore, no voltage is being induced in this loop and its other side (position 5). As the loop moves to position 2 (and 6), it begins cutting lines of force at an oblique angle, and a voltage is induced in the sides of the loop. If the ends of the loop were connected to a load, current would flow in the loop. The symbol $\otimes$ shown in position 2 indicates that electron current would flow into the page; the symbol $\odot$ in position 6 indicates that current would flow out of the page (as would be expected, because 2 and 6 are in series). As the loop

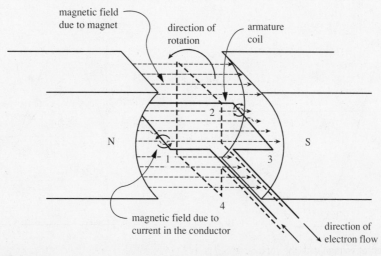

Figure 33–1. Simple electric generator.

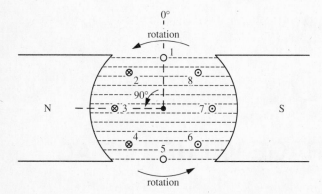

Figure 33–2. Rotation of the single loop in the magnetic field.

continues to rotate, the angle at which it cuts the lines of force becomes more nearly perpendicular until at position 3 and 7 the loop is crossing the lines of force at a right angle. Since the loop is assumed to be rotating at a constant speed, positions 3 and 7 are where the greatest number of lines of force are being cut per unit of time, and positions 1 and 5 are where the least lines of force are being cut per unit of time. It follows then that at positions 3 and 7 the highest voltage is being induced, and at positions 1 and 5 the lowest voltage is being induced (0 V in this case). Past position 3 the loop begins cutting the lines obliquely until at position 5 the loop is not cutting any lines.

If the voltage induced in the loop were plotted against the position of the loop within the magnetic field, the graph would resemble the one in Figure 33–3.

A complete rotation of the loop past the north and south poles is called a cycle. One cycle is equal to 360 electrical degrees. The positive or the negative half of each cycle is called an alternation.

The voltage induced in the loop will be a sine wave. If we call the peak of the instantaneous value of this sine wave V_M,

then the voltage V at any instant of time, when the loop has rotated θ degrees, is

$$V = V_M \sin \theta \qquad \textbf{(33–1)}$$

The loop represents only one coil of an ac generator (also called an *alternator*) armature. In a practical generator the loop ends are brought out and connected to slip rings. The typical ac generator also contains more than two poles. The armature can be rotated mechanically by a steam turbine, a gasoline engine, a water wheel, or even a hand crank. In the large ac generators used by power companies, the armature is stationary and built around the frame of the machine, and the magnetic poles are mounted on a shaft and rotated within the armature. The effect of wires cutting a magnetic field is the same whether the wire moves across a stationary field or a moving field cuts across a stationary wire. In both cases, a voltage is induced in the wire.

AC Voltage and Current

The generator in Figure 33–1 produces an ac voltage. When an ac voltage is connected to a resistive load, *alternating current* flows in the circuit. Because a cycle of ac voltage consists of one *positive* and one *negative* alternation, current in the circuit will flow in one direction during the positive alternation and in the opposite direction during the negative alternation. Ohm's and Kirchhoff's laws apply to a resistive ac circuit in the same way that they apply to dc circuits. However, in a dc circuit the applied voltage has a *single* value. In ac circuits the voltage is constantly changing in amplitude and polarity.

The effect of an ac voltage on current in a resistive circuit is shown in Figure 33–4. As the voltage $V_M \sin \theta$ increases from 0 to V_M, the current increases from 0 to I_M; that is, a maximum current occurs at the same time as a maximum voltage is reached. As the voltage drops from V_M back to 0, the current drops from I_M to 0. During the second (negative) alternation of V, the current through R reverses direction.

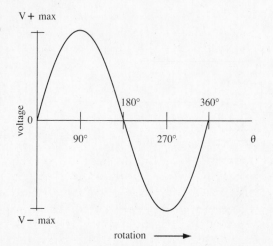

Figure 33–3. The variation of voltage with the rotation angle is a sine wave.

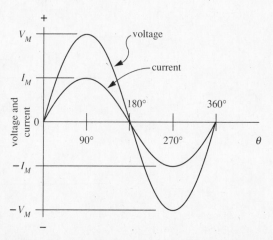

Figure 33–4. Voltage and current are in phase in a resistive circuit.

As the voltage reaches its maximum negative value ($-V_M$), the current also reaches its maximum value ($-I_M$); it then decreases to zero as the voltage decreases to 0. Thus, Figure 33–4 shows that current variations follow exactly the voltage variation in a resistive ac circuit. It is also evident that the shape of the current waveform is also a sine wave. The instantaneous values of current in an ac waveform can be given by the formula

$$I = I_M \sin \theta \qquad (33-2)$$

We can also say that in a resistive ac circuit, current and voltage are *in phase*. There are ac circuits that contain capacitors and inductors in which current and voltage are not in phase. These are considered in later experiments.

Peak, RMS, and Average Values

The amplitude of a dc voltage can be identified by a single value. Can we specify an ac voltage by a single value? There are actually three different values that can be used—namely, the *peak*, the *rms*, and the *average* value. Each identifies a different characteristic of the voltage, but they are all related. Given one value, the other two can be easily calculated.

Peak value. If we say that the peak value of a sinusoidal voltage is 100 V, we mean that the voltage reaches a maximum of +100 V on the positive alternation and –100 V on the negative alternation. From the peak value we can calculate the instantaneous level of voltage at any angle θ in the cycle using formula (33–1).

The peak value does *not* produce the same power as the same dc value, because the ac voltage varies constantly in amplitude, whereas the dc voltage maintains a constant level.

RMS value. There is a value of ac voltage that will produce the same power as the equivalent dc level. This is the rms, or *root mean square*, value. Thus, if we say that the rms value of an ac voltage is 100 V, we mean that it will deliver the same power as 100 V dc. The rms value is the *square root* of the *mean* (average) of the sum of the *squares* of the instantaneous values of voltage in an ac alternation. That is,

$$V_{rms} = \sqrt{\frac{V_1^2 + V_2^2 + \cdots + V_n^2}{n}}$$

where $V_1, V_2, \ldots, V_n$ are successive instantaneous values of $V_M \sin \theta$. It can be shown that

$$V_{rms} = 0.707 V_M \qquad (33-3)$$

and that

$$V_M = 1.414 V_{rms} \qquad (33-4)$$

So we have two unique but related values of V that describe the amplitude of an ac voltage. The rms values (also called *effective* values) are more popularly used than peak values in giving the amplitude of an ac voltage. That is why ac voltmeters are calibrated in rms rather than peak values.

Electric power is a square function of voltage (V^2/R) or current ($I^2 \times R$). That is why the rms value, which is derived from a square value (V_1^2, etc.), is comparable to the equivalent value of dc voltage.

Average value. There is yet another method of identifying an ac voltage, that is, by using the *average* value of the waveform. Voltage V_{ave} is also dependent on the peak value and the rms value and may be determined from these values by the formulas

$$V_{ave} = 0.636 V_M = 0.899 V_{rms} \qquad (33-5)$$

The peak voltage V_M can also be specified in terms of V_{rms} and V_{ave}. Thus,

$$V_M = 1.414 V_{rms} = 1.572 V_{ave} \qquad (33-6)$$

Alternating current is also identified in any one of three ways, peak (I_M), rms (I_{rms}), and average (I_{ave}). The formulas for current are identical to the formulas for voltage except that I replaces V.

Measurement of AC Voltages and Currents

In the design of ac circuits, voltage and current measurements are usually made in rms values. An analog meter responds to average values, but the ac scale is calibrated in rms values. Normally a VOM does not have an ac current function. Current measurements can be made with a digital VOM that has both ac voltage and current scales. These are also calibrated in rms values. Some meters have both rms and peak value scales, but these may be used only in the measurement of *sinusoidal* voltages or currents.

An oscilloscope is normally used for measuring peak ac voltages or peak-to-peak values.

SUMMARY

1. An ac generator produces a sinusoidal voltage.
2. The formula for the instantaneous value V of a sinusoidal voltage is

$$V = V_M \sin \theta$$

where V_M is the peak, or maximum voltage, of the waveform and θ is the angle the waveform has reached at the instant it is measured.
3. The amplitude of a sinusoidal ac waveform may be specified in three ways: (a) its peak value V_M; (b) its root mean square value V_{rms}; and (c) its average value V_{ave}.
4. The rms value is comparable to the equivalent value dc voltage because the *power* that each can produce is the same.

5. The three values of a sinusoidal voltage are interdependent, as their formulas indicate. Thus,

$$V_{rms} = 7.707 V_M$$
$$V_{ave} = 0.636 V_M = 0.899 V_{rms}$$

6. These formulas may also be written as

$$V_M = 1.414 V_{rms} = 1.572 V_{ave}$$

7. The ac scale of an analog voltmeter and ammeter is calibrated in rms values, although the meter movement responds to average values.
8. When an ac voltage is given without specifying whether it is rms, peak, or average, the rms value is meant.
9. The relationships among rms, peak, and average values of alternating current are the same as those for ac voltage, with I substituted for V in the voltage formulas.
10. Ohm's and Kirchhoff's laws apply to resistive ac circuits just as they do to dc circuits. Thus, the rms value of current in a resistor, across which V_{rms} is applied, may be calculated from the formula

$$I_{rms} = \frac{V_{rms}}{R}$$

Similarly, I_M may be calculated for the resistor R by substituting in the formula

$$I_M = \frac{V_M}{R}$$

SELF-TEST

Check your understanding by answering the following questions:

1. An ac voltage changes constantly in _____ and _____ .
2. The shape of the voltage waveform delivered by an ac generator is a _____ .
3. (True/False) The rms and average values of a sinusoidal voltage or current may be calculated from its peak value. _____
4. The V_M of a sinusoidal voltage is 25 V. The rms value of that voltage is _____ V.
5. The rms value of a sinusoidal current is 5 mA. The peak value of that current is _____ mA.
6. In question 5, the average value of current is _____ mA.
7. The rms voltage measured across a 47-Ω resistor is 15 V. The rms current in that resistor is _____ mA.
8. The peak current in the resistor in question 7 is _____ mA.
9. The ac voltmeter is usually calibrated in _____ (peak/rms/average) values.
10. The peak value of an ac voltage may be measured with a(n) _____ .

MATERIALS REQUIRED

Power Supplies:
- 120-V, 60-Hz power source
- Isolation transformer
- Variable-voltage autotransformer (Variac or equivalent)

Instruments:
- Sine-wave generator (AF range)
- DMM
- Oscilloscope
- AC ammeter (or additional DMM)

Resistors ($^1/_2$-W, 5%):
- 2   33-Ω
- 1   47-Ω
- 2   1000-Ω
- 1   1500-Ω

Miscellaneous:
- SPST switch
- Line cord set consisting of polarized plug, fuse, and on-off switch

PROCEDURE

CAUTION: You will be dealing with voltages that are potentially lethal. Pay strict attention to all electrical safety rules when performing this experiment.

A. 60-Hz Supply

A1. With the line cord unplugged and its switch off, connect the circuit of Figure 33–5. Set the variable voltage transformer to its lowest voltage; S_1 is **open**.

A2. Plug the line cord into a 120-V, 60-Hz outlet; turn the line switch **on**; **close** S_1. Adjust the output voltage of the variable transformer to 35 V. Maintain this voltage throughout Part A. Record this voltage under "Voltage, rms, Measured" in Table 33–1 (p. 235). (The voltmeter will read the voltage in rms units.)

A3. Measure the voltage across R_1, R_2, and R_3 with the voltmeter, and record the values under "Voltage, rms, Measured" in Table 33–1. **Open** S_1; line cord switch **off**.

A4. Connect an oscilloscope across the output of the variable-voltage transformer. Line cord switch **on**; S_1 **closed**.

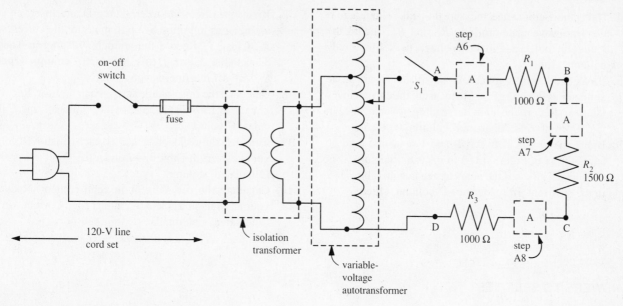

Figure 33-5. Circuit for procedure step A1.

Measure the peak voltage across the transformer output terminals with the oscilloscope and record the value in Table 33-1 under the "Voltage, peak, Measured" column.

A5. Repeat step A4, using the oscilloscope to measure the peak voltage across each of the resistors $R_1, R_2,$ and R_3. Turn **off** power whenever you change circuit connections. Record the values in Table 33-1. Switch line cord **off**; open S_1.

A6. With power to the resistors **off**, break the circuit between A and B and insert an ammeter in the circuit as shown in Figure 33-5 to measure the current through R_1. Set the range of the ac ammeter at 20 mA or above. Turn line cord switch **on**; **close** S_1. Record the ammeter reading in the "Current, rms, Measured" column in Table 33-1. S_1 **open**; line cord switch **off**.

A7. Break the circuit between B and C and insert an ammeter as in step A6 to measure the current through R_2. Line cord switch **on**; S_1 **closed**. Record the ammeter reading in Table 33-1 in the "Current, rms, Measured" column. S_1 **open**. Line cord switch **off**.

A8. Break the circuit between C and D and insert an ammeter as in step A6 to measure the current through R_3. Line cord switch **on**; S_1 **closed**. Record the ammeter reading in Table 33-1 in the "Current, rms, Measured" column. S_1 **open**; line cord switch **off**; line cord **unplugged** from outlet.

A9. Calculate the rms current that should be in each resistor. Record your answers in Table 33-1 in the "Current, rms, Calculated" column.

A10. Calculate the rms voltage across each resistor. Record your answers in Table 33-1 in the "Voltage, rms, Calculated" column.

A11. Calculate the peak voltage across each resistor.

Record your answers in Table 33-1 in the "Voltage, peak, Calculated" column.

B. 1000-Hz Supply

B1. With the sine-wave generator **off** and S_1 **open**, connect the circuit of Figure 33-6.

B2. Turn **on** the generator. Set the frequency to 1000 Hz. (The oscilloscope can be used to verify the value.) With an ac voltmeter connected across its output, adjust the generator output to 5 V (or its highest output voltage, but pick an even value to keep calculations simple). Record the value in Table 33-2 (p. 235) in the "Voltage, rms, Measured" column.

B3. With the voltmeter measure the rms voltage across each resistor, $R_1, R_2,$ and R_3. Record the values in Table 33-2 in the "Voltage, rms, Measured" column.

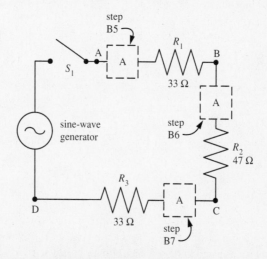

Figure 33-6. Circuit for procedure step B1. The ac ammeter is added in steps B5, B6, and B7.

B4. Using the oscilloscope, measure the peak voltage across the sine-wave generator, R_1, R_2, and R_3. Record the values in Table 33–2 in the "Voltage, peak, Measured" column. **Open** S_1.

B5. Break the circuit between A and B and insert an ac ammeter as in Figure 33–6 to measure the current in R_1. **Close** S_1. Record the ammeter reading in Table 33–2 in the "Current, rms, Measured" column. **Open** S_1.

B6. Break the circuit between B and C and insert an ac ammeter as in Figure 33–6 to measure the current in R_2. **Close** S_1. Record the ammeter reading in Table 33–2 in the "Current, rms, Measured" column. **Open** S_1.

B7. Break the circuit between C and D and insert an ac ammeter as in Figure 33–6 to measure the current in R_3. **Close** S_1. Record the ammeter reading in Table 33–2 in the "Current, rms, Measured" column. **Open** S_1; turn **off** the sine-wave generator.

B8. Calculate the rms voltage across each resistor. Record your answers in Table 33–2 in the "Voltage, rms, Calculated" column.

B9. Calculate the peak voltage across each resistor. Record your answers in Table 33–2 under the "Voltage, peak, Calculated" column.

B10. Calculate the rms current in each resistor. Record your answers in Table 33–2 under the "Current, rms, Calculated" column.

ANSWERS TO SELF-TEST

1. amplitude; polarity
2. sine wave
3. true
4. 17.7
5. 7.07
6. 4.50
7. 319
8. 451
9. rms
10. oscilloscope

Experiment 33 Name _____ Date _____

TABLE 33–1. 60-Hz Voltage and Current

	Voltage, rms, V		Voltage, peak, V		Current, rms, mA	
	Measured	Calculated	Measured	Calculated	Measured	Calculated
Variable transformer output		✕		✕	✕	✕
R_1						
R_2						
R_3						

TABLE 33–2. 1000-Hz Voltage and Current

	Voltage, rms, V		Voltage, peak, V		Current, rms, mA	
	Measured	Calculated	Measured	Calculated	Measured	Calculated
Sine-wave generator output		✕		✕	✕	✕
R_1						
R_2						
R_3						

QUESTIONS

1. Explain, in your own words, the relationship between peak, rms, and average values of an ac voltage.

_____

_____

_____

_____

2. Refer to your data in Table 33–1. Compare the measured peak voltage with the calculated value. Explain any differences.

_____

_____

_____

_____

3. Refer to your data in Table 33–1. Compare the measured rms voltage with the calculated value. Explain any differences.

_____

_____

_____

_____

Peak, RMS, and Average Values of AC

4. Formula 33–3 can be used to find V_{rms} when V_M is known:

$$V_{rms} = 0.707V_M$$

The constant 0.707 is, therefore, equal to the ratio V_{rms}/V_M. Using your measured data in Table 33–1 for V_{rms} and V_M, calculate the average value of V_{rms}/V_M for the three resistor measurements. Compare your answer with the constant 0.707; explain any difference. (Show all calculations.)

5. Repeat the process of Question 4 using your data in Table 33–2. Compare your answer with the constant 0.707; explain any difference. Also compare your calculated value with your value for Question 4. Explain any difference.

6. Do the data in the current column of Tables 33–1 and 33–2 verify the application of Ohm's law to resistive ac circuits? Refer to data to substantiate your answer.

_____

_____

_____

_____

_____

34

CHARACTERISTICS OF INDUCTANCE

OBJECTIVES

1. To observe experimentally the effect of an inductance on current in a dc and ac circuit
2. To verify experimentally the formula for inductive reactance:

$$X_L = 2\pi fL$$

BASIC INFORMATION

Inductance and Reactance of a Coil

Resistors limit the amount of current in dc as well as in ac circuits. In addition to resistors, reactive components, such as inductors and capacitors, impede currents in ac circuits.

In the circuit of Figure 34–1 a sine wave of voltage V produces a sine-wave current through the coil. This changing current creates a varying magnetic field about the coil. The periodic buildup and collapse of the magnetic field causes magnetic flux to cut across the windings of the coil, inducing a voltage V' in these windings. By Lenz' law this voltage is in opposition to the source voltage V and can therefore be termed a *counter emf*. When V is positive, V' is negative but less than V. The numerical value of the resultant emf that powers the circuit is $|V| - |V'|$, where $|V|$ and $|V'|$ are the absolute values of V and V', respectively. Because of the counter emf the alternating current in the coil is less than it would be if V were a dc source and a steady-state current were flowing. This property of the coil that opposes a changing current is termed the *inductance* of the coil.

The unit of inductance is the *henry* (abbreviated H), named in honor of the American physicist Joseph Henry. The number of henrys in an inductor can be measured by means of an instrument called an *inductance bridge*. The symbol used to denote inductance is L. The amount of opposition offered by an inductor is called *inductive reactance* and is denoted by the symbol X_L. Inductive reactance is measured in ohms.

The inductive reactance of a coil is not constant but is a variable quantity. It depends on the inductance L and the frequency f of the voltage source. Inductive reactance may be calculated from the formula

$$X_L = 2\pi fL \qquad \text{(34–1)}$$

where

■ π is the constant 3.14,
■ f is the frequency in hertz,
■ L is the inductance in henrys.

Some important facts may be deduced from formula (34–1). The first is that X_L varies directly with frequency—that is, it is directly proportional to f. This variation may be shown graphically by a specific example.

Problem. Draw a graph of X_L versus f for an inductor whose inductance is $L = 1.59$ H.

Solution. The inductive reactance X_L is found for each of a series of frequencies. Thus, when $f = 0$ Hz,

$$X_L = 2\pi fL = 2(3.14)(0)(1.59) = 0 \ \Omega$$

When $f = 100$ Hz,

$$X_L = 6.28(100)(1.59) = 1000 \ \Omega$$

Similarly, it can be shown that when

$$f = 200 \text{ Hz}, X_L = 2000 \ \Omega$$
$$f = 300 \text{ Hz}, X_L = 3000 \ \Omega$$
$$f = 400 \text{ Hz}, X_L = 4000 \ \Omega$$
$$f = 500 \text{ Hz}, X_L = 5000 \ \Omega$$

Figure 34–2 (p. 238) shows the variation of X_L versus f for $L = 1.59$ H. It is a straight-line graph, so we call the relationship between X_L and f a *linear* one.

A second fact that may be seen from formula (34–1) is that for direct current, that is, for $f = 0$, $X_L = 0$. This fact is consistent with the definition of inductance, as that characteristic

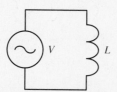

Figure 34–1. The sine wave of voltage will produce a current sine wave in the inductance, L.

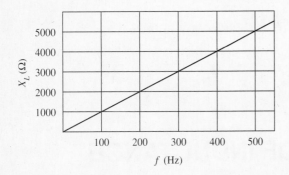

Figure 34–2. As frequency increases, inductive reactance increases.

of a coil that opposes a *change* in current. A steady direct current produces no change. Inductance has no effect on a direct current with a constant voltage; $X_L = 0$ for such a direct current.

Formula (34–1) also defines the relationship between X_L and L. If f is kept constant, X_L increases or decreases as L is increased or decreased, respectively. Again, if X_L is plotted against L, a straight line results.

Resistance of a Coil

An inductor consists of a number of turns of insulated wire wound around a core. There may be only several turns, or there may be thousands of turns. The more turns there are on a particular core, the greater is the inductance of the coil. The diameter of the wire used in winding a coil depends on the maximum current the coil will be required to carry. The larger the diameter of the wire, the greater the current capabilities of the coil. A smaller-diameter wire will carry less current. If more current is permitted to flow in a coil than that for which it is rated, the coil will overheat and may destroy the insulation around the windings, thus shorting windings together. Or an overheated coil may burn out, that is, the wire may open. In either event the coil will become defective.

Many circuits require relatively low currents. Very fine diameter wire is normally used in the construction of the coils used in these circuits. Since the resistance of a wire is directly proportional to its length, the resistance of inductors consisting of many turns of fine wire will be relatively high.

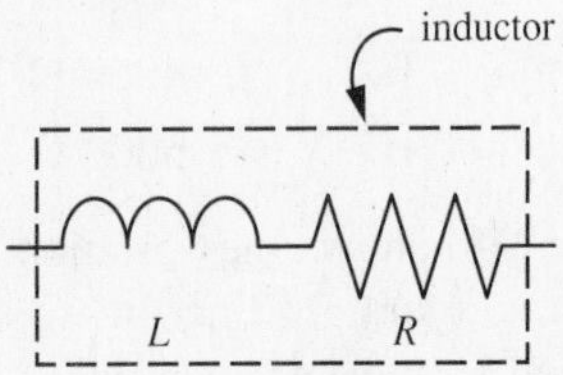

Thus, the characteristics of inductance *and resistance* are associated with an inductor. Although the resistance of a coil cannot be separated from the inductance, a schematic symbol of a coil with inductance L and resistance R (Figure 34–3) shows these two characteristics as though they were separate and acting in series. The resistance can be measured with an ohmmeter. For practical reasons in solving circuits problems a coil is usually represented by its series-equivalent form (Figure 34–3).

The resistance associated with the coil suggests that one method of troubleshooting an inductor is to place an ohmmeter across the two terminals and compare the measured with the rated resistance. If they are the same, it may be assumed that the winding is continuous—that is, it is neither open nor shorted. If the resistance is infinite, the inductor is open. If the resistance is very much lower than the rated value, say, 20 Ω instead of 500 Ω, it may be assumed that the coil winding is shorted. In practice, resistance checks of an inductor are made to determine if it is continuous or open. Special devices are used to check coils for shorted turns.

Measurement of X_L

The inductive reactance of a coil (X_L) may be determined by measurement. In the circuit of Figure 34–4 a sinusoidal voltage V causes a current I to flow. Ohm's law for ac circuits states that the voltage V_L across the coil is $V_L = I \times Z$. The symbol Z stands for *impedance*. This is the term used to express the total opposition to current in ac circuits. If the R of the coil is low compared with X_L, $Z = X_L$ and

$$V_L = I \times X_L \qquad \textbf{(34–2)}$$

This formula may be written in the form

$$X_L = \frac{V_L}{I} \qquad \textbf{(34–3)}$$

Formula (34–3) can be used to solve for X_L of a coil if frequency f is given. This is accomplished using an ac voltmeter to measure the voltage V_L across L and measuring the current I with an ac ammeter. The measured values of V_L and I are then substituted in formula (34–3), and X_L is calculated.

Note that a slight error is introduced by this procedure, for L has resistance in addition to inductance. However, if the

Figure 34–3. Schematic series-equivalent form of an inductor shows separate symbols for the resistive and inductive characteristics.

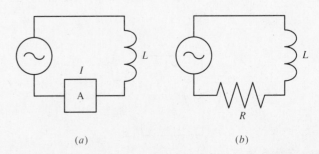

(a) $\qquad\qquad\qquad (b)$

Figure 34–4. Circuits used to determine X_L.

value of X_L is appreciably higher than R of the coil, R may be ignored.

An alternative method for measuring the current I does not require the use of an ammeter. Instead of the meter A, a resistor R of known value is placed in series with L, as in Figure 34–4(b). In a series circuit, current is the same everywhere. Hence, the current in R is the same as the current in L. The voltage V_R is then measured across R. Current is next determined by substituting the measured value of V_R and the known value of R in the formula

$$I = \frac{V_R}{R}$$ (34–4)

Next, V_L is measured and X_L calculated by substituting the values of V_L and I in formula (34–3).

SUMMARY

1. Inductance L is that characteristic of a coil that opposes a change in current. The unit of inductance is the henry (H).
2. The amount of opposition that an inductance offers to a changing current is called inductive reactance (X_L). Inductive reactance is measured in ohms.
3. The inductive reactance of a coil is not a constant but is directly proportional to the frequency f of the changing current. It is also directly proportional to the value of the inductance L.
4. Inductive reactance may be calculated from the formula

$$X_L = 2\pi fL$$

where

X_L is in ohms,
f is in hertz,
L is in henrys.

5. Every coil has some resistance R associated with it. Only the resistance of an inductance will limit direct current in a coil. The inductance of a coil will not affect direct current.
6. The continuity of a coil winding may be determined by measuring the resistance of the coil. If the resistance measured is infinite, the coil winding is open.
7. The ac voltage across a coil V_L is equal to the product of the alternating current in the coil and the inductive reactance of the coil; that is,

$$V_L = I \times X_L$$

8. If the ac voltage V_L across a coil and the alternating current I in the coil are known, the inductive reactance of the coil may be calculated from the formula

$$X_L = \frac{V_L}{I}$$

This formula assumes that the resistance R of the coil is low compared with the reactance of the coil.

SELF-TEST

Check your understanding by answering the following questions:

1. (True/False) Inductance of a coil is measured in ohms.
   _____
2. The _____ of a coil is directly _____ to the _____ of the voltage source, if the L of the coil remains constant.
3. If $L = 2.5$ H and $f = 1000$ Hz, $X_L = $ _____ Ω.
4. If the frequency f is doubled while L is held constant, X_L is _____ .
5. The graph of X_L versus f is a _____ if L is held constant.
6. The resistance R of a coil is associated with the resistance of the _____ of the coil.
7. If the ac voltage V_L across a coil is 6 V and the current I in the coil is 0.02 A, then $X_L = $ _____ Ω.
8. If $f = 60$ Hz and $X_L = 1000$ Ω, then $L = $ _____ H.

MATERIALS REQUIRED

Power Supplies:
- 120-V, 60-Hz source
- Variable-voltage autotransformer (Variac or equivalent)
- Isolation transformer
- Variable 0–15 V dc, regulated

Instruments:
- Oscilloscope
- Variable-frequency sine-wave generator
- DMM
- AC ammeter (25-mA range)
- DC ammeter (25-mA range)

Resistors:
- 1   1000-Ω (½-W, 5%)
- 1 resistor equal to the dc resistance of the choke

Inductors:
- 1 large choke (8 or 9 H) (Magnetek-Triad #C18A 8-H, 110-Ω, 300-mA; Magnetek-Triad #C10X 9-H, 250-Ω, 125-mA; or equivalent)
- 1   30-mH choke

Miscellaneous:
- SPDT switch
- SPST switch
- Polarized line cord set with on-off switch and fuse

PROCEDURE

A. Effect of Inductance on Direct and Alternating Currents

A1. Measure the resistance of the large choke (8 or 9 H). Record the value in Table 34–1 (p. 243). Also record the value of the resistor equal to the dc resistance of the choke.

A2. With the power supply **off**, S_1 **open**, and S_2 in the Ⓐ position, connect the circuit of Figure 34–5. Adjust the variable power supply V_{PS} to its lowest value. Set the meters to read dc voltage and current.

A3. Power **on**; **close** S_1. Increase V_{PS} until the ammeter A reads 20 mA. Do not change this voltage setting, because it will be used in the next step. Record the voltage in Table 34–1 under "Direct-Current Source."

A4. Throw S_2 to position Ⓑ. If necessary, adjust the voltage to equal that in step A3. Measure the current through the inductance and record this value as well as the voltage in Table 34–1 under "Direct-Current Source." **Open** S_1; turn V_{PS} **off**.

A5. With the line cord unplugged, the on-off switch **off**, S_1 **open**, and S_2 in position Ⓐ, connect the circuit of Figure 34–6. The choke and resistor are the same as used in the circuit of Figure 34–5. Set the meters to measure ac voltage and current. Adjust the autotransformer to its lowest value.

A6. Plug the line cord into a 120-V source and turn the line cord switch **on**. **Close** S_1. Increase the voltage of the autotransformer until the ammeter reads 20 mA. Maintain the setting of the autotransformer for the next step. Record the reading of the voltmeter under "Alternating-Current Source" in Table 34–1.

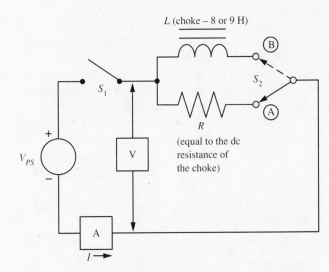

Figure 34–5. Circuit for procedure step A2.

A7. Throw S_2 to position Ⓑ. If necessary, adjust the voltage to equal that of step A6. Measure the current through the inductance and record this value and the voltage in Table 34–1 under "Alternating-Current Source" column.

A8. Increase the voltage of the autotransformer until the ammeter reads 25 mA. Record the voltage in Table 34–2 (p. 243). Decrease the voltage in steps and record it at 20 mA, 15 mA, and 10 mA in Table 34–2. **Open** S_1; turn line cord switch **off**; unplug cord.

A9. Calculate V/I for each voltage and current in Table 34–2. Record your answers in the table. Calculate the average V/I and record your answer in the table. Calculate X_L for the inductance of the choke you are using and 60-Hz frequency. Record your answer in the table.

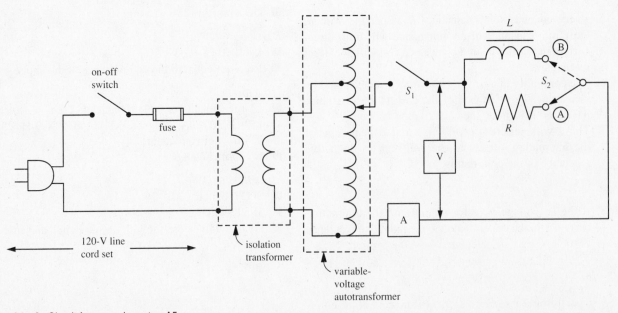

Figure 34–6. Circuit for procedure step A5.

B. Effect of Frequency on Inductance

B1. With S_1 **open** and the variable-voltage sine-wave generator **off**, connect the circuit of Figure 34–7. OSC is an oscilloscope calibrated to measure peak-to-peak voltage. Set the generator at its lowest output voltage.

B2. Examine Table 34–3 (p. 243). Using the oscilloscope, you will take a series of measurements of peak-to-peak voltage around the circuit at the frequencies noted in the table.

B3. **Close** S_1. Adjust the sine-wave generator to an output of 15 V p-p. (If the generator cannot produce a voltage that high, choose as high a voltage as possible but one that will make calculations simple.)

B4. Complete Table 34–3 by measuring and recording the output voltage of the generator, the voltage across the inductance, and the voltage across the resistor.

B5. Calculate the current in the circuit for each frequency using the formula

$$I = \frac{V_R}{R}$$

where

I is the current in the circuit,
V_R is the voltage across the resistor,
R is the measured resistance of the resistor.

Record your answers in Table 34–3.

B6. Calculate the inductive reactance for each of the frequencies, using your measured values for the voltage across the inductor V_L, the voltage across the resistor V_R, and the measured resistance of the resistor R. Since

$$X_L = \frac{V_L}{I_L} \quad \text{and} \quad I_L = I_R = \frac{V_R}{R}$$

$$X_L = \frac{V_L}{\dfrac{V_R}{R}} = \frac{V_L}{V_R} \times R$$

where

X_L is inductive reactance,
V_L is the voltage across the inductance,
I_L is the current in the inductance.

Record your answers in Table 34–3.

B7. Calculate the inductive reactance of the inductor using the formula

$$X_L = 2\pi f L$$

where

f is the frequency of the generator,
L is the rated value of the inductor (30 mH in this experiment).

Record your answers in Table 34–3.

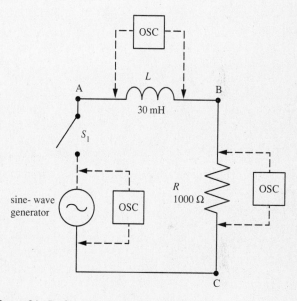

Figure 34–7. Circuit for procedure step B2.

ANSWERS TO SELF-TEST

1. false
2. inductive reactance; proportional; frequency
3. 15,700
4. doubled

5. straight line
6. winding
7. 300
8. 2.65

Name _____ Date _____

TABLE 34–1. Effect of Inductance on Direct and Alternating Current

Resistance (Measured), Ω	Position of S_2	Component in Circuit	Direct-Current Source		Alternating-Current Source	
			Voltage, V	Current, A	Voltage, V	Current, A
Resistor	Ⓐ	R		0.02		0.02
Choke	Ⓑ	L				

TABLE 34–2. Inductive Reactance X_L at 60 Hz

Current, A	0.025	0.020	0.015	0.010	Average $\dfrac{V}{I}$ at Line Frequency, Ω
Voltage, V					
$\dfrac{V}{I}$ Calculated					
Choke X_L Calculated (rated)					

TABLE 34–3. Effect of Frequency on Inductance

Frequency f, Hz	Sine-Wave Generator Voltage V, $V_{p\text{-}p}$	Voltage across Inductance V_L, $V_{p\text{-}p}$	Voltage across Resistor V_R, $V_{p\text{-}p}$	Current I Calculated, A	Inductive Reactance X_L, Ω	
					$\dfrac{V_L}{V_R} \times R$	$2\pi f L$
2,000						
3,000						
5,000						
7,000						
8,000						
9,000						
10,000						

QUESTIONS

1. Explain, in your own words, the effect of inductance on current in a dc circuit. Is the effect the same in an ac circuit? If not, explain the difference.

_____

_____

_____

_____

Characteristics of Inductance

2. Refer to your data in Table 34–1. Is there any significant difference between the effect of L and R on current in a dc circuit as compared to the effect of L and R on current in an ac circuit? Cite specific data in your table.

_____

_____

_____

_____

_____

_____

3. Refer to your data in Table 34–2. Is X_L the same for each calculated value of V/I? If not, explain any differences.

_____

_____

_____

_____

_____

_____

4. Refer to your data in Table 34–3. Are the calculated values of I peak values or rms values? Explain.

_____

_____

_____

_____

_____

5. On a separate $8\frac{1}{2} \times 11$ sheet of graph paper, plot a graph of current I_L through the inductance versus the voltage V_L across the inductance using your data in Table 34–2. The horizontal (x) axis should be V_L and the vertical (y) axis should be I_L. Label the graph and the axes.

6. On a separate $8\frac{1}{2} \times 11$ sheet of graph paper (do not use the sheet from Question 5), plot a graph of frequency f versus inductive reactance X_L using your data in Table 34–3. Choose an appropriate scale to show the relationship between X_L and f. The horizontal axis should be frequency and the vertical axis should be inductive reactance.

35

OSCILLOSCOPE FREQUENCY AND PHASE MEASUREMENTS

OBJECTIVES

1. To measure the frequency of an ac waveform using an oscilloscope with a calibrated horizontal sweep
2. To measure the phase angle between voltage and current in an inductance, using a dual-trace oscilloscope

BASIC INFORMATION

Measuring Frequency

In Experiment 32 the oscilloscope was used to measure ac and dc voltages. Oscilloscopes can also be used to measure frequency indirectly.

If the time base of the oscilloscope is calibrated in time units per division, then the number of horizontal divisions covered by a single cycle will represent the period of the cycle. In Figure 35–1 a single cycle of a sine wave spans 8 horizontal divisions. If the calibrated time base is set at 0.5 ms/div., the 8 divisions represent

$$t = 0.5 \text{ ms/div.} \times 8 \text{ div.}$$
$$= 4 \text{ ms}$$

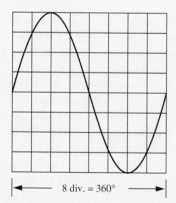

|←——— 8 div. = 360° ———→|

Figure 35–1. Sine wave displayed on the oscilloscope screen. The graticule grid consists of 8 × 8 divisions.

where t is the period of the waveform. Because the period is the reciprocal of frequency, the frequency of the wave can be calculated using the formula

$$f = \frac{1}{t} \qquad \textbf{(35–1)}$$

In the case of Figure 35–1, the frequency is

$$f = \frac{1}{4 \times 10^{-3}\text{s}} = 250 \text{ Hz}$$

The time base is provided by the horizontal sweep. The calibrated sweep is the rate at which the sweep travels horizontally across the face of the screen.

By accurately measuring the number of divisions spanned by a single cycle and multiplying the divisions by the time-base setting, the period of the wave can be found. Taking the reciprocal of the period will give the frequency, as in formula (35–1).

As we discussed in the experiment on oscilloscope voltage measurements, voltmeters can provide much more accurate measurements, although oscilloscopes do have important and unique characteristics for measuring voltages, especially when dealing with peak-to-peak values. Similarly, frequency measurements can be made quickly and accurately with electronic frequency meters or counters. Nevertheless, the oscilloscope is often the only instrument available, and the accuracy requirements may be well within those obtainable with the scope.

Phase Relationships

Ohm's law states that as the voltage across a resistor varies, the current through the resistor varies directly. That is, if the voltage is doubled, the current doubles; if the voltage decreases by a third, the current decreases by a third. This is true whether the source is dc or ac.

In the case of a sinusoidal alternating current, at any instant the voltage and current obey Ohm's law, and at the next instant the change in voltage is accompanied by a like change in current. When the instantaneous voltage is zero, the instantaneous current is zero. When v is maximum, i is

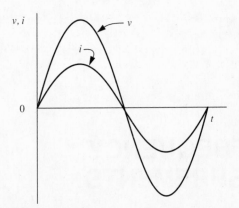

Figure 35–2. In a resistive circuit, voltage and current are in phase.

maximum (where v and i are the instantaneous values of voltage and current, respectively). Figure 35–2 shows this relationship graphically. In the case of ac resistive circuits, we say the voltage and current are *in phase*.

When an inductor is connected in an ac circuit, the voltage across the inductor and the current through the inductor are not in phase. For a pure inductor (that is, a device having only inductance) the voltage and current are out of phase by 90°. By convention we say the current *lags* the voltage by 90°. For example, when the voltage reaches its maximum instantaneous value, the current is zero. Not until one-fourth time period later will the current reach its maximum instantaneous value, and by that time the voltage is equal to zero. In this experiment the oscilloscope will be used to show graphically, as in Figure 35–3, how the voltage and current sine waves are shifted in time.

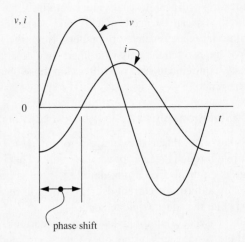

Figure 35–3. In a nonresistive circuit, voltage and current are not in phase.

Dual-Trace Oscilloscope

A dual-trace oscilloscope is, in effect, two separate scopes capable of simultaneously showing two traces on the same screen. The dual-trace scope contains two separate input channels and two separate time-base (horizontal) and Volts/Div. (vertical) controls.

Every oscilloscope manufacturer has its own design for the functions and controls of the dual-trace scope it produces. Whereas vertical amplifiers and calibrated horizontal sweeps are features of all oscilloscopes, the actual arrangement, labeling, and specifications of the controls differ from manufacturer to manufacturer. It is always good practice to have the operating manual immediately available when using a scope.

Modern scopes include the capability to display multiple traces (more than two), to display numerical quantities of voltage and phase angles directly on the screen, and to process waveform information digitally. Through the use of special interfaces, the personal computer can be used to display waveforms on its video monitor.

Dual-trace scopes are most often used to observe two waveforms of the same frequency or derived from the same frequency. However, many scopes are capable of displaying two unrelated frequencies if necessary.

Phase Measurement

A dual-trace oscilloscope can be used to display the phase relationship between two signals. One signal is established as a reference through the Channel 1 input. The leading signal can be used as the trigger source, or, if the signal comes from a line source, the line can be the trigger source. Once the positioning controls center the reference waveform, these controls are not adjusted again. The second signal is input in Channel 2. The Volts/Div. control can be used to adjust the height of the second waveform (it need not be the same height as the reference waveform). Engaging the dual-trace switch will display both signals simultaneously. The dual-trace switch is labeled differently by different manufacturers: DUAL, A & B, ALT, and CHOP are examples of the dual-trace mode-switch labels to be found on popular modern scopes. Figure 35–4 (p. 247) shows how two waves, one representing voltage across an inductor v_L and the other current through the inductor i_L, will appear on the oscilloscope screen. Although the two waves represent different quantities, the vertical amplifiers were adjusted so that the waveforms of both channels have the same height.

Notice that the two waveforms are offset from one another by 2 div. That is, the maximum positive value of v_L occurs at B; 2 div. later, at C, i_L reaches its maximum positive value. The actual difference, the angle by which i_L lags v_L can be calculated as follows.

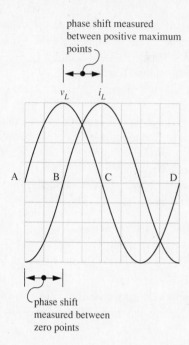

Figure 35 – 4. Using the dual waveforms to measure the phase-shift angle.

The distance AD represents one cycle of 360°. This distance covers 8 div. of the graticule. Therefore, each division is equal to

$$\text{Degrees/division} = \frac{360°}{8} = 45°/\text{div}.$$

Because 2 div. separate the two positive maximum values in the display of Figure 35–4, the current is shown lagging the voltage by

$$\text{Amount of lag} = 2 \text{ divisions} \times 45°/\text{div}.$$
$$= 90°$$

This verifies the fact that in a pure inductor the current

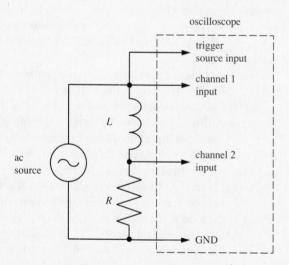

Figure 35 – 5. Circuit used to measure the phase-shift between voltage and current in an inductive circuit.

through the inductor lags the voltage across the inductor by 90°.

In introducing the example, we mentioned the reference waveform as the basis of our measurements. From a practical standpoint a reference waveform can be set up using the circuit of Figure 35–5. This circuit will behave almost as a pure inductance provided X_L is much greater then R. If X_L is 10 times greater than R, the error in phase angle will be approximately 6°; that is, the angle between v_L and i_L will be about 84° rather than 90°. As X_L increases greater than $10\,R$, the error decreases and the angle approaches 90°.

In this circuit, Channel 1, to which the leading waveform is connected, is triggered by the source supplying the circuit (if the source is 60 Hz, the line can be the trigger source). In the series circuit of Figure 35–5 the current is the same throughout. It is in phase with the voltage across R. Because the circuit acts as a pure L, the voltage across the source leads the current and therefore the voltage across R. Thus, the waveforms across R and across the source, in effect, represent the current and voltage waveforms of an inductance.

SUMMARY

1. Triggered oscilloscopes can be used to measure the frequency of an ac waveform, provided the scope has a calibrated time base, or horizontal sweep.
2. To measure the frequency of a waveform, a single cycle is displayed on the screen of the scope. The waveform will extend across the screen over a number of horizontal graticule divisions. The setting of the time base control in time units per division is then multiplied by the number of divisions to give the period t of the waveform. Frequency is the reciprocal of the period.

$$f = \frac{1}{t} \text{ Hz}$$

where t is the period in seconds.

3. A dual-trace oscilloscope can be used to measure the phase difference or displacement of two waves with the same frequency or derived from the same frequency.
4. A dual-trace oscilloscope is, in effect, two scopes capable of displaying their input waveforms on a single screen.
5. Dual-trace scopes have two separate time-base functions and two separate vertical amplifiers, each serving separate input channels.
6. To measure phase difference the scope must be triggered from the source of the leading waveform or from the line, if the circuit is fed from the line.
7. Phase difference between two waveforms is calculated by multiplying the number of divisions on the graticule between the comparable maximum or zero points of each waveform by the degrees/division factor.

SELF-TEST

Check your understanding by answering the following questions:

1. A single cycle of a sine-wave voltage measures 4 divisions along the horizontal axis of the graticule of an oscilloscope. The time-base control is set at 2 ms/div. The period of the waveform is _____ s.
2. The frequency of the waveform in questions 1 is _____ Hz.

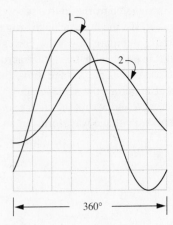

Figure 35–6. Circuit for question 4.

3. A series circuit containing resistance and inductance may be considered a pure inductance if the inductive reactance is at least _____ times _____ (greater/less) than the resistance.
4. In Figure 35–6, waveform 1 _____ (leads/lags) waveform 2 by _____ °.
5. (True/False) It is common practice to use a dual-trace oscilloscope to find the phase difference between two waveforms of different frequencies that are not derived from the same frequency. _____
6. The trigger source for a circuit operated from a 1000-Hz supply should be _____ Hz.

MATERIALS REQUIRED

Power Supply:
- AF sine-wave generator (oscillators of known but disguised frequency may be used in place of the generator for part A)

Instrument:
- Dual-trace oscilloscope

Resistors ($\frac{1}{2}$-W, 5%):
- 2   1000-Ω

Inductor:
- 1   30-mH

PROCEDURE

NOTE: Oscilloscopes vary widely from manufacturer to manufacturer. Check the operating manual of the scope you will be using to find the comparable controls and functions mentioned in the following steps. Check with your instructor if in doubt about the proper settings or adjustments of the scope.

A. Frequency Measurements

A1. Your instructor will provide you with an ac source of unknown frequency. Connect the output of the source to Channel 1 of the oscilloscope.
A2. Turn **on** the scope; turn **on** the ac source. Adjust the scope so that one cycle of the waveform is displayed on the screen.
A3. Measure the number of divisions spanned by one cycle and record the value in Table 35–1 (p. 251).
A4. Record the Time-base/Div. setting in Table 35–1.
A5. Calculate the period of the waveform using your data from steps A3 and A4. Record your answer in Table 35–1.
A6. Calculate the frequency of the ac source and record your answer in Table 35–1. Verify your frequency measurement with your instructor. If your answer is not correct within an acceptable range of error, repeat part A.

B. Phase Measurements

B1. With the oscilloscope and AF sine-wave generator **off**, connect the circuit of Figure 35–7. Channel 1 is connected to the output of the generator, as is the trigger source connection. On some scopes Channel 1 can also serve as the trigger source, and the extra connection to EXT trigger is unnecessary. Channel 2 is connected across resistor R_2.
B2. Turn **on** the scope and generator. Adjust the generator to 50 kHz. Switch to Channel 1; this will be the reference signal channel. Adjust the scope and output level of the generator until a single stationary sine wave is displayed on the screen about the entire width of the screen and 6 divisions peak-to-peak. Center the waveform about the horizontal and vertical centerlines of the graticule.
B3. Switch to Channel 2. Adjust the Volts/Div. control until a single cycle of a sine wave is displayed. Adjust the height of the sine wave until it is 4 divisions peak-to-peak. Adjust the vertical position control until the waveform is centered on the horizontal centerline of the graticule. *Do not adjust or change the horizontal positioning control.*

B4. Switch to the dual-trace mode that permits displaying both the Channel 1 and Channel 2 signals on the screen simultaneously. Check the operating manual of your scope to find the proper control or controls. Two sine waves should be displayed. Draw the waveforms in Table 35–2 (p. 251). Label the Channel 1 waveform v and the Channel 2 waveform i. Turn the oscilloscope and generator **off**.

B5. Replace R_1 (1000-Ω resistor) with a 30-mH coil as in Figure 35–8.

B6. Turn **on** the scope and generator. Adjust the generator to 100 kHz. Check to see that the waveform of Channel 1 is still 6 divisions peak-to-peak. Adjust the generator output and Volts/Div. scope control if necessary. Center the single cycle waveform as in step B2.

B7. Switch to Channel 2. Adjust the Volts/Div. control until the displayed waveform is 6 divisions peak-to-peak. Center the waveform vertically on the horizontal centerline of the graticules. *Do not adjust or change the horizontal positioning control.*

B8. Switch to the dual-trace mode that permits displaying both the Channel 1 and Channel 2 signals on the screen simultaneously. Two sine waves should be displayed on the screen, but their maximum points as well as their zero points should be displaced. Draw the two waveforms in Table 35–3 (p. 251). Label the Channel 1 signal v and the Channel 2 signal i.

B9. Note the positions of the maximum positive points and the zero points of each waveform. Measure the distance, in divisions, between the maximum points. Verify the measurement by measuring the divisions between zero points. (They should be the same.) Record the value in Table 35–3. Calculate the phase displacement of the two waves and record your answer in Table 35–3. Turn the scope and generator **off** and disconnect the circuit.

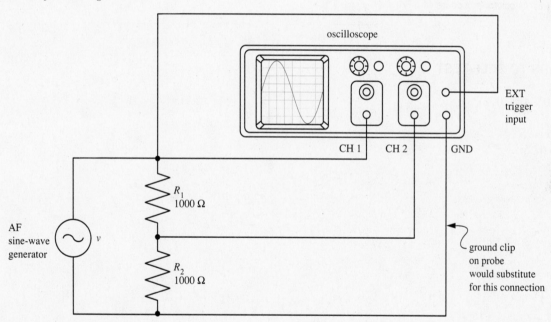

Figure 35–7. Circuit for procedure step B1.

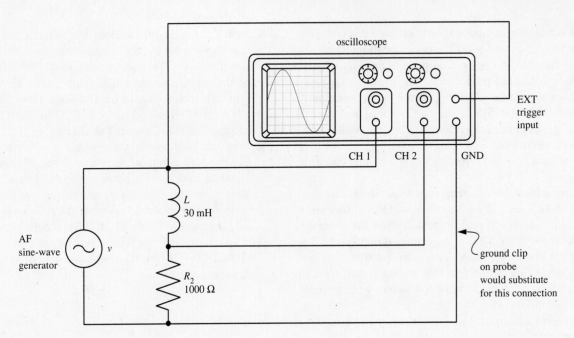

Figure 35–8. Circuit for procedure step B5.

ANSWERS TO SELF-TEST

1. 8×10^{-3}, or 0.008
2. 125
3. 10; greater
4. leads; 67.5°
5. false
6. 1000

Experiment 35 Name ______________________ Date ____________

TABLE 35–1. Frequency Measurement

Width of 1 Cycle, div.	*Time-Base Setting,* time units/div.	*Period of Wave, t,* s	*Frequency of Wave, f,* Hz

TABLE 35–2. Phase Relationship in a Resistive Circuit

TABLE 35–3. Phase Relationship in an Inductive Circuit

Distance between maximum points: ____________ div.

Phase displacement ________ °

QUESTIONS

1. Explain, in your own words, the general process for measuring the frequency of an ac waveform using an oscilloscope. Discuss any special conditions and restrictions connected with the process.

2. Explain, in your own words, the general process for measuring the phase angle between voltage and current in an inductance. Discuss any special conditions and restrictions connected with the process.

3. Were the conditions and restrictions, if any, mentioned in your answer to Question 2 met in this experiment? Explain.

_____

_____

_____

_____

_____

_____

4. How could the result of procedure step B8 indicate whether or not the two waveforms were properly centered? Refer to your waveform in Table 35–3.

_____

_____

_____

_____

_____

_____

5. Did your results in this experiment verify that voltage and current are in phase in a resistive circuit? Explain.

_____

_____

_____

_____

_____

_____

6. Did your results in this experiment verify that voltage and current are 90° out of phase in a pure inductive circuit? Explain.

_____

_____

_____

_____

_____

36

INDUCTANCES IN SERIES AND PARALLEL

OBJECTIVES

1. To investigate the effect of direct current on an inductance
2. To verify experimentally that the total inductance L_T for two inductors L_1 and L_2 connected in series, where no mutual coupling exists, is

$$L_T = L_1 + L_2$$

3. To verify experimentally that L_T for inductors connected in parallel, where no mutual coupling exists, is

$$\frac{1}{L_T} = \frac{1}{L_1} + \frac{1}{L_2}$$

BASIC INFORMATION

Effect of Direct Current on the Inductance of an Iron-Core Choke

In electronics, iron-core inductors are often referred to as *chokes*, or *choke coils*. They are so called because they are used to impede the flow of alternating current. They thereby serve as current-smoothing elements in power supplies. The inductance of these chokes is dependent on the direct current present in them. A filter choke rated 8 H at 50 mA direct current means that L is 8 H when a 50-mA direct current flows in the choke. In the absence of the direct current, the choke will not measure 8 H. If there are fewer than 50 mA dc, the choke will measure more than 8 H; when there are more than 50 mA dc, the choke will measure less than 8 H.

In this experiment you will verify the effect of direct current on the inductance of an iron-core choke by means of the experimental circuit that can switch alternately from a pure ac source to another source containing both alternating and direct current. The pure ac circuit consists of an inductor connected in series with a resistor. There is no direct current in the circuit. The value of inductance is determined by finding the alternating current I in the circuit and the ac voltage across the inductor. The inductance can be calculated from the formula

$$X_L = \frac{V_L}{I} \quad \text{and} \quad L = \frac{X_L}{2\pi f}$$

A dc supply is then connected in series with V_{ac}. The dc voltage is adjusted to the rated dc level of the choke. Again the current and voltage are measured. The inductance can be calculated from these values for comparison with the previous value of L.

Inductance in Series

Just as in the case of resistors, inductors may be connected in series. The effect of series-connected inductors is to offer greater opposition to alternating currents than an individual inductor can, acting alone. Measurement shows that series-connected inductors, where no mutual coupling exists, act like an equivalent single inductor whose inductance L_T is the sum of the individual inductances. That is,

$$L_T = L_1 + L_2 + L_3 + \cdots + L_n \quad \text{(36-1)}$$

Verification of the total inductance of series-connected inductors, as given by formula (36–1), may be obtained by measuring L_T directly on an inductance bridge. Inductance may also be measured indirectly, as in this experiment.

Consider the circuit of Figure 36–1. The ac source voltage V_{ac} is supplying current to R and L. The value of X_L may be determined, as in Experiment 35, by measuring I and the voltage V_L across L and substituting these values in the formula

$$X_L = \frac{V_L}{I} \quad \text{(36-2)}$$

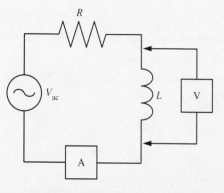

Figure 36-1. A series *RL* circuit may be used to determine the inductive reactance, X_L, of an inductor.

After X_L has been calculated, L may be found from the formula

$$L = \frac{X_L}{2\pi f} \qquad (36-3)$$

where

L is the required inductance in henrys,
X_L is the calculated value of inductive reactance in ohms,
f is the frequency of the ac source in hertz.

In the circuit of Figure 36–1 the current I may be measured directly with an ac ammeter. Another method is to measure the resistance of R and the voltage V_R across R. The current may then be calculated from the formula

$$I = \frac{V_R}{R} \qquad (36-4)$$

This method for finding inductance can be applied to a single inductor or to a number of inductors connected in series. Naturally, the same current will be in each inductor and in the series resistor.

Inductance in Parallel

Inductors may be connected in parallel in the same way resistors are connected in parallel. Two or more inductors connected in parallel, where no mutual coupling exists, act as an equivalent single inductor whose inductance L_T is given by the general formula

$$\frac{1}{L_T} = \frac{1}{L_1} + \frac{1}{L_2} + \frac{1}{L_3} + \cdots + \frac{1}{L_n} \quad (36-5)$$

Thus, the total inductance of parallel-connected inductances is less than the smallest inductance in the circuit.

Two Parallel Inductors

Two inductors connected in parallel are equivalent to a single inductor L_T whose value can be found using the formula

$$L_T = \frac{L_1 \times L_2}{L_1 + L_2} \qquad (36-6)$$

Similar Inductors Connected in Parallel

Two or more parallel inductors with the same inductance rating and with no mutual coupling can be replaced by an equivalent inductor whose value can be found using the formula

$$L_T = \frac{L}{n} \qquad (36-7)$$

where

L_T is the total inductance,
L is the value of the individual inductances that make up the parallel combination,
n is the number of parallel inductances.

Measuring Parallel Inductances

Verification of the total inductance L_T of parallel-connected inductors as given by formulas (36–5), (36–6), and (36–7) may be made by measuring across the parallel combination with an inductance bridge.

Inductance can also be measured indirectly by measuring the total current delivered to the parallel circuit and the voltage across the parallel-connected inductors. Inductive reactance can be found using the formula

$$X_L = \frac{V_L}{I}$$

After X_L has been calculated, L_T can be found from the formula

$$L = \frac{X_L}{2\pi f} \qquad (36-8)$$

where

L_T is the total inductance in henrys,
X_L is the inductive reactance in ohms,
f is the frequency of the source in hertz.

In the circuit of Figure 36–2 the current delivered to the parallel combination is being measured by the ammeter A. The voltage across the combination is V_L. The voltage across the resistor is V_R. The formula

$$I = \frac{V_R}{R}$$

gives the value of the total current drawn by the parallel inductances. Application of formulas (36–2) and (36–8) will result in a value for L_T.

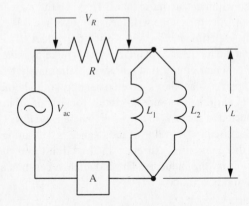

Figure 36–2. Parallel-connected inductances.

SUMMARY

1. The total inductance L_T of series-connected inductors, where no mutual coupling exists, is given by the formula

$$L_T = L_1 + L_2 + L_3 + \cdots + L_n$$

where $L_1, L_2, L_3, \ldots, L_n$ are inductors connected in series. This is the same as the formulas for R_T of resistors connected in series.

2. The inductance of iron-core inductors, or chokes, is rated for a particular level of direct current. Thus, a choke rated 8 H at 50 mA dc will measure more than 8 H when there are fewer than 50 mA dc and less than 8 H when there are more than 50 mA dc.

3. The inductance of coils and chokes may be measured directly with an inductance bridge.

4. The inductance of coils and chokes may also be determined indirectly by measuring the ac voltage V_L across the coil and the alternating current I in the coil. From these measurements X_L can be calculated, for $X_L = V_L/I$. Knowing X_L, it is possible to calculate L by the formula

$$L = \frac{X_L}{2\pi f}$$

5. The total inductance L_T of two or more parallel-connected inductances with no mutual coupling is given by the formula

$$\frac{1}{L_T} = \frac{1}{L_1} + \frac{1}{L_2} + \frac{1}{L_3} + \cdots + \frac{1}{L_n}$$

which is the same as the formula for parallel-connected resistors.

6. The total inductance of parallel-connected inductances can be measured directly using an inductance bridge.

7. The total inductance can be measured indirectly by measuring the current delivered to the parallel combination and the voltage across the combination. The inductive reactance can be calculated from these measurements using the formula

$$X_L = \frac{V_L}{I}$$

Once X_L is determined, the value of L_T can be found using the formula

$$L = \frac{X_L}{2\pi f}$$

SELF-TEST

Check your understanding by answering the following questions:

1. Three chokes, 4.2, 2.5, and 8 H, are connected in series. No mutual coupling exists. Their total inductance is _____ H.

2. The ac voltage V_L across a coil L is 22 V. The frequency is 60 Hz. The alternating current in the coil is 0.025 A.
   (a) $X_L =$ _____ Ω
   (b) $L =$ _____ H

3. A choke L and a 1-kΩ resistor R are connected in series as in Figure 36–3. The voltage across the resistor is 50 V, and the voltage across the choke is 40 V. The frequency of the applied voltage is 60 Hz.
   (a) The current I in the circuit = _____ A
   (b) X_L of the choke = _____ Ω
   (c) L of the choke = _____ H

4. Three chokes, 4.2, 2.5, and 8 H, are connected in parallel with no mutual coupling. The total inductance is $L_T =$ _____ H.

5. Four inductors having no mutual coupling are connected in parallel. Each inductor is 4.2 H. The total inductance is $L_T =$ _____ H.

6. In the circuit of Figure 36–4, $V_R = 20$ V, $V_L = 30$ V, $f = 1000$ Hz, and $R = 10$ kΩ.
   (a) The current delivered by the voltage source is _____ A
   (b) The inductive reactance of the circuit is _____ Ω
   (c) The total inductance is $L_T =$ _____ H

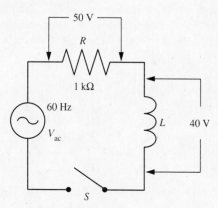

Figure 36–3. Circuit for question 3.

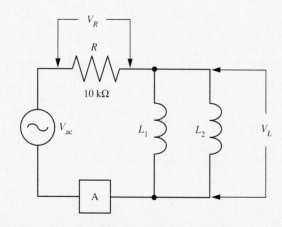

Figure 36–4. Circuit for question 6.

MATERIALS REQUIRED

Power Supplies:
- Variable 0–15 V dc, regulated
- Isolation transformer
- Variable-voltage autotransformer (Variac or equivalent)

Instruments:
- VOM or DMM
- Oscilloscope
- 0–100-mA dc milliammeter

Resistors:
- 1 10,000-Ω, $^{1}/_{2}$-W, 5%
- 1 500-Ω, 5-W

Inductors:
- 2 large iron-core chokes (8 or 9 H, as in Experiment 34)

Miscellaneous:
- SPDT switch
- SPST switch
- Polarized line cord with on-off switch and fuse

PROCEDURE

A. Effect of DC on Inductance

A1. With the line cord unplugged, its switch **off**, and S_1 **open**, connect the circuit of Figure 36–5. Set the autotransformer to its lowest voltage and S_2 to position Ⓐ.

A2. **Close** S_1. With the DMM and oscilloscope connected across the resistor, increase the ac voltage until the voltage across the resistor, V_R, is 1 V. Calculate the ac current in the circuit and record your answer in row 1 of Table 36–1 (p. 259) under "AC Current."

A3. Verify that V_{ac} = 1 V; adjust the autotransformer, if necessary. Connect the DMM and oscilloscope across the inductor. Measure the voltage across the inductor V_L. Record the value in row 1 of Table 36–1. Open S_1. Calculate the inductive reactance and inductance of the inductor and record your answers in row 1 of Table 36–1.

A4. Set S_2 to position Ⓑ. **Close** S_1. Adjust the dc power supply until the dc ammeter measures 15 mA in the circuit. The ac voltage across the resistor should still be 1 V; if it is not, adjust the autotransformer. The

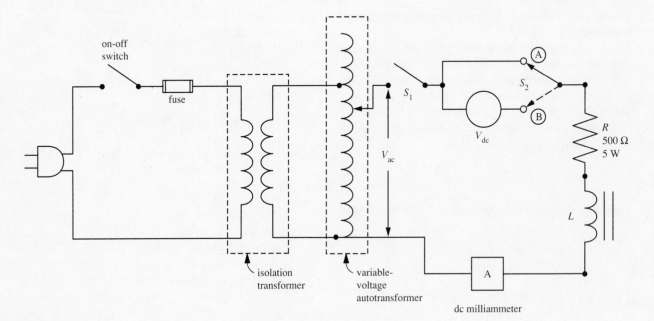

Figure 36–5. Circuit for procedure step A1.

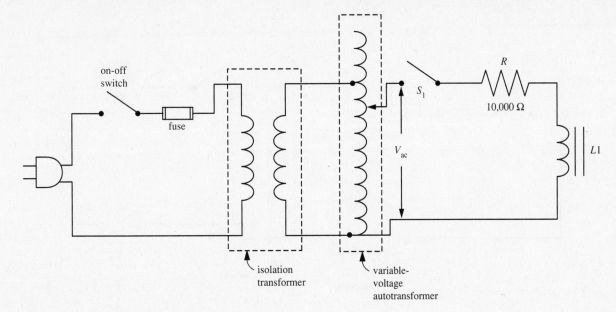

Figure 36 – 6. Circuit for procedure step B1.

ammeter must be measuring a constant 15 mA. Using both the DMM and the oscilloscope, measure the voltage across the inductor. Record the value in row 2 of Table 36–1. Calculate the ac current, inductive reactance, and inductance and record your answers in row 2 of Table 36–1.

A5. Repeat step A4 with 30 mA dc in the circuit. Record all data in row 3 of Table 36–1.

A6. Repeat step A4 with the rated dc current of the choke. Record all data in row 4 of Table 36–1. **Open** S_1. Turn line cord switch **off**. The ac circuit will be reused in Part B.

A7. In row 5 of Table 36–1, record the rated inductance, dc current, and resistance values of the inductor and any other pertinent nameplate data.

B. Behavior of Inductors Connected in Series

B1. The circuit of step A6 should be connected as shown in Figure 36–6. (The dc power supply, S_2, and the dc ammeter will not be used in this part.) The inductance is one of two being used in this experiment.

B2. **Close** S_1. Adjust the ac voltage to 5 V. Using the DMM and oscilloscope, measure the rms voltages across inductor 1 and the resistor. Record the values in Table 36–2 (p. 259). **Open** S_1. Disconnect inductor 1.

B3. Calculate the current in the circuit and the inductance of inductor 1. Record your answers in Table 36–2.

B4. Connect inductor 2 in the circuit of Figure 36–6. Check to verify that $V_{ac} = 5$ V. If necessary, adjust the volt-

age. Repeat step B2. Record the measured values for V_{L2} and V_R in Table 36–2. **Open** S_1, but do not disconnect inductor 2.

B5. Repeat step B3 for inductor 2.

B6. Break the circuit of Figure 36–6 and connect inductor 1 in series with inductor 2 as in Figure 36–7 (p. 258). (Separate $L1$ and $L2$ as much as possible to avoid coupling action between the chokes.)

B7. Measure the voltage across the combination of the two inductors $L1$ and $L2$—that is, across A and B, and across the resistor. Record the values in Table 36–2. **Open** S_1; leave the line cord switch **on** for part C.

B8. Calculate the current in the circuit and the inductance of the series combination of $L1$ and $L2$. Record your answers in Table 36–2.

C. Behavior of Inductors Connected in Parallel

C1. With S_1 **open** and the circuit from step B7, disconnect $L2$ from the circuit and reconnect it as in Figure 36–8 (p. 258). (Separate $L1$ and $L2$ far enough to prevent coupling.)

C2. Measure the voltage across the parallel combination of $L1$ and $L2$ (across A and B) and the voltage across the resistor V_R. Record the values in Table 36–2 for the parallel combination. **Open** S_1, turn line cord switch **off**; unplug the line cord.

C3. Calculate the total current in the circuit and the total inductance L_T of the parallel combination of $L1$ and $L2$. Record your answers in Table 36–2.

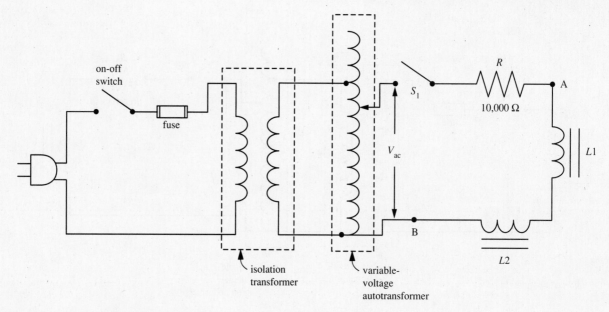

Figure 36 – 7. Circuit for procedure step B6.

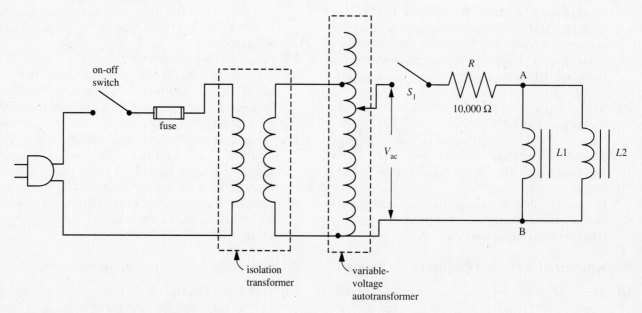

Figure 36 – 8. Circuit for procedure step C1.

ANSWERS TO SELF-TEST

1. 14.7
2. (a) 880; (b) 2.33
3. (a) 0.05; (b) 800; (c) 2.12
4. 1.31
5. 1.05
6. (a) 2 m; (b) 15,000; (c) 2.39

Experiment 36　　　　Name _____ Date _____

TABLE 36–1. Effect of DC on the Inductance of an Iron-Core Choke

Row	Position of S_2	DC Current I_{dc}, mA	Voltage across Resistor, V_R, V	AC Current I_{ac}, A	Voltage across Inductor V_L, V	Inductive Reactance X_L (Calculated), Ω	Inductance L (Calculated), H
1	(A)	0	1				
2	(B)	15	1				
3	(B)	30	1				
4	(B)		1				
5	Rated values of inductor:						

TABLE 36–2. Determining the Total Inductance of Chokes in Series and in Parallel

Inductor	Voltage across Inductor(s) V_L, V	Voltage across Resistor V_R, V	Total Current in Circuit I, A	Inductance L, H	Total Inductance L_T, H
1					
2					
1 and 2 in series					
1 and 2 in parallel					

QUESTIONS

1. Explain, in your own words, the effect of direct current on the value of inductance of an inductor.

_____

_____

_____

_____

_____

_____

_____

2. Explain the process for finding the inductance of an iron-core choke experimentally.

_____

_____

_____

_____

_____

_____

_____

_____

_____

3. Are the values of inductance determined from procedure steps B3 and B5 equal to the rated values of the two inductors? If not, explain the discrepancy.

_____

_____

_____

_____

_____

_____

_____

_____

_____

4. Using Formulas (36–1) and (36–5) and the values of $L1$ and $L2$ from Table 36–2, calculate L_T. Compare this with the value of L_T determined from V_L and I in Table 36–2. Explain any difference.

_____

_____

_____

_____

_____

_____

_____

_____

_____

_____

5. Compare the formulas for calculating the total inductance of series and parallel inductances with the formulas for calculating the total resistance of series and parallel resistances.

_____

_____

_____

_____

_____

_____

_____

_____

37

RC TIME CONSTANTS

OBJECTIVES

1. To determine experimentally the time it takes a capacitor to charge through a resistance
2. To determine experimentally the time it takes a capacitor to discharge through a resistance

BASIC INFORMATION

Capacitance and Capacitors

Capacitance is created when two conductors are separated by a nonconductor, or *dielectric*. This description would appear to be applicable to many electrical conditions, and, in fact, it is. For example, two wires separated only by air (a dielectric) exhibit capacitance that under certain situations can create problems in a circuit. The symbol for capacitance is *C*.

The unit of capacitance is the *farad*, named after the English scientist Michael Faraday. The abbreviation for farad is F. Since the farad is such a large unit of measurement a device with 1 F of capacitance would be very big, so farads are usually expressed in smaller units using metric prefixes. Typical capacitance values used in electronic circuits are given in millionths of a farad (a microfarad, or μF). Other typical units are picofarads (pF) and nanofarads (nF).

Charging and Discharging a Capacitor

In electric circuits capacitors are used for many purposes. For example, they are used to store energy, to pass alternating current while blocking direct current, and to shift the phase relationship between current and voltage. They are also used as components in filter and resonant circuits. In this experiment they will be used in timing circuits.

A capacitor can store a charge of electrons over a period of time. The process of building up the charge of electrons in the capacitor is known as *charging*. To charge a capacitor it is necessary to have a voltage across the capacitor. Figure 37–1(*a*) (p. 262) shows a circuit containing a dc voltage source and a capacitor. A three-position switch connects the capacitor to either a series circuit containing a voltage source and a resistor or a series circuit containing

the capacitor and resistor alone. A neutral position disconnects the capacitor from both circuits. In Figure 37–1(*a*) the switch is in neutral position 0.

If the switch is moved to position c, as in Figure 37–1(*b*), a charge of electrons will flow from the voltage source to the bottom plate of the capacitor. At the same time, an equal charge of electrons will flow to the voltage source from the top plate of the capacitor. This process continues until the capacitor is fully charged. This point is reached when the voltage applied to the capacitor is equal and opposite to the voltage across the plates of the capacitor. The polarity of the charged capacitor is shown in Figure 37–1(*b*). Since the voltage across the capacitor is equal to the voltage applied to the capacitor by the source, there is no further movement of charges. If the switch in the circuit is now moved to the neutral position 0, the capacitor will remain fully charged.

With the capacitor fully charged, the switch is moved to position d as shown in Figure 37–1(*c*). This places the capacitor in a complete circuit containing the capacitor and a resistor. The charge of electrons can now travel from the bottom plate through the resistor to the top plate of the capacitor until the charges on both plates are equal. At this point the capacitor is said to be fully discharged. A voltmeter placed across the capacitor would read 0 V. Because current is defined as the movement of charges, the charging and discharging of a capacitor involves the flow of current. As noted in Figure 37–1(*b*) and (*c*), the direction of current flow during the charging process is opposite to the direction of flow during the discharging process.

The Charge on a Capacitor

The size of a capacitor, given in farads (more commonly in microfarads or picofarads), is known as its capacitance.

The relationship between the charge *Q* on a capacitor and the capacitance of the capacitor *C* is given by the formula

$$Q = C \times V \qquad (37\text{--}1)$$

where

Q is the charge in coulombs,
C is the capacitance in farads,
V is the voltage across the capacitor in volts.

Time Required to Charge a Capacitor

The circuit of Figure 37–1(*a*) will be used to demonstrate another characteristic of the charging and discharging process. Figure 37–2(*a*) indicates two voltages of particular interest in these processes, V_R, the voltage across the resistor R, and V_C, the voltage across the capacitor C.

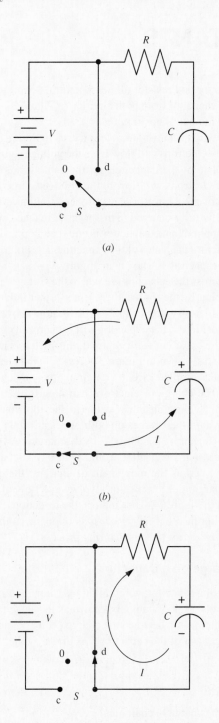

Figure 37–1. Charging and discharging a capacitor.

The capacitor is in a completely discharged state, and the switch is in position 0. Now the switch is moved to position c and the voltmeter across C is observed. At the instant the switch is moved to position c, the meter reads 0 V, but as C was charging, the meter indicated an increasing voltage. This demonstrates that a capacitor does not charge instantaneously; it takes *time* to charge.

Figures 37–2(*b*) through (*e*) show how current and voltage in the circuit of Figure 37–2(*a*) behave. At the instant the switch is moved to position c, the charge of electrons rushes to the bottom plate of the capacitor. The only object impeding this flow of current is the resistor. At that instant, the capacitor appears merely as a short circuit. Therefore, at that instant, the resistor develops the full voltage drop equal to the applied voltage V. The value of current at that instant is equal to $(V/R) = I_c$. Since C appears as a short circuit, the voltage across C, V_C, is zero.

As C charges, the voltage across C increases and opposes the voltage of the source V. This reduces I_c, and as a result, V_R decreases. Finally, when C is fully charged, its voltage is equal and opposite to V; therefore, $I_C = 0$, and $V_R = 0$.

Figure 37–2(*b*) shows that the voltage across AB is a constant V once the switch is moved to position c. In Figure 37–2(*c*), an almost instantaneous inrush of current I_C flows when the switch is moved to position c. However, as the capacitor charges, I_C decreases until it is zero, when C is fully charged.

Figure 37–2(*c*) shows the effects of the inrush of I_C and its decrease to zero. The voltage drop across R, V_R, goes from V to zero, as in Figure 37–2(*d*).

The voltage across C as seen in Figure 37–2(*e*) builds from zero until at its fully charged state it is equal to V.

The active voltage V_A causing current to flow is the difference between the applied voltage V and the voltage to which the capacitor has been charged to that point. This relationship is expressed by the formula

$$V_A = V - V_C \qquad \textbf{(37–2)}$$

When $V = V_C$, the capacitor is fully charged, and $V_A = 0$. Figures 37–2(*b*) through (*e*) show these conditions.

The circuit shown in Figures 37–1(*a*) and 37–2(*a*) is known as an *RC* (resistor-capacitor) circuit. The time it takes the capacitor to charge to any particular level is given in a unit expressed in time *constants*. Time constants can be calculated from the formula

$$t = R \times C \qquad \textbf{(37–3)}$$

where

t = time constant in seconds

R = resistance in ohms

C = capacitance in farads

It can be shown (though the derivation is beyond the scope of this book) that in one time constant, a capacitor charges to approximately 63.2 percent of the voltage applied across it.

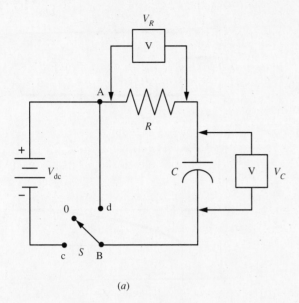

(a)

Figure 37–2. Capacitor *C* charging through an *RC* circuit.

Thus, in an *RC* circuit containing $R = 1$ MΩ and $C = 1$ μF, the time constant will be

$$
\begin{aligned}
t &= R \times C \\
&= 1 \times 10^6 \times 1 \times 10^{-6} \\
&= 1 \text{ s}
\end{aligned}
$$

This means that after 1 s the capacitor will have developed 63.2 percent of the applied voltage. If the applied voltage was 100 V, a voltmeter across *C* would read approximately 63.2 V 1 s after the voltage was applied.

Charge Rate of a Capacitor

Figure 37–3, curve A, is a graph showing the rise of voltage across a capacitor during charging. The capacitor is charging through an *RC* circuit. The horizontal axis in this graph is calibrated in time constants, whereas the vertical axis is calibrated in percent of full charge.

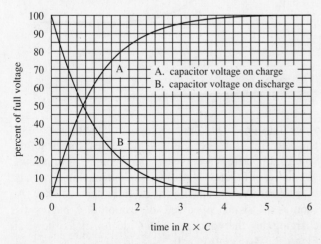

Figure 37–3. Universal graph for the charge and discharge of a capacitor.

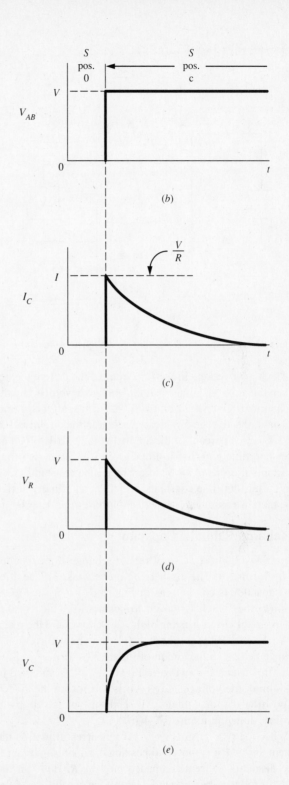

At one time constant, the graph shows that the capacitor has reached 63.2 percent of full charge; at two time constants, the capacitor has reached 86 percent of full charge; three time constants bring the charge to 95 percent, and four time constants yield 98 percent of full charge. At five time constants the capacitor has reached over 99 percent of full charge and for all practical purposes can be considered fully charged.

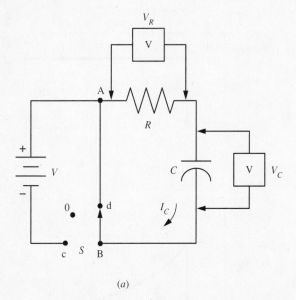

(a)

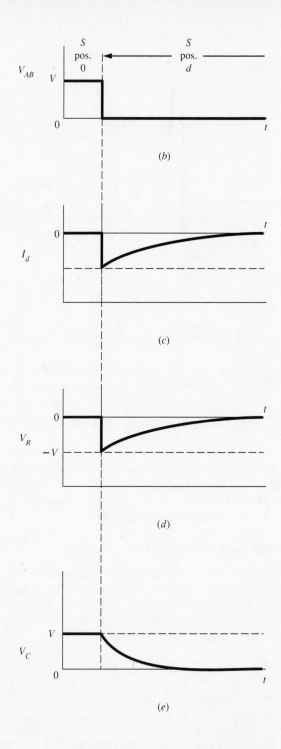

Figure 37-4. Capacitor *C* discharging through an *RC* circuit.

The active voltage in the *RC* circuit is the difference between the applied voltage and the voltage developed by the capacitor, as given by formula (37–2). Thus, after one time constant, the voltage producing current in the circuit is 100 – 63.2 = 26.8 percent of the applied voltage. If $V = 100$ V, after one time constant it drops to 26.8 V; after two time constants it drops to 14 V; after three time constants, $V_A = 5$ V; after four time constants, $V_A = 2$ V; and finally at five constants we assume for practical purposes that $V_A = 0$.

Discharge Rate of a Capacitor

Figure 37–3, curve B, shows the discharge of a capacitor in an *RC* circuit. The discharging process takes place after the capacitor is charged, as in Figure 37–2(a). In Figure 37–4(a) the switch is moved to position d, which results in a series circuit consisting of the capacitor *C* and the resistor *R*. Figures 37–4(b) through (e) show how current and voltage behave in the circuit of Figure 37–4(a).

At the instant the switch in Figure 37–4(a) is moved to position d, the voltage across AB is *V*. Closing the switch at position d short-circuits AB, and the voltage drops to zero, as shown in Figure 37–4(b).

When the switch is in position 0, no current flows. At the instant the switch is moved to position d, an inrush current I_d (the discharge current), impeded only by *R*, flows in the circuit produced by voltage V_C across the capacitor. As the capacitor discharges, V_C decreases and I_d decreases, as is shown in Figure 37–4(c). Note that this curve shows current flows in the direction opposite to the direction of I_C during the charging process.

With the switch in position 0, no current flows through *R*, and therefore the voltage drop across *R* is zero. With the switch in position d, current begins flowing at its maximum

value, thus producing the maximum voltage drop across *R* equal to V_C. As V_C decreases, I_d decreases; therefore, the voltage drop across *R* decreases, as Figure 37–4(d) shows.

Finally, in Figure 37–4(e) we see the maximum voltage $V_C = V$ across the capacitor at its fully charged state. As soon as the switch is moved to position d, *C* begins discharging and the voltage across *C* decreases as shown in Figure 37–4(e) until $V_C = 0$.

SUMMARY

1. The charge in coulombs Q on a capacitor is equal to the capacitance C in farads times the voltage V in volts to which the capacitor has charged. Thus, $Q = C \times V$.
2. The time required to charge a capacitor to 63.2 percent (approximately) of the applied voltage is called a time constant.
3. When a capacitor of C farads is charging through a resistance of R ohms, the time constant t in seconds of the charging circuit is $t = R \times C$.
4. At any instant of time while a capacitor is charging, the active voltage V_A equals the difference between the charging source voltage V and the voltage across the capacitor; that is,

$$V_A = V - V_C$$

5. It takes five time constants for a capacitor to charge to 99+ percent (approximately) of the applied voltage. For practical purposes we say that a capacitor is fully charged after five time constants.
6. When a capacitor of C farads is discharging through a resistor of R ohms, the discharge time constant t in seconds is $t = R \times C$.
7. When discharging, a capacitor will lose 63.2 percent of its charge in one time constant.
8. In each succeeding time constant the capacitor will lose an additional 63.2 percent of the voltage still across it.
9. At five time constants, we say the capacitor is fully discharged.
10. The graph of the charge and discharge of a capacitor is shown in Figure 37–3, which represents the universal chart for the charge and discharge of a capacitor.
11. A time constant is a *relative*, not an absolute, measure of time.

SELF-TEST

Check your understanding by answering the following questions:

1. A 0.25-µF capacitor is charging through a 2.2-MΩ resistor in series with an applied voltage of 20 V. The circuit has a time constant of _____ s.
2. In one time constant the capacitor in question 1 will have charged to _____ V.
3. In _____ s the capacitor in question 1 will have charged to 17.2 V.
4. A 0.05-µF capacitor charged to 100 V is permitted to discharge through a 220-kΩ resistor. At the end of _____ s, the voltage across the capacitor will have dropped to 37 V (approximately).
5. The discharge time constant in question 4 is _____ s.
6. It takes _____ time constants to discharge a capacitor almost completely.

MATERIALS REQUIRED

Power Supply:
■ Variable 0–15 V dc, regulated

Instrument:
■ Electronic analog voltmeter or multitester (the meter must not be heavily damped)

Resistors (½-W, 5%):
■ 1   1-MΩ
■ 1   10-kΩ

Capacitors:
■ 1   1-µF

Miscellaneous:
■ 2 SPST switches
■ Timer (can be watch with sweep second hand or digital watch with seconds readout)

PROCEDURE

A. Discharging of a Capacitor

A1. With power **off** and S_1 and S_2 **open**, connect the circuit of Figure 37–5 (p. 266). Set the power supply to its lowest voltage.
A2. **Close** S_1 and increase the power supply voltage until the electronic meter reads 12 V.
A3. Without changing the setting of the power supply, measure the output voltage of the supply. Record this value in Table 37–1 (p. 269).

A4. The circuit of Figure 37–5 is a simple series circuit consisting of a 1-MΩ resistor in series with the internal resistance of the electronic meter R_{in}. Using Ohm's law, the value of R_{in} can be calculated.

$$R_{in} = \frac{12}{V_{PS} - 12} \times 1 \text{ M}\Omega$$

where 12 is the reading of the EVM in step 2 and V_{PS} is the voltage across the power supply from step A3.

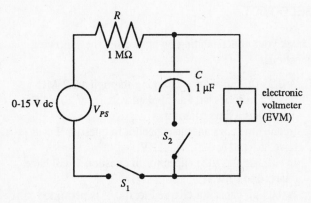

Figure 37–5. Circuit for procedure step A1.

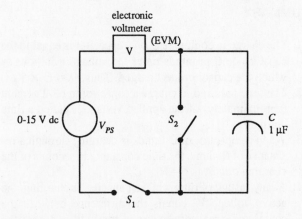

Figure 37–6. Circuit for procedure step B1.

Calculate R_{in} for your meter and record your answer in Table 37–1. Check the meter specifications or nameplate to determine the rated value for R_{in}. Record the value in Table 37–1.

A5. Using the rated values for R_{in} and C, calculate the time constant of a circuit containing only R_{in} and C. Record your answer in Table 37–1, under "Discharge Time Constant $R_{in}C$, Calculated, s."

Table 37–1 has a column of time-constant multiples from 1 through 5 and 10. Calculate the elapsed time in seconds for each time constant. For example, if $RC = 10$ s, then after one time constant 10 s will have passed; after two time constants, another 10 s will have passed, for a total of 20 s elapsed time; after three time constants, 30 s, and so on. Record these times in the "Discharge Time, Seconds" column.

A6. Charge the capacitor by **closing** S_2 (S_1 is still **closed** from step A2). At the instant S_2 is closed, the capacitor acts as a short, and the meter reading will drop. As the capacitor charges, the meter reading will increase until it stabilizes at 12 V, indicating the capacitor is fully charged.

NOTE: In the following steps the behavior of the capacitor during discharge will be recorded. Three discharge trials will be run. The voltage across the capacitor will be measured at intervals of the time constant as the capacitor discharges.

A7. Open S_1. This causes the capacitor to discharge through the meter. Using a timer that measures seconds accurately, measure the voltage across the capacitor at 1, 2, 3, 4, and 5 time constants, and at 10 time constants. Record the values in Table 37–1, under "1st Trial."

A8. Close S_1. The capacitor will recharge until the meter reads 12 V, at which point the capacitor is fully charged.

A9. Repeat step A7; record the voltages for the six time intervals under "2d Trial."

A10. Repeat step A8 to recharge the capacitor.
A11. With the capacitor fully charged, repeat step A7; record the voltages for the six time intervals under "3d Trial." **Open** S_1 and S_2; turn **off** power.
A12. Calculate the average measured voltage for each of the six time constants and record your answers in the "Average" column of Table 37–1. Also, calculate the voltage for each multiple of the calculated time constant $R_{in}C$. Record your answers in Table 37–1 under the "Calculated" column.

B. Charging of a Capacitor

B1. With power **off** and S_1 and S_2 **open**, connect the circuit of Figure 37–6. Set the power supply to its lowest voltage.

B2. Close S_1 and S_2 and increase the power supply voltage until the EVM measures 12 V. The capacitor is fully discharged at this point.

B3. Complete the "Charge Time" column in Table 37–2 (p. 269) by entering the values in the "Seconds" column from the same column in Table 37–1. The discharge time constant is the same as the charge time constant, and both are equal to $R_{in}C$.

NOTE: In the following steps the behavior of the capacitor during charge will be recorded. Three charge trials will be run. For each trial the voltage measured by the meter (that is, the voltage across R_{in}) will be recorded at each multiple of the time constant.

B4. Open S_2. This causes the capacitor to charge through the meter. At the instant S_2 is opened, C acts as a short circuit, and the total voltage is across the meter. As the capacitor charges, the meter readings will decrease. Record the voltmeter measurements for each of the six time intervals in Table 37–2, under "1st Trial." At 10 times the time constant, the capacitor can be considered fully charged.

B5. **Close** S_2; this discharges the capacitor. When the meter measures 12 V, the capacitor is fully discharged.

B6. Repeat step B4. Record the values in Table 37–2, under "2d Trial."

B7. Repeat step B5 to discharge the capacitor fully.

B8. Repeat step B4. Record the values in Table 37–2, under "3d Trial." **Open** S_1 and S_2; turn **off** power.

B9. Calculate the voltage across the capacitor for each trial and each time interval. The voltage across the capacitor will be

$$V_C = V_{PS} - V_{R_{in}}$$

where

V_C is the voltage across the capacitor,

V_{PS} is the voltage of the power supply (12 V in this experiment),

$V_{R_{in}}$ is the voltage read by the electronic voltmeter.

Record your answers in Table 37–2.

B10. Calculate the average voltage across the capacitor for each time interval. Record your answer in the "Average" column.

B11. For each multiple of the time constant, calculate the voltage across the capacitor. Record your answers in Table 37–2 under "Calculated."

ANSWERS TO SELF-TEST

1. 0.55
2. 12.6
3. 1.1
4. 0.011
5. 0.011
6. 5

TABLE 37–1. Discharging of a Capacitor

Power Supply Voltage V_{PS}, V	Internal Resistance of Meter R_{in}, Ω		Discharge time Constant $R_{in}C$, Calculated, s	Voltage Across Capacitor V_C, V
	Calculated	Rated		
				12

Discharge Time		Voltage across Capacitor V_C, V				
Time Constants	Seconds	1st Trial	2d Trial	3d Trial	Average	Calculated
1						
2						
3						
4						
5						
10						

TABLE 37–2. Charging of a Capacitor

Charge Time		Voltmeter Reading $V_{R_{in}}$, V			Voltage across Capacitor V_C, V				
Time Constants	Seconds	1st Trial	2d Trial	3d Trial	1st Trial	2d Trial	3d Trial	Average	Calculated
1									
2									
3									
4									
5									
10									

QUESTIONS

1. Explain, in your own words, the charging and discharging processes of a capacitor.

_____

_____

_____

_____

_____

_____

_____

_____

_____

　　　　　RC Time Constants　　**269**

2. Explain, in your own words, what the time constant of an *RC* circuit is.

_____

_____

_____

_____

_____

_____

_____

3. Discuss the key factors that probably limited the accuracy of your measurements in this experiment.

_____

_____

_____

_____

_____

_____

_____

_____

4. On a separate $8^{1}/_{2} \times 11$ sheet of graph paper, plot the following time constants versus voltage graphs using the same set of axes. The time constants should be on the horizontal (x) axis and the voltages across C should be on the vertical (y) axis. Label all axes and graphs.
   (a)   Time constant versus the average measured value of the voltage across the capacitor from Table 37−1.
   (b)   Time constant versus the calculated voltage across the capacitor from Table 37−1.
   (c)   Time constant versus the average measured value of the voltage across the capacitor from Table 37−2.
   (d)   Time constant versus the calculated voltage across the capacitor from Table 37−2.

5. Do the graphs plotted in Question 4 agree with the universal time-constant graph as shown in Figure 37−3? Discuss any differences.

_____

_____

_____

_____

_____

_____

_____

_____

38

REACTANCE OF A CAPACITOR (X_c)

OBJECTIVE

To verify experimentally the formula for capacitive reactance

$$X_C = \frac{1}{2\pi f C}$$

BASIC INFORMATION

Reactance of a Capacitor

The capacitive reactance X_C of a capacitor is the amount of opposition it offers to current in an ac circuit. The unit of capacitive reactance is the ohm. However, like the X_L of a coil, the X_C of a capacitor cannot be measured with an ohmmeter. Rather, capacitive reactance must be measured indirectly from its effect on current in an ac circuit.

Capacitive reactance is dependent on frequency and is given by the formula

$$X_C = \frac{1}{2\pi f C} \qquad \textbf{(38–1)}$$

where

X_C is in ohms,
C is the capacitance in farads,
f is the frequency in hertz.

The value of C in microfarads (μF) may be substituted directly in the formula

$$X_C = \frac{10^6}{2\pi f C} \qquad \textbf{(38–2)}$$

Remember that the C in formula (38–2) is expressed directly in microfarads. For example, to find the reactance of a 0.1-μF capacitor at a frequency of 1000 Hz, substitute in formula (38–2):

$$X_C = \frac{10^6}{(6.28)(1000)(0.1)} = 1592 \ \Omega$$

Formulas (38–1) and (38–2) show that the higher the frequency, the lower the value of the reactance of a capacitor, and the lower the frequency, the higher the reactance. For

direct current, $f = 0$; thus, X_C is infinite. This means that a capacitor appears as an open circuit to dc, and no direct current will flow through a capacitor.

The reactance of a capacitor may be determined by measurement. In the circuit of Figure 38–1 a sinusoidal voltage V_{ac} causes a current I to flow in the circuit. Ohm's law for ac circuits states that

$$I = \frac{V}{Z} \qquad \textbf{(38–3)}$$

where

I is in amperes,
V is in volts,
Z is in ohms.

In Figure 38–1, if R in the circuit is very small compared with X_C, the reactance of C can be considered the impedance of the circuit. That is,

$$X_C = Z \qquad \textbf{(38–4)}$$

Hence, the capacitive current I is given by the formula

$$I = \frac{V}{X_C} = \frac{V_C}{X_C} \qquad \textbf{(38–5)}$$

Formula (38–5) may be written in the form

$$X_C = \frac{V_C}{I} \qquad \textbf{(38–6)}$$

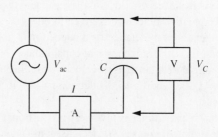

Figure 38 – 1. A sinusoidal voltage V causes current I to flow in the circuit. Since R in the circuit is very low compared to X_C, $Z = X_C$ and $V_C = V$.

We use formula (38–6) in determining the reactance of a capacitor at a given frequency f by measuring the voltage V_C across C with an ac voltmeter and measuring the current in the circuit with an ac ammeter. The values of V_C and I are then substituted in formula (38–6), and X_C is calculated. Note in the experimental circuit of Figure 38–1 that the voltmeter V is connected directly across C and not across the voltage source V. Thus, we measure V_C and not V. The reason is that the ammeter A has resistance, across which there will be a voltage drop. If the resistance of the ammeter is high enough, its effect on impeding current would have to be considered if the applied voltage V were measured. By measuring V_C, formula (38–6) can be used to find X_C.

An alternative method for measuring alternating current does not require the use of an ac ammeter.

Consider the circuit of Figure 38–2. This is a series circuit. Thus, the current flowing through R and C is the same. With this circuit arrangement it is possible to determine the value of X_C experimentally by measuring the voltages V_C across C and V_R across R. To determine the current I in amperes in the circuit, the known value of R in ohms and the measured value V_R in volts are substituted in Ohm's law.

$$I = \frac{V_R}{R} \qquad \text{(38–7)}$$

Knowing I, it is possible to find X_C by substituting for I and V_C in formula (38–6). It is also possible to find X_C without calculating I by combining formulas (38–6) and (38–7) and solving for X_C. We get

$$X_C = \frac{V_C}{V_R} \times R \qquad \text{(38–8)}$$

By substituting the measured values V_C, V_R, and R in formula (38–8), we can determine the value of X_C. Having found X_C experimentally, we can verify approximately the validity of the formula for X_C by comparing the experimental value of X_C with the formula value. This method requires that the capacitance C be known.

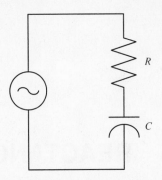

Figure 38–2. Series *RC* circuit.

5. The capacitive reactance of a capacitor C in a circuit of frequency f may be calculated from the formula

$$X_C = \frac{1}{2\pi f C}$$

where X_C is in ohms, f is in hertz, and C is in farads. When C is given in microfarads (μF), the formula can be written as

$$X_C = \frac{10^6}{2\pi f C \text{ (in } \mu\text{F)}}$$

6. Capacitive reactance cannot be measured directly—only by its effect in an ac circuit. Thus, in Figure 38–1, we measure the voltage V_C across C and the current I in the circuit. Then, using Ohm's law for ac circuits, we can find X_C by substituting V_C and I in the formula

$$X_C = \frac{V_C}{I}$$

7. Another method of determining X_C is illustrated by the circuit of Figure 38–2. Here the voltages V_R and V_C are measured across R and C, respectively. Then X_C is calculated by substituting V_R and V_C in the formula

$$X_C = \frac{V_C}{V_R} \times R$$

SUMMARY

1. The amount of opposition that a capacitor offers to current in an ac circuit is called capacitive reactance, X_C. Capacitive reactance is given in ohms.
2. The capacitive reactance of a capacitor is not constant but varies with frequency and capacitance.
3. Capacitive reactance is inversely proportional to frequency. As f increases, X_C decreases; as f decreases, X_C increases.
4. Capacitive reactance is inversely proportional to capacitance. As C increases, X_C decreases; as C decreases, X_C increases.

SELF-TEST

Check your understanding by answering the following questions:

1. The unit of capacitance is the _____ . The unit of capacitive reactance is the _____ .
2. The reactance of a capacitor depends on the _____ of the voltage source and the _____ of the capacitor.
3. The X_C of a capacitor whose value is 0.1 μF in a circuit where a 60-Hz signal is applied is _____ Ω.

4. If the frequency of the signal in question 3 is changed to 600 Hz, the X_C of the 0.1-μF capacitor is _____ Ω.

5. The voltage across a capacitor is 9.5 V, and the current in the capacitor is 0.01 A. The capacitive reactance of the capacitor is _____ Ω.

6. A 1000-Ω resistor and a 0.05-μF capacitor are connected in series. The voltage measured across the resistor is 5 V, and the voltage measured across the capacitor is 15 V. The X_C of the capacitor is _____ Ω.

7. The frequency of the applied voltage in question 6 is _____ Hz.

MATERIALS REQUIRED

Power Supplies:
- Isolation transformer
- Variable-voltage autotransformer (Variac or equivalent)

Instruments:
- Electronic voltmeter (EVM) or DMM
- Oscilloscope
- 0–5-mA ac ammeter
- Capacitance bridge or meter

Resistor:
- 1 5600-Ω, ½-W, 5%

Capacitors (nonelectrolytic):
- 1 0.5-μF or 0.47-μF, 25-WV dc
- 1 0.1-μF, 100-WV dc

Miscellaneous:
- SPST switch
- Polarized line cord with on-off switch and fuse

PROCEDURE

1. Measure the capacitance of the two test capacitors using a capacitance bridge or meter. Record the values in Table 38–1 (p. 275) in the "Measured Value" column.

2. With the line cord unplugged, the line switch **off** and S_1 **open**, connect the circuit of Figure 38–3 (p. 274). Adjust the autotransformer to its lowest voltage setting.

3. Plug in the line cord; turn line switch **on** and **close** S_1. Increase the output voltage of the autotransformer until the ammeter measures 2 mA. Measure the voltage across the capacitor, V_C, and record the value in Table 38–1 for the 0.5-μF capacitor.

4. Repeat step 3 for a current of 3 mA.

5. Repeat step 3 for a current of 4 mA. After measuring V_C, open S_1 and disconnect the 0.5-μF capacitor from the circuit. Adjust the autotransformer to its lowest setting. The remaining circuit will be reused in step 7.

6. Calculate the capacitive reactance of the capacitor for each of the three currents (2 mA, 3 mA, and 4 mA). First, calculate X_C using the voltmeter and ammeter measurements and applying Ohm's law for ac circuits. Next, calculate X_C using the formula

$$X_C = \frac{1}{2\pi f C}$$

Record your answers in Table 38–1 for the 0.5-μF capacitor in the "Voltmeter-Ammeter Value" and "Reactance Formula Value" columns.

7. Connect a 0.1-μF capacitor in the circuit of Figure 38–3. Repeat steps 3 through 6, but use 1, 2, and 3 mA. Record all values in Table 38–1 for the 0.1-μF capacitor. After taking all measurements, **open** S_1. This circuit will be reused in step 9.

8. Use an ohmmeter to measure the resistance of the 5600-Ω resistor. Record your answer in Table 38–2 (p. 275).

9. With the circuit of step 7 and S_1 **open**, connect the 5600-Ω resistor in series with the 0.1-μF capacitor as shown in Figure 38–4 (p. 274).

10. **Close** S_1. Increase the output voltage of the autotransformer to 10 V rms. With the oscilloscope calibrated for peak-to-peak voltage measurements, connect the hot lead of the scope to point A. Measure the peak-to-peak voltage across the capacitor and across the resistor. Record the values in Table 38–2 for the 0.1-μF capacitor. **Open** S_1. Disconnect the capacitor. This circuit will be reused in step 12.

11. Calculate the capacitive reactance using the voltage-ratio formula (38–8). Record your answer in Table 38–2 in the "Capacitive Reactance (Calculated)" column.

12. With S_1 **open**, connect a 0.5-μF capacitor in the circuit of step 10.

13. Repeat step 10. Record the measurements of V_C and V_R in Table 38–2 for the 0.5-μF capacitor. After completing the measurements, **open** S_1, turn line switch **off**; unplug line cord.

14. Repeat step 11.

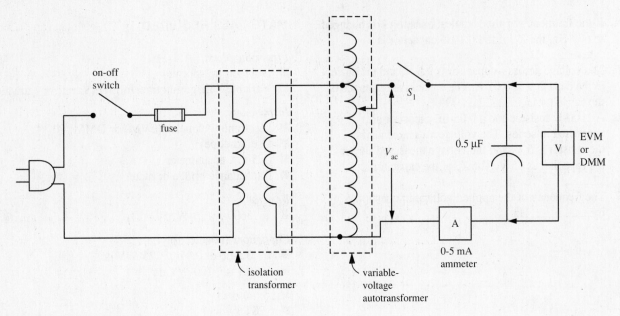

Figure 38-3. Circuit for procedure step 2.

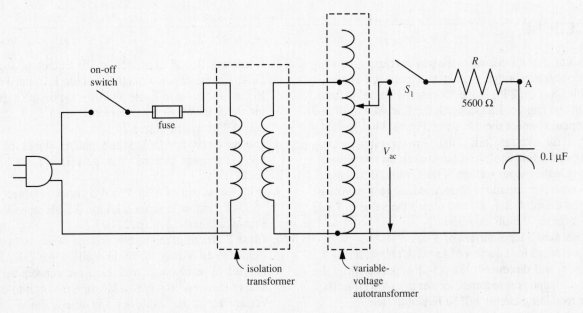

Figure 38-4. Circuit for procedure step 9.

ANSWERS TO SELF-TEST

1. farad; ohm
2. frequency; capacitance
3. 26,500
4. 2650
5. 950
6. 3000
7. 1060

Name _____ Date _____

TABLE 38-1. Voltmeter-Ammeter Method for Determining Capacitive Reactance

Capacitance C, μF		Current I, mA	Voltage across Capacitor V_C, V	Capacitive Reactance X_C (Calculated), Ω	
Rated Value	Measured Value			Voltmeter-Ammeter Value	Reactance Formula Value
0.5		2			
		3			
		4			
0.1		1			
		2			
		3			

TABLE 38-2 Voltage-Resistance Method for Determining Capacitive Reactance

Capacitor Rated Value, μF	Resistor R, Ω		Voltage across Capacitor V_C, V_{P-P}	Voltage across Resistor V_R, V_{P-P}	Capacitive Reactance X_C (Calculated) Voltage-Ratio Formula, Ω
	Rated Value	Measured Value			
0.1	5600				
0.5	5600				

QUESTIONS

1. Explain, in your own words, two methods that can be used to determine capacitive reactance experimentally. Discuss any special conditions or restrictions in using these methods.

_____

_____

_____

_____

_____

_____

_____

_____

_____

_____

_____

_____

_____

_____

_____

_____

_____

Reactance of a Capacitor (X_C)   **275**

2. Refer to your data in Table 38–1. Should all values of X_C obtained using the voltmeter-ammeter measurements be the same? If so, explain why. If not, explain the difference.

_____

_____

_____

_____

_____

_____

_____

_____

3. Refer to your data in Table 38–2. Should the two values of X_C calculated using the voltage-ratio formula be the same? If not, explain the difference.

_____

_____

_____

_____

_____

_____

_____

4. Compare the value of capacitive reactance calculated using the reactance formula with the experimental values in Tables 38–1 and 38–2. Should they be the same? If not, explain the reasons for the differences.

_____

_____

_____

_____

_____

_____

_____

_____

CAPACITORS IN SERIES AND PARALLEL

OBJECTIVE

1. To verify experimentally that the total capacitance C_T of capacitors connected in series is

$$\frac{1}{C_T} = \frac{1}{C_1} + \frac{1}{C_2} + \frac{1}{C_3} + \cdots + \frac{1}{C_n}$$

2. To verify experimentally that the total capacitance C_T of capacitors connected in parallel is

$$C_T = C_1 + C_2 + C_3 + \cdots + C_n$$

BASIC INFORMATION

Total Capacitance of Series-Connected Capacitors

In the series-connected circuit of Figure 39–1 the same line current I flows through C_1 and C_2. The effect of connecting these capacitors in series is to increase the total capacitive reactance and thus reduce the line current that would flow if either capacitor were in the circuit alone.

The total capacitive reactance of the series circuit X_{CT} is the sum of the capacitive reactances of C_1, and C_2:

$$X_{C_T} = X_{C_1} + X_{C_2}$$

Substituting the formula for capacitive reactance gives

$$\frac{1}{2\pi f C_T} = \frac{1}{2\pi f C_1} + \frac{1}{2\pi f C_2}$$

The term $2\pi f$ cancels, leaving the formula for capacitances in series:

$$\frac{1}{C_T} = \frac{1}{C_1} + \frac{1}{C_2}$$

which in its general form is

$$\frac{1}{C_T} = \frac{1}{C_1} + \frac{1}{C_2} + \frac{1}{C_3} + \cdots + \frac{1}{C_n} \quad \textbf{(39–1)}$$

Because the reactance of the series combination is greater than the reactance of either capacitor considered alone, the total capacitance C_T must be less than the capacitance of any of the individual series capacitances.

The total capacitance C_T of capacitor combinations may be determined experimentally in a number of ways. The simplest is to connect the required combination and measure C_T with a capacitance meter. If a laboratory-standard capacitance bridge is used, C_T can be measured to a high degree of accuracy.

A second method, used in this experiment, is to determine the capacitive reactance X_{C_T} of the combination. Once X_{C_T} is determined, C_T may be found from the formula

$$C_T = \frac{1}{2\pi f X_{C_T}} \quad \textbf{(39–2)}$$

The reactance of the capacitor combination may be found by measuring the current I_C and the voltage V_C across the combination. The total capacitive reactance X_{C_T} is then calculated from the formula

$$X_{C_T} = \frac{V_C}{I_C} \quad \textbf{(39–3)}$$

The capacitive current may be measured with an ac ammeter. Or the ac voltage V_R across a series-connected resistor can be measured, and the series current, which is also the capacitive current I_C can be calculated from the formula

$$I_C = \frac{V_R}{R} \quad \textbf{(39–4)}$$

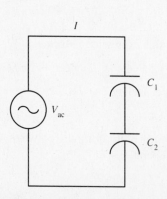

Figure 39–1. Series-connected capacitors.

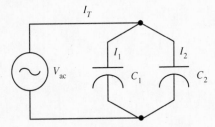

Figure 39-2. Parallel-connected capacitors.

Total Capacitance of Parallel-Connected Capacitors

Parallel-connected capacitors are frequently used in power and electronics circuits. The circuit in Figure 39-2 has two capacitors, C_1 and C_2, connected in parallel across the voltage source V_{ac}. Thus, the voltage across each capacitor is the same. The current supplied to this circuit divides, so that I_1 flows to C_1 and I_2 flows to C_2. The total current is therefore $I_T = I_1 + I_2$. The two capacitors could be replaced by a single equivalent capacitor drawing the total current I_T.

The flow of current is impeded by the capacitive reactance of each capacitor, which is given by the formula

$$I_1 = \frac{V_{ac}}{X_{C_1}}; \quad I_2 = \frac{V_{ac}}{X_{C_2}} \qquad (39-5)$$

The total current, in terms of V_{ac} and X_C, is

$$I_T = \frac{V_{ac}}{X_{C_1}} + \frac{V_{ac}}{X_{C_2}} \qquad (39-6)$$

$$= \frac{V_{ac}}{\dfrac{1}{2\pi f C_1}} + \frac{V_{ac}}{\dfrac{1}{2\pi f C_1}} = V_{ac}(2\pi f C_1) + V_{ac}(2\pi f C_2)$$

Dividing by V_{ac}, we obtain

$$\frac{I_T}{V_{ac}} = 2\pi f C_1 + 2\pi f C_2 \qquad (39-7)$$

But

$$\frac{I_T}{V_{ac}} = \frac{1}{X_{C_T}} = 2\pi f C_T$$

Substituting this expression in formula (39-7) gives

$$2\pi f C_T = 2\pi f C_1 + 2\pi f C_2$$

which, upon canceling $2\pi f$, becomes

$$C_T = C_1 + C_2$$

The general form of this formula for two or more capacitors connected in parallel is

$$C_T = C_1 + C_2 + C_3 + \cdots + C_n \qquad (39-8)$$

SUMMARY

1. In series-connected capacitors the current is the same in each of the capacitors. The total reactance X_{C_T} of series-connected capacitors is the sum of the reactances of the series capacitors in the circuit.
2. The total current I_T in a circuit consisting of series-connected capacitors is less than the current would be if any one of the capacitors were in the circuit by itself.
3. In a circuit consisting of two or more series-connected capacitors, the total capacitance C_T of the combination is

$$\frac{1}{C_T} = \frac{1}{C_1} + \frac{1}{C_2} + \frac{1}{C_3} + \cdots + \frac{1}{C_n}$$

4. The total capacitance of *series-connected capacitors* is calculated in the same way as the total resistance R_T of *parallel-connected resistors*.
5. One method to experimentally determining the total capacitance C_T of series-connected capacitors is to measure the C_T of the combination with a capacitance bridge or meter.
6. Another method to determine C_T is to measure the total ac current I_T of the combination and the voltage V_C across the combination, and to calculate X_{C_T} using the formula

$$X_{C_T} = \frac{V_C}{I_T}$$

Knowing X_{C_T} and frequency f of the source, calculate C_T by substituting X_{C_T} and f in the formula

$$C_T = \frac{1}{2\pi f X_{C_T}}$$

7. In a circuit consisting of two or more capacitors connected in parallel, the total current I_T is equal to the sum of the branch currents.
8. The total current in a parallel circuit is greater than the branch current in any one of the capacitors in the circuit.
9. In a circuit consisting of two or more parallel-connected capacitors, the total capacitance C_T is equal to the sum of the individual branch capacitors,

$$C_T = C_1 + C_2 + C_3 + \cdots + C_n$$

This is similar to the formula for finding the total resistance of series-connected resistors.

SELF-TEST

Check your understanding by answering the following questions:

1. In the circuit of Figure 39-1 $C_1 = 0.40\ \mu F$ and $C_2 = 0.05\ \mu F$. The total capacitance C_T of this combination is _____ μF.

2. If the frequency f of the applied voltage V is 100 Hz, then in the circuit of question 1,

(a) X_{C_1} = _____ Ω

(b) X_{C_2} = _____ Ω

(c) X_{C_T} = _____ Ω

3. In the circuit of Figure 39–3 the frequency of the source is 1 kHz, and the capacitances C_1, C_2, and C_3, are equal. The voltage across the resistor V_R = 4 V, and the voltage across the three series-connected capacitors V_{C_T} = 6 V. Find the value of each of the capacitors: $C_1 = C_2 = C_3 =$ _____ μF.

4. In Figure 39–2 the current I_1 in C_1 is 0.04 A, and the current I_2 in C_2 is 0.02 A. The total current I_T in the circuit is _____ A.

5. In Figure 39–2 C_1 = 0.5 μF, and C_2 = 1.0 μF. The total capacitance C_T is _____ μF.

6. The applied voltage V_{ac} in Figure 39–2 is 12 V. The total current in the circuit is 0.01 A. Find the total reactance and the total capacitance of the circuit at a frequency of 60 Hz.

(a) X_{C_T} = _____ Ω

(b) C_T = _____ μF

MATERIALS REQUIRED

Power Supplies:
■ Isolation transformer
■ Variable-voltage autotransformer (Variac or equivalent)

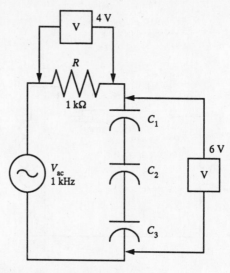

Figure 39–3. Circuit for question 3.

Instruments:
■ Oscilloscope
■ Capacitance bridge or meter
■ AC milliammeter or DMM
■ Electronic voltmeter (EVM) or a second DMM

Resistors ($\frac{1}{2}$-W, 5%):
■ 1   56-kΩ

Capacitors:
■ 1   0.05-μF or 0.047-μF, 25-WV dc
■ 2   0.1-μF, 25-WV dc
■ 1   0.5-μF or 0.47-μF, 25-WV dc

Miscellaneous:
■ SPST switch
■ Polarized line cord with on-off switch and fuse

PROCEDURE

If they are not already marked, label the capacitors 1 through 4 for identification during this experiment.

A. Total Capacitance of Capacitors in Series

A1. Using a capacitance bridge or meter, measure the capacitance of each of the capacitors in this experiment and record the values in Table 39–1 (p. 283).

A2. Connect the capacitors in series in the five combinations listed in Table 39–2 (p. 283). Measure the total capacitance of each of the combinations using a capacitance bridge or meter. Record the values in Table 39–2 in the "Measured" column.

A3. Using the measured values of capacitance from Table 39–1, calculate the total capacitance of each of the five combinations in Table 39–2. Record your answers in the table in the "Calculated" column.

A4. With the line cord unplugged, line switch **off**, and S_1 open, connect the circuit of Figure 39–4(a) (p. 280). Set the autotransformer to its lowest output voltage. In this circuit a 0.5-μF capacitor is in series with a 0.1-μF capacitor. This is the $C_3 + C_4$ series combination in Table 39–2.

A5. Plug in the line cord, turn the line switch **on and close** S_1. Increase the output of the autotransformer until V_{ac} = 10 V rms.

A6. Using the EVM, measure the voltage across the resistor V_R and across the two capacitors (that is, across AB), V_{C_T}. Use the oscilloscope, calibrated to measure rms voltage, to verify your measurements. (Check with your instructor if there are significant differences in the measurements.) Record the values in Table 39–2 in the V_R and V_{C_T} columns. **Open** S_1; turn **off** line cord switch. This circuit will be reused in step A8.

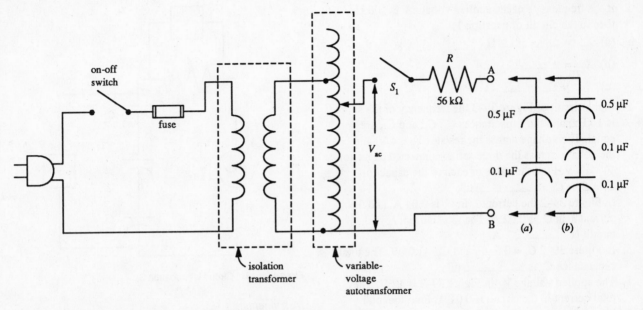

Figure 39–4. (a) Circuit for procedure step A4. (b) Circuit for procedure step A8.

A7. Calculate the current I_T in the circuit using Ohm's law. Use the measured values of voltage and resistance. Record your answer in Table 39–2 in the "I_T" column. Calculate capacitive reactance using your calculated I_T and the measured V_{C_T}. Record your answer in Table 39–2 in the "X_{C_T}" column. Calculate C_T using your calculated value of X_{C_T}. The line frequency f is = 60 Hz. Record your answer in Table 39–2 in the "C_T" column.

A8. Add a 0.1-μF capacitor in series in the circuit of step A4, as shown in Figure 39–4(b). The new circuit has a 0.5-μF and two 0.1-μF capacitors in series. This is the $C_2 + C_3 + C_4$ combination of Table 39–2.

A9. **Close** S_1. Increase the voltage until V_{ac} = 10 V rms. Measure V_R and V_{C_T} (across AB) as in A6. Record your values in Table 39–2. Open S_1. Turn **off** line cord switch. This circuit will be adapted for Part B.

A10. Calculate I_T, X_{C_T}, and C_T as in step A7. Record all answers in Table 39–2.

B. Total Capacitance of Capacitors in Parallel

B1. With S_1 **open** and the autotransformer set at its lowest output voltage, connect the circuit of Figure 39–5. In this case a 0.5-μF capacitor is in parallel with a 0.1-μF capacitor. This is the $C_3 + C_4$ parallel combination of Table 39–3 (p. 283).

B2. **Close** S_2 and increase V_{ac} to 10 V. Measure the voltage across the parallel combination and the total current in the circuit. Record the values in Table 39–3. **Open** S_1; turn **off** line switch.

B3. Applying Ohm's law for reactance, calculate X_{C_T} using the measured values of I_T and V_{C_T}. Record your answer in Table 39–3. Calculate total capacitance using your calculated value of X_{C_T}. The line frequency f is 60 Hz. Record your answer in Table 39–3, in the "Voltmeter-Ammeter Method" column. Applying the formula $C_T = C_1 + C_2 + \cdots + C_n$, calculate C_T using the measured values of C_3 and C_4 from Table 39–1. Record your answer in Table 39–3 in the "Formula Value" column.

B4. Add a third capacitor, 0.1-μF, in parallel as shown in Figure 39–5. In this case there are three capacitors connected in parallel: 0.5-μF and two 0.1-μF capacitors. This is the $C_2 + C_3 + C_4$ parallel combination of Table 39–3. Repeat step B2. After recording the measurements in Table 39–3, **open** S_1, turn **off** the line switch, and unplug the line cord.

B5. Repeat step B3 for the three-capacitor combination.

B6. With the $C_3 + C_4$ parallel combination disconnected from the circuit, measure the total capacitance with a capacitance bridge or meter. Record the value in Table 39–3 in the "Measured Value" column.

B7. Repeat step B6 with the $C_2 + C_3 + C_4$ parallel combination.

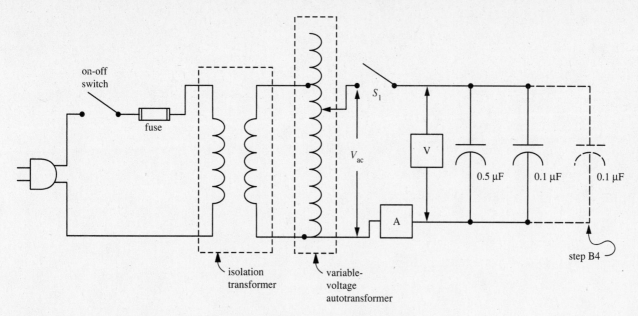

Figure 39–5. Circuit for procedure step B1. The third parallel capacitor is aded in step B4.

ANSWERS TO SELF-TEST

1. 0.044
2. (a) 3980; (b) 31,800; (c) 35,780
3. 0.32
4. 0.06
5. 1.5
6. (a) 1200; (b) 2.2

Name _____   Date _____

TABLE 39–1. Measured Values of Capacitance

Capacitor Number	Rated Value, μF	Measured Value, μF
1	0.05	
2	0.1	
3	0.1	
4	0.5	

TABLE 39–2. Determining Total Capacitance of Capacitors in Series

Series Combination	Method 1		Method 2				
	Total Capacitance C_T, μF		Voltage across Resistor V_R, V	Voltage across Series Combination V_{C_T}, V	Current I_T, mA	Total Capacitive Reactance X_{C_T}, Ω	Total Capacitance C_T, μF
	Measured	Calculated					
$C_1 + C_2$							
$C_1 + C_2 + C_3$							
$C_1 + C_2 + C_3 + C_4$							
$C_3 + C_4$							
$C_2 + C_3 + C_4$							

TABLE 39–3. Determining Total Capacitance of Capacitors in Parallel

Parallel Combination	Total Current I_T, A	Voltage across Parallel Combination V_{C_T}, V	Total Capacitive Reactance (Calculated) X_{C_T}, Ω	Total Capacitance C_T, μF		
				Voltmeter-Ammeter Method	Formula Value	Measured Value
$C_3 + C_4$						
$C_2 + C_3 + C_4$						

QUESTIONS

1. Explain, in your own words, the effect of adding more capacitors in parallel on the total current of the parallel circuit, on the current through each branch capacitor, and on the voltage across each branch capacitor. Assume a constant-voltage source.

_____

_____

_____

_____

_____

_____

_____

Capacitors in Series and Parallel **283**

2. Explain, in your own words, the effect of adding more series capacitors in a series circuit on the total current of the series circuit, the current through each capacitor, and the voltage across each series capacitor. Assume a constant-voltage source.

_____

_____

_____

_____

_____

3. Refer to your data for Method 1 in Table 39–2. Compare the measured values with the calculated values for the various combinations of series capacitors. Discuss any differences between measured and calculated values.

_____

_____

_____

4. Refer to your data for Method 2 in Table 39–2. Compare the measured values with the calculated values for the two series combinations used. Discuss any differences between measured and calculated values.

_____

_____

_____

_____

5. Refer to Table 39–3. Which method appears to be most accurate for determining the value of capacitance? Discuss your reasons.

_____

_____

_____

_____

6. Compare the formulas for finding total capacitance for capacitors in series and in parallel with the formula for finding total resistance for resistors in series and in parallel.

_____

_____

_____

_____

EXPERIMENT
40

THE CAPACITIVE VOLTAGE DIVIDER

OBJECTIVES

1. To show that the voltage V_1 across a capacitor C_1 in a series-connected capacitive voltage divider is given by the formula

$$V_1 = V \times \frac{C_T}{C_1}$$

where V is the voltage applied across all the capacitors and C_T is the total capacitance of the series-connected capacitors
2. To verify experimentally the formula in objective 1

BASIC INFORMATION

AC Voltage Across a Capacitor

Ohm's law applied to ac circuits states that the current I in a circuit is equal to the applied voltage V divided by the total opposition to alternating current. The total opposition to current in ac is called *impedance*. The symbol for impedance is Z. Thus,

$$I = \frac{V}{Z} \qquad (40-1)$$

In a circuit containing only capacitance C, such as in Figure 40–1, the total opposition to alternating current is the reactance X_C of capacitor C. Therefore, in the circuit of Figure 40–1, Z and X_C are the same, and

$$I = \frac{V}{X_C} \qquad (40-2)$$

From formula (40–2) we see that the voltage drop V_C across capacitor C is equal to the product of the current I in the capacitor and the reactance X_C of the capacitor.

$$V_C = I \times X_C \, V_C = I \times X_C \qquad (40-3)$$

Capacitive Voltage Divider

If an ac voltage is applied across capacitors connected in series, as in Figure 40–2, there is a voltage drop across each capacitor. This voltage drop is equal to the current times the capacitive reactance. In a series circuit the current I is the same everywhere in the circuit, so the voltage across each capacitor is

$$V_1 = I \times X_{C_1} \qquad (40-4)$$

$$V_2 = I \times X_{C_2}$$

The smaller the capacitor, the greater the reactance and, therefore, the greater the voltage drop across it in a capacitve voltage divider. In Figure 40–2 the reactance of a 0.05-μF capacitor is twice the reactance of a 0.1-μF capacitor; thus, $V_1 = 2V_2$. If the applied voltage V were 18 V in that circuit, then V_1 would be = 12 V and V_2 would be = 6 V; the sum of V_1 and V_2 would equal the applied voltage V.

Formula (40–4) can be written in terms of the respective capacitances:

$$V_1 = I \times X_{C_1} = I \times \frac{1}{2\pi f C_1} \qquad (40-5)$$

$$V_2 = I \times X_{C_2} = I \times \frac{1}{2\pi f C_2}$$

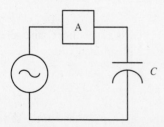

Figure 40–1. The total opposition to ac in this circuit is the capacitive reactance X_C caused by C.

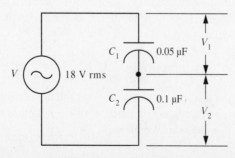

Figure 40–2. Voltage drops in a capacitive voltage divider.

The ratio of voltage drops and capacitances can be found by dividing V_1 by V_2.

$$\frac{V_1}{V_2} = \frac{\dfrac{1}{2\pi f C_1}}{\dfrac{1}{2\pi f C_2}} = \frac{2\pi f C_2}{2\pi f C_1}$$

The $2\pi f$ terms cancel, leaving

$$\frac{V_1}{V_2} = \frac{C_2}{C_1} \qquad \textbf{(40–6)}$$

Using similar reasoning for series circuits with more than two capacitors, it can be shown that, in general,

$$\frac{V_a}{V_b} = \frac{C_b}{C_a}$$

where V_a is the voltage across capacitor C_a in series with capacitor C_b, and V_b is the voltage across capacitor C_b.

Now consider the total reactance X_{C_T} of series-connected capacitors,

$$V = I \times X_{C_T} \qquad \textbf{(40–7)}$$

That is, the applied voltage V equals the product of I and the total reactance of the series-connected capacitors.

Since $V_1 = I \times X_{C_1}$, dividing V_1 by V leads to

$$\frac{V_1}{V} = \frac{I \times X_{C_1}}{I \times X_{C_T}} = \frac{X_{C_1}}{X_{C_T}} = \frac{\dfrac{1}{2\pi f C_1}}{\dfrac{1}{2\pi f C_T}}$$

and

$$\frac{V_1}{V} = \frac{C_T}{C_1}$$

Therefore,

$$V_1 = V \times \frac{C_T}{C_1} \qquad \textbf{(40–8)}$$

Formula (40–8) states that the voltage drop across any capacitor C_1 in a series-connected capacitive voltage divider equals the product of the applied voltage and the ratio of the total capacitance C_T to the capacitance of C_1. Note that this is similar to the formula for voltage distribution in a series-connected resistive voltage divider, with the exception that in a resistive divider,

$$V_1 \text{ (across } R_1) = V \times \frac{R_1}{R_T}$$

That is, in a series capacitive divider the positions of C_1 and C_T are reversed, as compared with the positions of R_1 and R_T in a series-resistive divider.

It is now possible to apply formula (40–8) to the circuit of Figure 40–2. In that circuit

$$\frac{1}{C_T} = \frac{1}{C_1} + \frac{1}{C_2} = \frac{1}{0.05} + \frac{1}{0.1}$$

$$C_T = \frac{1}{30} = 0.0333 \ \mu F$$

Substituting $C_T = 0.0333$, $C_1 = 0.05$, and $V = 18$ V in formula (40–8) gives

$$V_1 = 18 \times \frac{0.0333}{0.05} = 12 \text{ V}$$

Similarly,

$$V_2 = 6 \text{ V}$$

These are the same results as were obtained in our earlier solution of this circuit.

We can now also verify formula (40–6), for according to that formula,

$$\frac{V_1}{V_2} = \frac{C_2}{C_1}$$

and

$$\frac{12}{6} = \frac{0.1}{0.05} = \frac{2}{1}$$

as expected.

SUMMARY

1. The voltage drop V_1 across a capacitor C_1 in a capacitive voltage divider is given by the formula $V_1 = I \times X_{C_1}$, where I is the current in the capacitor and X_{C_1} is the reactance of the capacitor.
2. In a series-connected capacitive voltage divider

    $$\frac{V_a}{V_b} = \frac{C_b}{C_a}$$

    where C_a and C_b are the capacitances of series-connected capacitors with voltage drops V_a and V_b, respectively.
3. The total current I in a capacitive voltage divider where the applied voltage is V and the total capacitance is C_T is

    $$I = \frac{V}{X_{C_T}}$$

4. The voltage V_1 across a capacitor C_1 in a series-connected voltage divider with applied voltage V is

    $$V_1 = V \times \frac{C_T}{C_1}$$

    where C_T is the total capacitance of the series-connected capacitors.

SELF-TEST

Check your understanding by answering the following questions:

1. In the circuit of Figure 40–1, $V = 10$ V, and $X_C = 500\ \Omega$. The current I in the circuit is _____ A.
2. In the circuit of Figure 40–2 $X_{C_1} = 1500\ \Omega$. Therefore,
   (a) $X_{C_2} =$ _____ Ω
   (b) $X_{C_T} =$ _____ Ω
   (c) $I =$ _____ A
3. In the capacitive voltage divider of Figure 40–3, $C_1 = 1.0\ \mu$F, $C_2 = 5.0\ \mu$F, $C_3 = 2.5\ \mu$F, and $V = 12$ V.
   (a) $V_1 =$ _____ V
   (b) $V_2 =$ _____ V
   (c) $V_3 =$ _____ V

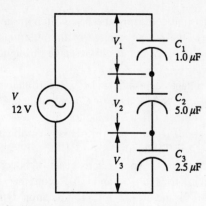

Figure 40–3. Circuit for question 3.

MATERIALS REQUIRED

Power Supplies:
- Isolation transformer
- Variable-voltage autotransformer (Variac or equivalent)

Instruments:
- Electronic voltmeter (EVM) or DMM
- Capacitance bridge or meter

Capacitors:
- 1 0.5-μF or 0.47-μF, 25-WV dc
- 2 0.1-μF, 25-WV dc
- 1 0.05-μF or 0.047-μF, 25-WV dc

Miscellaneous:
- SPST switch
- Polarized line cord with on-off switch and fuse

PROCEDURE

If they are not already labeled, mark the capacitors with numbers 1 through 4 for identification.

1. Using a capacitance bridge or meter, measure the capacitance of each of the four capacitors used in this experiment. Record the values in Table 40–1 (p. 289).

2. With the line cord unplugged, the line switch **off** and S_1 **open**, connect the circuit of Figure 40–4. Set the autotransformer to its lowest output voltage.

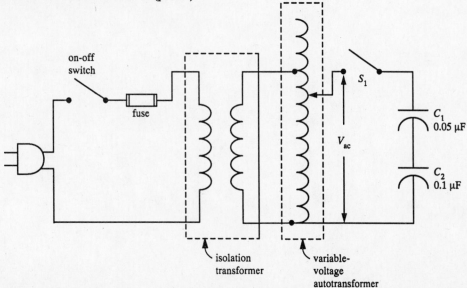

Figure 40–4. Circuit for procedure step 2.

3. Plug in the line cord; turn the line switch **on** and **close** S_1. Increase the output voltage of the autotransformer until $V_{ac} = 10$ V. Maintain this voltage throughout this experiment.
4. Measure voltages across C_1 and C_2. Record the values in Table 40–2 (p. 289) for the $C_1 + C_2$ combination. **Open** S_1.
5. Connect another 0.1-μF capacitor in series with the 0.05- μF and 0.1-μF capacitors. This is combination of $C_1 + C_2 + C_3$ in Table 40–2.
6. **Close** S_1. Measure the voltage across each of the capacitors connected in series (C_1, C_2, and C_3). Record the values in Table 40–2 for the $C_1 + C_2 + C_3$ combination. **Open** S_1.
7. Replace C_1 with C_4 so that the series combination consists of 0.1-μF, 0.1-μF, and 0.5-μF capacitors (combination $C_2 + C_3 + C_4$ in Table 40–2).

8. **Close** S_1; check V_{ac} and adjust to 10 V if necessary. Measure the voltage across each capacitor in the series combination. Record the values in Table 40–2 for the $C_2 + C_3 + C_4$ combination. **Open** S_1; line switch **off**; unplug line cord.
9. Using the measured values for C_1, C_2, C_3, and C_4, calculate the total capacitance for each of the three series combinations in Table 40–2. Record your answers in the table.
10. Using your calculated values for C_T, calculate the voltage across each capacitor using the voltage-ratio formula. Record your answers in Table 40–2 in the "Calculated Voltage" columns.

ANSWERS TO SELF-TEST

1. 0.02
2. (a) 750; (b) 2250; (c) 0.008;
3. (a) 7.5; (b) 1.5; (c) 3.0

TABLE 40–1. Measured Values of Capacitance

Capacitor Number	1	2	3	4
Rated value, μF	0.05	0.1	0.1	0.5
Measured value, μF				

TABLE 40–2. Verifying the Voltage-Ratio Formula for Series-Connected Capacitors

Series Combination	V	Measured Voltages, V				Total Series Cap. (Cal.) C_T, μF	Calculated Voltage, V			
		V_1	V_2	V_3	V_4		V_1	V_2	V_3	V_4
$C_1 + C_2$	10			✕	✕				✕	✕
$C_1 + C_2 + C_3$	10				✕					✕
$C_2 + C_3 + C_4$	10	✕					✕			

1. Explain, in your own words, the relationship between the voltage across a capacitor in a capacitive voltage divider and the voltage applied to the voltage divider.

_____

_____

_____

_____

_____

_____

_____

_____

_____

2. Refer to your data in Table 40–2. Compare the measured voltages V_1 through V_4 with the calculated voltages. Explain any differences.

_____

_____

_____

_____

_____

_____

_____

_____

_____

_____

    The Capacitive Voltage Divider   **289**

3. Refer to your data in Table 40–2. Do the measured values verify the relationship stated in Question 1? Cite specific data to support your answers.

_____

_____

_____

_____

_____

_____

_____

_____

_____

_____

4. What conclusion can you make about the sum of the voltages V_1 through V_4 in a series-connected capacitive voltage divider? Refer to specific data in Table 40–2 to support your answer.

_____

_____

_____

_____

_____

_____

_____

_____

_____

_____

_____

_____

IMPEDANCE OF A SERIES *RL* CIRCUIT

OBJECTIVES

1. To verify experimentally that the impedance Z of a series *RL* circuit is given by the formula

$$Z = \sqrt{R^2 + X_L^2}$$

2. To study the relationship between impedance, resistance, inductive reactance, and phase angle

BASIC INFORMATION

Impedance of a Series *RL* Circuit

The total opposition to alternating current in an ac circuit is called *impedance* Z. Ohm's law applied to ac circuits states that

$$I = \frac{V}{Z}$$

$$V = I \times Z$$

$$Z = \frac{V}{I}$$

Consider the circuit of Figure 41–1. If we assume that the inductance L through which alternating current flows has zero resistance, the current is impeded only by X_L. That is, $Z = X_L$. In this case, if $L = 8$ H and $f = 60$ Hz,

$$X_L = 2\pi f L = 6.28(60)(8) = 3014 \ \Omega$$

How much current I will there be in the circuit if $V= 10$ V? Applying Ohm's law, we obtain

$$I = \frac{V}{X_L} \qquad \textbf{(41–1)}$$

where I is in amperes, V is in volts, and X_L is in ohms. Therefore,

$$I = \frac{10}{3014} = 3.32 \ \text{mA}$$

If there is resistance R associated with the inductance L or if L is in series with a resistor of, say, 3000 Ω (Figure 41–2), the current will be less than 3.32 mA. How much current will flow, assuming the same X_L as previously computed? If an ammeter were placed in the circuit of Figure 41–2, it would measure 2.351 mA. That means the impedance of the circuit is

$$Z = \frac{V}{I} = \frac{10}{0.002351} = 4253 \ \Omega$$

Obviously this is less than the arithmetic sum of R and X_L (which is 6014 Ω). Even though they have the same unit—ohms—resistance and reactance cannot be added arithmetically.

The current in R is out of phase with the voltage in X_L by 90°. Time-varying (ac) currents and voltages are often

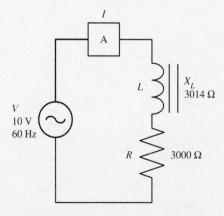

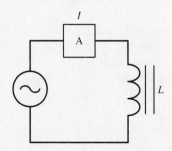

Figure 41–1. The current I in a circuit containing only inductance is limited by the inductive reactance X_L caused by L.

Figure 41–2. The impedance Z of a circuit containing R in series with L is greater than that of a circuit with L alone.

represented by *phasors*. Because I and V are related to Z we use the term phasor also when discussing Z.

The phasor diagram in Figure 41–3 shows R and X_L as phasors 90° apart. By convention R is shown at 0° (the horizontal line to the right). Inductive reactance X_L is represented by the vertical line in the upward direction, and capacitive reactance is represented by a vertical line in the downward direction. The impedance Z of a circuit is the *phasor sum* of R and X. Thus, in Figure 41–3 the phasor sum of R and X_L is the line Z. The angle θ is called the *phase angle* between R and Z.

The phasors can be added graphically by drawing R and X_L to some convenient scale and then completing the rectangle so that two sides are equal to R and two sides are equal to X_L. The diagonal of the rectangle is equal to Z. If this line is measured using the same scale as R and X_L, the value of Z can be found directly.

Note that the phasors R, X_L, and Z form a right triangle. In a right triangle, the sides are related by the following rule (referred to in mathematics texts as the pythagorean theorem):

The sum of the squares of the two sides of a right triangle is equal to the square of the longest side (called the *hypotenuse* of the triangle).

In terms of the R, X_L, Z triangle in Figure 41–3, this relationship can be written as the formula

$$Z^2 = R^2 + X_L^2$$

To find Z, we take the square root of both sides of the formula.

$$Z = \sqrt{R^2 + X_L^2} \qquad \textbf{(41–2)}$$

If we go back to the circuit of Figure 41–2, formula (41–2) can be used to find the circuit Z and verify the measured I.

$$Z = \sqrt{3000^2 + 3014^2}$$

$$Z = 4253 \ \Omega$$

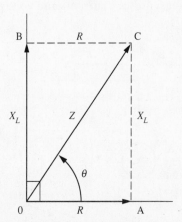

Figure 41–3. Phasor diagram of an *RL* series circuit. Impedance Z is the phasor sum of R and X_L.

With 10 V applied, the current in the circuit of Figure 41–2 is

$$I = \frac{V}{Z} = \frac{10}{4253}$$

$$I = 0.002351 \ \text{A, or } 2.351 \ \text{mA}$$

which verifies the measured values previously given.

Formula (41–2) and the illustrative problem point up a very important difference between the mathematics of ac circuits and that of dc circuits. In a dc circuit consisting of series-connected resistors, the total opposition to current R_T is simply the arithmetic sum of R_1, R_2, etc. In an ac circuit consisting of a series-connected resistor and inductance, the total opposition to current, the impedance Z, is *not* the arithmetic sum of R and X_L, but is the *phasor sum* of R and X_L.

Alternative Method for Solving Series *RL* Circuits

Certain relationships between the sides of a right triangle and its angles can be used to solve problems in ac circuits.

If the two sides of the triangle representing R and X_L are known, the third side of the triangle, Z, can be found without using formula (41–2). The angle between R and Z is known as the phase angle. The quotient of X_L divided by R is known as the *tangent* of the phase angle, written as

$$\tan \theta = \frac{X_L}{R}$$

In the previous problem, $X_L = 3014 \ \Omega$ and $R = 3000 \ \Omega$. Thus,

$$\tan \theta = \frac{3014}{3000} = 1.005$$

A table of trigonometric functions or the trigonometric functions on a calculator can then be used to find the value of θ. The calculator method makes use of the tan⁻¹ key. Once 1.005 is keyed into the calculator, press the tan⁻¹ key or sequence of keys (tan⁻¹ requires a two-key operation on most scientific calculators). The answer displayed will be the value of the angle θ. In the problem given

$$\theta = \tan^{-1}(1.005)$$

The keystroke operation on most calculators will be

(On some calculators the f, or function key, is marked "shift.") The displayed answer will be 45.143°. This is the value of the phase angle between R and Z.

The quotient of R divided by Z is known as the *cosine* of the phase angle:

$$\cos \theta = \frac{R}{Z}$$

or solving for Z,

$$Z = \frac{R}{\cos \theta} = \frac{3000}{\cos 45.143}$$

To find Z, the keystroke operation on most calculators will be

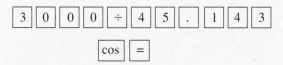

The displayed answer will be 4253.

Remember that for practical purposes the answers given by the calculator can be rounded to three decimal places.

The following problem is solved using a calculator throughout without rounding until the final answer.

Problem. A series RL circuit has $R = 40 \ \Omega$ and $X_L = 25 \ \Omega$. Find the impedance of the circuit.

Solution. Draw the phasor diagram (Figure 41–4).

$$\theta = \tan^{-1} \left(\frac{X_L}{R} \right) = \tan^{-1} \left(\frac{25}{40} \right)$$

$\theta = \boxed{2} \boxed{5} \boxed{\div} \boxed{4} \boxed{0} \boxed{=} \boxed{f} \boxed{\tan^{-1}} = 32.0053832$

$Z = \boxed{\cos} \boxed{1/x} \boxed{\times} \boxed{4} \boxed{0} \boxed{=} = 47.16990566$

$$Z = 47.170 \ \Omega$$

Using formula (41–2)

$$Z = \sqrt{40^2 + 25^2} = 47.170 \ \Omega$$

which agrees with the previous answer.

It should be noted that many scientific calculators can also solve formula (41–2) with just a few keystrokes. The student should refer to the instruction manual supplied with the calculator being used.

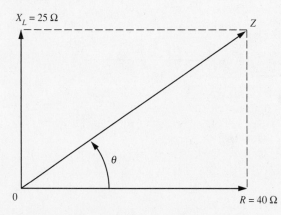

Figure 41–4. Phasor diagram for series *RL* problem.

SUMMARY

1. In an ac circuit the total opposition to current is called the impedance of the circuit. The symbol for impedance is Z. The unit of impedance is the ohm.
2. Ohm's law extended to ac circuits states that the current I equals the ratio of the applied voltage V and impedance Z. Thus,

$$I = \frac{V}{Z}$$

3. In a series RL circuit, the impedance Z is the phasor sum of R and X_L, where X_L is out of phase with R by an angle of 90° (Figure 41–3).
4. The numerical value of Z can be found from the impedance right triangle in Figure 41–3 using the formula

$$Z = \sqrt{R^2 + X_L^2}$$

5. In Figure 41–3 if θ is the phase angle between Z and R, then

$$\theta = \tan^{-1} \left(\frac{X_L}{R} \right)$$

6. Impedance Z may be calculated if θ and R are known. Thus,

$$Z = \frac{R}{\cos \theta}$$

where the angle θ is calculated from the formula in item 5 above.

SELF-TEST

Check your understanding by answering the following questions:

1. In the series RL circuit (Figure 41–2), $X_L = 100 \ \Omega$ and $R = 200 \ \Omega$. $Z =$ _____ Ω.
2. In the circuit of question 1, $\theta =$ _____ °.
3. In the circuit of question 1, $\dfrac{R}{\cos \theta} =$ _____ Ω.
4. The ratio R/cos θ, where θ is the phase angle between Z and R, gives the _____ of the circuit of Figure 41–2.
5. In a series RL circuit, $R = 45 \ \Omega$, $X_L = 45 \ \Omega$, and the applied voltage $V = 10$ V. The current I in the circuit is _____ A.
6. In the circuit of Figure 41–2, $V_R = 15$ V. The current in the inductor L is _____ A.

MATERIALS REQUIRED

Power Supplies:
- Isolation transformer
- Variable-voltage autotransformer (Variac or equivalent)

Instruments:
- Electronic voltmeter (EVM) or DMM
- 0–25-mA ac ammeter

Resistors:
- 1   2700-Ω, $^1/_2$,-W 5%
- 1   5100-Ω, 2-W

Inductor:
- 1 large iron-core choke (8 or 9 H) (Magnetek-Triad #C18A, 8 H, 100 Ω, 300 mA; Magnetek-Triad #C10X 9 H, 250 Ω, 125 mA; or equivalent)

Miscellaneous:
- Polarized line cord with on-off switch and fuse

PROCEDURE

1. Using an ohmmeter, measure the resistance of the 5100-Ω and 2700-Ω resistors. Record the values in Table 41–1.
2. With the line cord unplugged, line switch **off,** and S_1 **open,** connect the circuit of Figure 41–5. Set the autotransformer to its lowest output voltage.
3. Plug in the line cord; line switch **on; close** S_1. Increase the output voltage of the autotransformer until the ammeter measures 15 mA. Measure the voltage across AB, V_{AB}, the resistor V_R, and the inductor V_L. Record the values in Table 41–1 for the 5100-Ω resistor. **Open** S_1; disconnect the 5100-Ω resistor. This circuit will be reused in step 5.
4. Calculate the inductive reactance of the inductor using I and V_L. Calculate the circuit impedance using I and V_{AB}. Calculate current using V_R and R (measured). Record all answers in the 5100-Ω row of Table 41–1.

5. Connect a 2700-Ω resistor in series with the inductor in the circuit of Figure 41–5.
6. **Close** S_1. Repeat the measurements of step 3. Record the values in the 2700-Ω row of Table 41–1. After all measurements have been completed, **open** S_1, turn line switch **off,** and unplug line cord.
7. Perform the calculations of step 4 using the data for the 2700-Ω resistor. Record all answers in the 2700-Ω row of Table 41–1.
8. Examine Table 41–2. Using the values of impedance (calculated from V_{AB} and I) and inductive reactance (calculated from V_L and I) in Table 41–1, calculate the phase angle θ and impedance using the phase-angle relationships. Complete Table 41–2 for the 5100-Ω and 2700-Ω circuits.

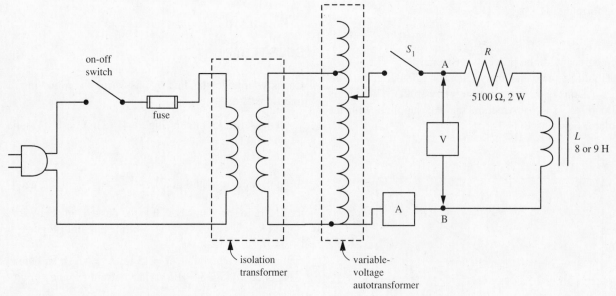

Figure 41–5. Circuit for procedure step 2.

ANSWERS TO SELF-TEST

1. 223.6
2. 26.6
3. 223.6
4. impedance
5. 0.157
6. 0.005

Experiment 41   Name _____ Date _____

TABLE 41–1. Verifying the Impedance Formula for an *RL* Circuit

Resistance R, Ω		Current I, mA	Voltage across AB, V_{AB}, V	Voltage across Resistor V_R, V	Voltage across Inductor V_L, V	Inductive Reactance (Calculated) Ohm's Law Formula, Ω	Circuit Impedance (Calculated) Ohm's Law Formula, Ω	Circuit Impedance (Calculated) R-X_L Formula, Ω	Current (Calculated) V_R/R, mA
Rated Value	Measured Value								
5100		15							
2700		15							

TABLE 41–2. Determining Phase Angle and Impedance

Resistance R, Ω		Inductive Reactance (from Table 41–1), Ω	$\tan \theta = \dfrac{X_L}{R}$	Phase Angle θ, degrees	Impedance $Z = \dfrac{R}{\cos \theta}$, Ω
Rated Value	Measured Value				
5100					
2700					

QUESTIONS

1. Explain, in your own words, the relationship between resistance, inductive reactance, and impedance in a series *RL* circuit.

_____

_____

_____

_____

_____

_____

_____

2. Explain, in your own words, how increases and decreases in resistance, with a constant inductive reactance, affect the phase angle of a series *RL* circuit.

_____

_____

_____

_____

_____

_____

_____

_____

Impedance of a Series RL Circuit   **295**

3. Refer to your data in Table 41–1. Are the two calculated values of impedance for the 5100-Ω resistor equal? Are the two calculated values of impedance for the 2700-Ω resistor equal? Explain any differences from expected results.

_____

_____

_____

_____

_____

_____

4. Refer to your data in Table 41–1. Are the calculated values of current equal to 15 mA? Should they be? Explain.

_____

_____

_____

_____

_____

_____

5. What factors might introduce errors in this experiment? Include both human and mechanical factors.

_____

_____

_____

_____

_____

_____

6. Refer to your data in Tables 41–1 and 41–2. Have the results in this experiment verified Formula (41–2)? Cite specific data to support your answer.

_____

_____

_____

_____

_____

_____

42

VOLTAGE RELATIONSHIPS IN A SERIES *RL* CIRCUIT

OBJECTIVES

1. To measure the phase angle θ between the applied voltage *V* and the current *I* in a series *RL* circuit
2. To verify experimentally that the relationships among the applied voltage *V*, the voltage V_R across *R*, and the voltage V_L across *L* are described by the formulas

$$V = \sqrt{V_R^2 + V_L^2}$$

$$V_R = V \times \frac{R}{Z}$$

$$V_L = V \times \frac{X_L}{Z}$$

BASIC INFORMATION

Phasors

A *phasor* is a quantity that is identified by two characteristics, *amplitude* and *direction*. In electricity we refer to sinusoidal voltages and currents as *phasor* quantities. Because it is related to sinusoidal voltages and currents, impedance will also be referred to as a phasor quantity.

Sinusoidal voltages or currents of the same frequency may be added and subtracted when they are represented by phasors.

Phase Angle Between Applied Voltage and Current in a Series *RL* Circuit

Because current is the same in every part of a series circuit, current *I* is shown as the reference phasor in considering the phase relations among V_R, V_L, and *V* in the series *RL* circuit [Figure 42–1(*a*)] (p. 298). Current in an inductance lags the voltage across the inductance by 90°. The phasor V_L in Figure 42–1(*b*) leads *I* by 90°; therefore, V_L leads V_R by 90°.

Phasor Sum of V_R and V_L Equals Applied Voltage *V*

Voltages V_R and V_L are, respectively, the voltages across *R* and *L* in the series *RL* circuit [Figure 42–1(*a*)]. Does the arithmetic sum of V_R and V_L equal the applied voltage *V*? No, because V_R and V_L are 90° out of phase. But *V* is the phasor sum of V_R and V_L. This fact is shown in Figure 42–1(*b*), where V_R and V_L are the legs of a right triangle and *V* is their phasor sum. Applying the pythagorean theorem, we have

$$V = \sqrt{V_R^2 + V_L^2} \qquad \textbf{(42–1)}$$

What is the phase relationship between the applied voltage *V* and the current *I* in the inductive circuit? In Figure 42–1(*b*), *I* is seen to lag the applied voltage *V* by an angle θ. This angle θ is the same as the angle θ between *Z* and *R* in the impedance phasor diagram (Figure 42–2, p. 298).

Figure 42–2(*a*) is Figure 42–1(*b*) redrawn, with the composition of each of the voltages as shown. Thus, V_R is the product of *I* and *R*, V_L is the product of *I* and X_L, and *V* is the product of *I* and *Z*. Since *I* is a common factor in each of these products, it may be canceled, leaving the impedance diagram [Figure 42–2(*b*)]. It is evident, therefore, that the angle θ is the same in Figures 42–1(*b*) and 42–2(*b*).

Further study of Figure 42–1(*b*) shows the relationship between *V*, V_R, V_L, and the phase angle θ. From the voltage phasor diagram

$$\frac{V_R}{V} = \cos\theta \qquad \textbf{(42–2)}$$

But from the impedance triangle we know that

$$\cos\theta = \frac{R}{Z}$$

Therefore,

$$\frac{V_R}{V} = \frac{R}{Z}$$

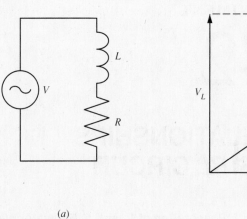

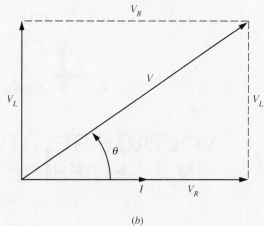

(a)   (b)

Figure 42–1. Phase relations in a series *RL* circuit.

or the voltage across the resistor is

$$V_R = V \times \frac{R}{Z} \qquad \textbf{(42–3)}$$

From the voltage triangle [Figure 42–1(*b*)],

$$\frac{V_L}{V_R} = \tan \theta = \frac{X_L}{R}$$

Therefore,

$$V_L = V_R \times \frac{X_L}{R}$$

Substituting V_R from formula (42–3), we obtain

$$V_L = V \times \frac{R}{Z} \times \frac{X_L}{R}$$

$$V_L = V \times \frac{X_L}{Z} \qquad \textbf{(42–4)}$$

Formulas (42–3) and (42–4) may be used to calculate the voltages V_R and V_L in a series *RL* circuit where the applied voltage V, the resistance R, and the inductive reactance X_L are known.

Problem. If 15 V is applied across a circuit consisting of a 40-Ω resistor in series with an inductance whose reactance is 25 Ω, what are the values of the phase angle θ between V and I, the voltage V_R across R, and the voltage V_L across L?

Solution. Using formulas (42–3) and (42–4), we obtain the solutions using a scientific calculator and the following keystrokes:

$$\theta = \boxed{2}\,\boxed{5}\,\boxed{\div}\,\boxed{4}\,\boxed{0}\,\boxed{=}\,\boxed{f}\,\boxed{\tan^{-1}} = 32.005°$$

$$Z = \boxed{\cos}\,\boxed{\div}\,\boxed{4}\,\boxed{0}\,\boxed{=}\,\boxed{1/x} = 47.170\ \Omega$$

(Answers have been rounded to three decimal places.)

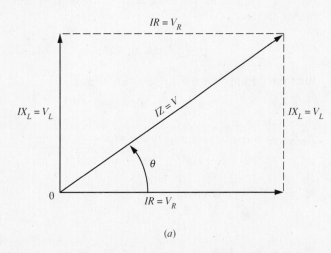

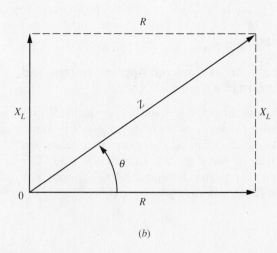

(a)   (b)

Figure 42–2.   The phase angle of the voltage phasor diagram is the same as the phase angle of the impedance phasor diagram.

298   *Experiment 42*

$$V_R = V \times \frac{R}{Z}$$

$$= 15 \times \frac{40}{47.170} = 12.720 \text{ V}$$

$$V_L = V \times \frac{X_L}{Z}$$

$$= 15 \times \frac{25}{47.170} = 7.950 \text{ V}$$

We can check the solutions using formula (42–1). By substituting the calculated values of V_R and V_L in the formula, we find that

$$V = \sqrt{12.720^2 + 7.950^2}$$

$$= 15.000 \text{ V}$$

The calculated value of V is the same as the given applied voltage V. Our solutions are therefore verified.

SUMMARY

1. In a series RL circuit the current I lags the applied voltage V by an angle θ, called the phase angle.
2. The phase angle θ between V and I is the same as the angle θ between Z and R in the impedance phasor diagram of the RL circuit. The angle θ is also the same as the angle between V and V_R in Figure 42–1(b).
3. The value of θ depends on the relative values of X_L, R, and Z and may be calculated from the following formula:

$$\theta = \tan^{-1}\left(\frac{X_L}{R}\right)$$

4. In a series RL circuit the voltage drop V_L across the inductance leads the voltage drop V_R across the resistor by 90°.
5. Voltages V_R and V_L are added using their phasors to obtain the applied voltage V in the circuit. That is, V is the phasor sum of V_R and V_L.
6. The relationship between the applied voltage V, the voltage drop V_R across R, and the voltage drop V_L across L is given by the formula

$$V = \sqrt{V_R^2 + V_L^2}$$

7. If the applied voltage V and R, X_L, and Z are known in a series RL circuit, then V_R and V_L may be calculated from the formulas

$$V_R = V \times \frac{R}{Z}$$

$$V_L = V \times \frac{X_L}{Z}$$

SELF-TEST

Check your understanding by answering the following questions:

1. In the circuit of Figure 42–1 the measured value of $V_R = 30$ V, and $V_L = 20$ V. The applied voltage V must then equal _____ V.
2. (True/False) The phase angle between V and V_R is the same as the phase angle between the applied voltage V and the current I through the resistor in a series RL circuit. _____
3. The phase angle θ between V and I in question 1 is _____ °.
4. In the circuit of Figure 42–1, $X_L = 68$ Ω, $R = 82$ Ω, and $V = 12$ V. In this circuit
   (a) $Z =$ _____ Ω
   (b) $\theta =$ _____ °
   (c) $V_R =$ _____ V
   (d) $V_L =$ _____ V

MATERIALS REQUIRED

Power Supplies:
■ Isolation transformer
■ Variable-voltage autotransformer (Variac or equivalent)

Instruments:
■ Dual-trace oscilloscope
■ EVM or DMM

Resistors (½-W, 5%):
■ 1   1200-Ω
■ 1   4700-Ω

Inductor:
■ Large iron-core choke (see Experiment 41 for type)

Miscellaneous:
■ SPST switch
■ Polarized line cord with on-off switch and fuse

PROCEDURE

1. Using an ohmmeter, measure the resistance of the 4700-Ω and 1200-Ω resistors. Record the values in Table 42–1 (p. 303).

2. With the line cord unplugged, line switch **off,** and S_1 **open,** connect the circuit of Figure 42–3. The oscilloscope should be turned off and the autotransformer should be set at its lowest output voltage. Channel 1 and Channel 2 input jacks should be connected as shown. The EXT TRIG jack should be connected to the choke as shown. Set the trigger switch to EXT.

    NOTE: Although the choke does have resistance, its value is so much less than the series 4700-Ω resistor that it can be ignored.

2. Plug in the line cord, turn line switch **on,** and **close** S_1. Increase the output of the autotransformer to 10 V rms.

3. Channel 1 is the voltage reference channel. Turn on the oscilloscope. Adjust the controls of the scope so that a single sine wave, about 6 divisions peak-to-peak, fills the width of the screen. Use the horizontal and vertical controls to center the waveform on the screen.

4. Switch to Channel 2. This is the current channel. Adjust the controls so that a single sine wave, about 4 divisions peak-to-peak, fills the width of the screen. Use the vertical control to center the waveform vertically. *Do not use the horizontal control.*

5. Switch the oscilloscope to dual-channel mode. The Channel 1 and Channel 2 signals should appear together. Note where the curves cross the horizontal (x) axis. These are the zero points of the two sine waves. With a centimeter scale, accurately measure the horizontal distance d between the two positive or two negative peaks of the sine waves. Verify your measurement by measuring the distance between the corresponding zero points of the two waves (see Figure 42–4). Record the measurement in Table 42–1, in the 4700-Ω row. Also measure the distance D from 0 to 360° for the voltage sine wave. Record the value in Table 42–1 for the 4700-Ω resistor. Turn **off** the scope; **open** S_1; disconnect the 4700-Ω resistor.

6. Using the formula in Figure 42–4 calculate the phase angle θ between the voltage and current in the circuit of Figure 42–3. Record your answer in Table 42–1.

7. Replace the 4700-Ω resistor with a 1200-Ω resistor in the circuit of Figure 42–3.

8. **Close** S_1. Repeat steps 3, 4, and 5 for the 1200-Ω resistor. After taking the final measurement, turn **off** the scope; **open** S_1; and disconnect the scope from the circuit.

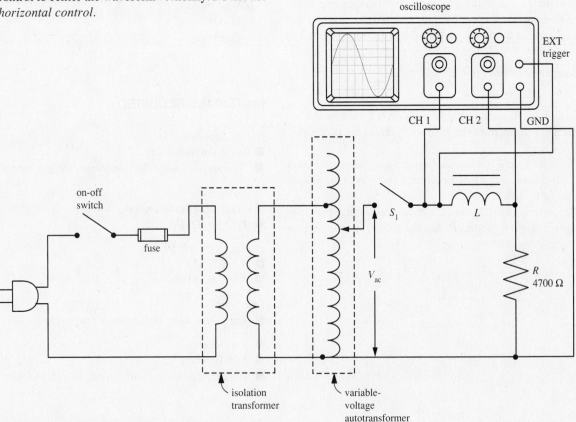

Figure 42–3. Circuit for procedure step 2.

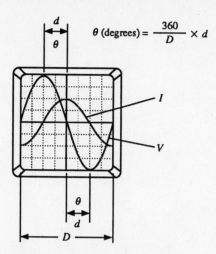

$$\theta \text{ (degrees)} = \frac{360}{D} \times d$$

Figure 42–4. Determining the phase angle between the voltage and current in an *RL* circuit using oscilloscope waveforms.

9. Repeat step 6 for the 1200-Ω resistor. The circuit of Figure 42–3 with the choke and 1200-Ω resistor connected in series will be used to verify the formula

$$\theta = \tan^{-1}\left(\frac{X_L}{R}\right)$$

where θ is the phase angle whose tangent is X_L/R. (The oscilloscope is not used in this part.)

10. **Close** S_1. The applied voltage V_{ac} should still be set to 10 V. Measure the voltage across the resistor V_R and across the inductor V_L. Record the values in Table 42–2 (p. 303), in the 1200-Ω row. **Open** S_1; disconnect the 1200-Ω resistor.
11. Calculate the current in the circuit using Ohm's law with the measured values of V_R and R. Record your answer in Table 42–2 for the 1200-Ω resistor.
12. Calculate the inductive reactance X_L of the inductor using Ohm's law for inductors with the measured value of V_L and the calculated value of I. Record your answer in Table 42–2.
13. Using the calculated value of X_L from step 12 and the measured value of R, calculate the phase angle θ

$$\theta = \tan^{-1}\left(\frac{X_L}{R}\right)$$

Record your answer in Table 42–2 for the 1200-Ω resistor.
14. Repeat steps 10 through 13 for the 4700-Ω resistor.
15. Using the measured values of V_R and V_L for the 1200-Ω resistor, calculate V_{ac} with the square root formula

$$V = \sqrt{V_R^2 + V_L^2}$$

Record your answer in the "Applied Voltage (Calculated)" column in Table 42–2. Repeat the calculation for V_R and V_L with the 4700-Ω resistor. Record your answer in Table 42–2.

ANSWERS TO SELF-TEST

1. 36.1
2. true
3. 33.7
4. (a) 106.5; (b) 39.7; (c) 9.24; (d) 7.66

Experiment 42   Name _____ Date _____

TABLE 42–1. Using the Oscilloscope to Find the Phase Angle θ in a Series *RL* Circuit

Resistance R, Ω		Width of Sine Wave D, cm	Distance Between Zero Points d, cm	Phase Angle θ, degrees
Rated Value	Measured Value			
4700				
1200				

TABLE 42–2. Phase Angle θ and Voltage Relationships in a Series *RL* Circuit

Resistor Rated Value, Ω	Applied Voltage V_{ac}, V	Voltage across Resistor V_R, V	Voltage across Inductor V_L, V	Current (Calculated) I, A	Inductive Reactance X_L, (Calculated), Ω	Phase Angle θ (Cal. from X_L and R), degrees	Applied Voltage (Calculated) V_{ac}, V
4700							
1200							

QUESTIONS

1. Explain, in your own words, the relationship between the voltage across a resistance, the voltage across an inductance, and the applied voltage in a series *RL* circuit.

_____

_____

_____

_____

_____

_____

_____

_____

2. Refer to your data in Tables 42–1 and 42–2. Is the phase angle θ found using the oscilloscope the same (for comparable resistors) as the phase angle found using resistance/inductive reactance formulas? Should the two angles be the same? Explain your answer.

_____

_____

_____

_____

_____

_____

_____

_____

_____

3. Refer to your data in Table 42–2. Do the results of this experiment verify Formulas (42–2), (42–3), and (42–4)? Support your answer with reference to your data.

_____

_____

_____

_____

_____

_____

_____

_____

_____

_____

_____

_____

_____

_____

_____

4. Do the results of this experiment verify Formula (42–1)? Refer to specific data to support your answer.

_____

_____

_____

_____

_____

_____

_____

_____

_____

_____

_____

_____

_____

_____

43

IMPEDANCE OF A SERIES *RC* CIRCUIT

OBJECTIVES

1. To verify experimentally that the impedance Z of a series *RC* circuit is given by the formula

$$Z = \sqrt{R^2 + X_C^2}$$

2. To study the relationships among impedance, resistance, capacitive reactance, and phase angle

BASIC INFORMATION

The impedance of a series *RC* circuit can be found the same way as the impedance of a series *RL* circuit. The same formulas can be used except that X_C replaces X_L. Thus,

$$Z = \sqrt{R^2 + X_C^2} \qquad \textbf{(43–1)}$$

Both R and X_C are phasor quantities and must be added using phasors to obtain Z.

Figure 43–1(*a*) is a series *RC* circuit whose impedance diagram is shown in Figure 43–1(*b*). Note that the phasor representing X_C is a vertical line in the downward direction (recall that the X_L phasor was a vertical line in the upward direction).

Problem 1. If in Figure 43–1(*a*), $R = 300 \ \Omega$, $X_C = 400 \ \Omega$, and $V = 25$ V, find Z and *I*.

Solution. We can find the impedance Z using formula (43–1):

$$Z = \sqrt{R^2 + X_C^2}$$
$$= \sqrt{300^2 + 400^2} = \sqrt{250,000}$$
$$Z = 500 \ \Omega$$

From Ohm's law

$$I = \frac{V}{Z} = \frac{25}{500}$$
$$I = 0.05 \text{ A or } 50 \text{ mA}$$

Alternative Method for Calculating Impedance

As with seri.. *RL* circuits, the $\tan^{-1}$ function on a scientific calculator may be used to find the phase angle θ between R and Z when R and X_C are known.

$$\theta = \tan^{-1}\left(\frac{X_C}{R}\right) \qquad \textbf{(43–2)}$$

The keystroke procedure is to divide X_C by R, find the answer, then press the f key and $\tan^{-1}$ key in sequence. The display will show the value of θ.

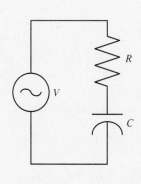

(*a*)   (*b*)

Figure 43–1. Phase relations in a series *RC* circuit.

The impedance can then be found using this value of θ and the formula

$$Z = \frac{R}{\cos \theta} \qquad (43-3)$$

Again, we can use the scientific calculator to produce the answer by using the following keystrokes: Enter the value of R, press the divide key, enter the value of θ, and then press the cos key and the = key. The display will show the value of Z. An example demonstrates the procedure to be used.

Problem 2. The values of problem 1 will be used. In addition to finding Z and I, the phase angle θ is also required.

Solution.

$$\theta = \tan^{-1}\left(\frac{X_C}{R}\right)$$

$$= \tan^{-1}\left(\frac{400}{300}\right)$$

The keystrokes are as follows:

$$\boxed{4}\,\boxed{0}\,\boxed{0}\,\boxed{\div}\,\boxed{3}\,\boxed{0}\,\boxed{0}\,\boxed{=}\,\boxed{f}\,\boxed{\tan^{-1}} = 53.130°$$

$$Z = \frac{R}{\cos \theta} = \frac{300}{\cos 53.130}$$

With 53.130 still in the display, press the following keys:

$$Z = \boxed{\cos}\,\boxed{\div}\,\boxed{3}\,\boxed{0}\,\boxed{0}\,\boxed{=}\,\boxed{1/x} = 500 \ \Omega$$

which is the same as the answer in problem 1. Continue the keystrokes to find I using the formula $I = V/Z$. With 500 in the display, press the following keys:

$$I = \boxed{\div}\,\boxed{2}\,\boxed{5}\,\boxed{=}\,\boxed{1/x} = 0.05 \ A$$

Using Ohm's Law to Verify *I*, *V*, and *Z*

One form of Ohm's law for ac circuits is

$$Z = \frac{V}{I} \qquad (43-4)$$

This formula can be used to verify the relationships among R, X_C, and Z. In the series circuit of Figure 43–1(*a*), the voltage V across the combination of R and C can be measured as can the current I in the circuit. The impedance Z can then be calculated using formula (43–4) and the measured value of V and I. If the value calculated in this way is the same as that calculated using formulas (43–1) and (43–3), the relationships among θ, X_C, and R have been verified.

Problem 3. In the series circuit of Figure 43–1(*a*), $R = 50 \ \Omega$, $X_C = 120 \ \Omega$, and $V = 10$ V. An ac ammeter connected into the circuit measures 77 mA. Verify the relationships among R, X_C, and θ.

Solution. The impedance can be found using formula (43–1)

$$Z = \sqrt{R^2 + X_C^2}$$

$$= \sqrt{50^2 + 120^2}$$

$$Z = 130 \ \Omega$$

Using Ohm's law to find Z, we have

$$Z = \frac{V}{I} = \frac{10}{0.077}$$

$$Z = 129.870, \text{ or } Z = 130 \ \Omega$$

This verifies the value found using formula (43–1). The phase angle θ can be found from formula (43–2):

$$\theta = \tan^{-1}\left(\frac{120}{50}\right)$$

The calculator solution is as follows:

$$\theta = \boxed{1}\,\boxed{2}\,\boxed{0}\,\boxed{\div}\,\boxed{5}\,\boxed{0}\,\boxed{=}\,\boxed{f}\,\boxed{\tan^{-1}}$$

$$= 67.380°$$

Also,

$$Z = \frac{R}{\cos \theta} = \frac{50}{\cos 67.380}$$

Proceeding from the previous display of 67.380°, Z can be found using the calculator and formula (43–3), as follows:

$$Z = \boxed{\cos}\,\boxed{\div}\,\boxed{5}\,\boxed{0}\,\boxed{=}\,\boxed{1/x} = 130 \ \Omega$$

Again, the relationships among R, X_C, and θ lead to the same answer.

SUMMARY

1. In a series-connected *RC* circuit, the impedance Z is the phasor sum of R and X_C, where X_C lags R by 90° [Figure 43–1(*b*)].
2. From the impedance right triangle in Figure 43–1(*b*), the numerical value of Z can be found from the formula

$$Z = \sqrt{R^2 + X_C^2}$$

3. In Figure 43–1(*b*), if θ is the phase angle between Z and R, then

$$\theta = \tan^{-1}\left(\frac{X_C}{R}\right)$$

4. The impedance Z may also be calculated if θ and R are known. Thus,

$$Z = \frac{R}{\cos \theta}$$

where θ is the phase angle calculated from the formula in item 3 above.

5. The impedance Z of an RC circuit may be found experimentally if the applied voltage V and the current I in the circuit are known. Z is calculated using Ohm's law:

$$Z = \frac{V}{I}$$

6. If the value of Z is substantially the same for each of the methods used, the right-triangle relationship among Z, X_C, R, and θ has been verified.

SELF-TEST

Check your understanding by answering the following questions.

1. In the series RC circuit [Figure 43–1(a)], $R = 300\ \Omega$, $X_C = 120\ \Omega$, and $Z = $ _____ Ω.
2. In the circuit of question 1, $\theta = $ _____ °.
3. In the circuit of question 1, $R/\cos \theta = $ _____ Ω.
4. The ratio $R/\cos \theta$, where θ is the phase angle between Z and R, gives the _____ of the circuit of Figure 43–1(a).

5. In a series RC circuit, $R = 120\ \Omega$, $X_C = 150\ \Omega$, and the applied voltage $V = 5$ V, the current $I = $ _____ A.
6. In the circuit of question 1, $V_R = 45$ V. The current I in the capacitor C is _____ A.

MATERIALS REQUIRED

Power Supplies:
- Isolation transformer
- Variable-voltage autotransformer (Variac or equivalent)

Instruments:
- EVM or DMM
- Multirange ac ammeter with 5- and 25-mA ranges or second multimeter with these ranges

Resistors:
- 1   5100-Ω, 1-W
- 1   22,000-Ω, ½-W, 5%

Capacitors:
- 1   0.5-μF or 0.47-μF
- 1   0.1-μF

Miscellaneous:
- SPST switch
- Polarized line cord with on-off switch and fuse

PROCEDURE

1. With the line cord unplugged, the line switch **off,** and S_1 **open,** connect the circuit of Figure 43–2 (p. 308). Set the autotransformer to its lowest output voltage. Set the ammeter at the 25-mA range and the voltmeter at the 100-V range.
2. **Close S_1.** Increase the output of the autotransformer until the ammeter measures 10 mA (0.01 A).
3. Measure the voltage across the series RC combination V_{AB} and across the resistor V_R. Record the values in row 1 of Table 43–1 (p. 309). After taking the measurements, **open** S_1 and disconnect the resistor and capacitor. The circuit will be reused in step 6.
4. Calculate the total impedance of the circuit using the measured values of V_{AB} and I. Record your answer in Table 43–1. Calculate the capacitive reactance of the capacitor using the rated value of C and the formula

$$X_C = \frac{1}{2\pi f C}$$

The line frequency is 60 Hz. (Check with your instructor if your area has a different line frequency.) Record your

answer in Table 43–1, row 1. Using the R-X_C formula (43–1) with the calculated value of X_C and the rated value of R, calculate the total impedance of the circuit. Record your answer under "Total Impedance (Calculated) R-X_C Formula" in row 1 of Table 43–1.

5. With the measured value of V_R and the rated value of the resistor, calculate the line current I. Record your answer in Table 43–1 in row 1.
6. Connect a 22,000-Ω resistor in series with a 0.1-μF capacitor and connect the combination across the output of the autotransformer as in Figure 43–2. Set the autotransformer to its lowest output voltage. Set the ammeter on the 5-mA range.
7. **Close S_1.** Increase the output of the autotransformer until the ammeter measures 2 mA (0.002 A).
8. Repeat steps 3 through 5. Record the measured values in row 2 of Table 43–1. After completing your measurements, **open** S_1, turn **off** the line switch, and unplug the line cord. Complete row 2 of Table 43–1 with the calculated values of Z, X_C, and I.

9. Examine Table 43–2 (p. 309). For the two *RC* circuits in Table 43–1, calculate the phase angle and the total impedance using the phase-angle formula (43–3). Use the values of X_C from Table 43–1 and the rated values of resistance in your calculations. Record your answers in Table 43–2.

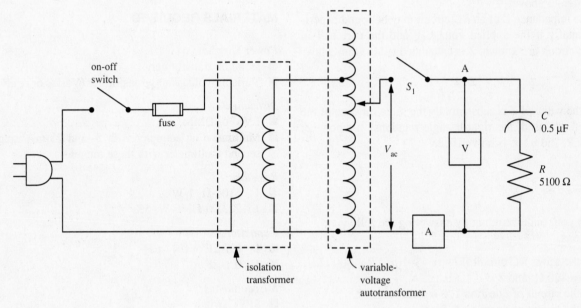

Figure 43–2. Circuit for procedure step 1.

ANSWERS TO SELF-TEST

1. 323
2. 21.8
3. 323
4. impedance
5. 0.026
6. 0.15

Experiment 43   Name _____ Date _____

TABLE 43–1.   Determining the Impedance of a Series *RC* Circuit

Row	Resistance (Rated Value) R, Ω	Capacitance (Rated Value) C, μF	Current (Measured) I, mA	Supply Voltage (Measured) V_{AB}, V	Voltage across R (Measured) V_R, V	Total Impedance (Calculated) Ohm's Law Z, Ω	Capacitive Reactance (Calculated) X_C, Ω	Total Impedance (Calculated) R-X_C Formula, Z, Ω	Line Current (Calculated) I, mA
1	5100	0.5	10						
2	22,000	0.1	2						

TABLE 43–2.   Determining Phase Angle and Impedance of a Series *RC* Circuit

Resistance (Rated Value) R, Ω	Capacitance (Rated Value) C, μF	Phase Angle (Calculated) θ, degrees	Total Impedance (Calculated) Phase Angle Formula Z, Ω
5100	0.5		
22,000	0.1		

QUESTIONS

1. Explain, in your own words, the relationship between resistance, capacitive reactance, and impedance in a series *RC* circuit.

_____

_____

_____

_____

_____

_____

_____

_____

2. Explain, in your own words, how increases and decreases in capacitive reactance affect the phase angle of a series *RC* circuit with a fixed resistance.

_____

_____

_____

_____

_____

_____

_____

_____

3. Refer to your data in Table 43–1. Are the two calculated values of impedance for the 5100-Ω circuit equal? Are the two calculated values of impedance for the 22,000-Ω circuit equal? If so, what can you conclude about Formula (43–1)? If not, explain the differences.

_____

_____

_____

_____

_____

_____

_____

_____

4. Refer to Tables 43–1 and 43–2. Compare the calculated value of impedance in Table 43–2 with the two values in Table 43–1 for comparable resistor-capacitor combinations. Are the comparable impedances equal? Should they be? Discuss your conclusions with specific reference to your data.

_____

_____

_____

_____

_____

_____

_____

_____

5. What factors might introduce errors in this experiment? Include both human and mechanical factors.

_____

_____

_____

_____

_____

_____

_____

_____

44

VOLTAGE RELATIONSHIPS IN A SERIES *RC* CIRCUIT

OBJECTIVES

1. To measure the phase angle θ between the applied voltage V and the current I in a series *RC* circuit
2. To verify experimentally that the relationships among the applied voltage V, the voltage V_R across R, and the voltage V_C across C are described by the formulas

$$V = \sqrt{V_R^2 + V_C^2}$$

$$V_R = V \times \frac{R}{Z}$$

$$V_C = V \times \frac{X_C}{Z}$$

BASIC INFORMATION

The relationships among R, X, Z, and θ in *RC* and *RL* circuits are very similar. The difference is that in an *RL* circuit, current I lags the applied voltage V, whereas in an *RC* circuit, current I *leads* the applied voltage V.

Phase Relationships Between the Applied Voltage and Current in a Series *RC* Circuit

The current I is common in every part of the *RC* series circuit of Figure 44–1(*a*). Therefore, current is used as the reference phasor in the phasor diagram showing V_R and V_C. Because current and voltage in a resistor are in phase, the voltage phasor V_R is shown in Figure 44–1(*b*) on the same line as the current phasor. But *current* in a *capacitor leads the voltage* across the capacitor by 90°. Therefore, the phasor V_C is shown lagging the current I and V_R by 90°.

Phasor V_R is the voltage across R, and V_C is the voltage across C in the series *RC* voltage divider [Figure 44–1(*a*)]. As in the case of an *RL* circuit, the applied voltage V is the phasor sum of V_R and V_C, as shown in Figure 44–1(*b*). We also see that V is the hypotenuse of a right triangle of which V_R and V_C are the legs. Therefore, applying the pythagorean formula, we obtain

$$V = \sqrt{V_R^2 + V_C^2} \tag{44–1}$$

Figure 44–1(*b*) also shows the phase relationship between the applied voltage V and the current I in the series *RC* circuit. The current I is seen to lead the applied voltage V by the angle θ.

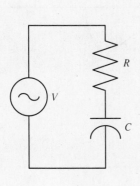

(*a*)

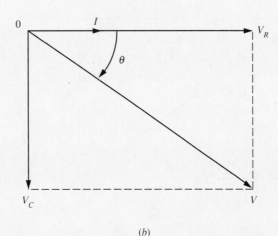

(*b*)

Figure 44–1. Phase relations in a series *RC* circuit.

The angle θ by which current leads the applied voltage in a series RC circuit is the same as the angle θ between the impedance phasor Z and the resistance phasor R. Figure 44–2 is Figure 44–1(b) redrawn showing the phase relationships among V, V_R, and V_C. Voltage V_R is the product of I and Z. Because I is a common factor in each of these products, it may be canceled, leaving the impedance diagram[Figure 44–2(b)]. This diagram shows that the angle θ is the same in Figures 44–1(b) and 44–2(b).

Further study of Figure 44–1(b) shows the relationship among V, V_R, V_C, and the phase angle θ. From the voltage phasor diagram

$$\frac{V_R}{V} = \cos \theta \qquad (44–2)$$

But from the impedance triangle

$$\cos \theta = \frac{R}{Z}$$

Therefore,

$$\frac{V_R}{V} = \frac{R}{Z}$$

or the voltage across the resistor is

$$V_R = V \times \frac{R}{Z} \qquad (44–3)$$

From the voltage triangle [Figure 44–1(b)]

$$\frac{V_C}{V_R} = \tan \theta = \frac{X_C}{R}$$

Therefore,

$$V_C = V_R \times \frac{V_C}{R}$$

Substituting V_R from formula (44–3), we have

$$V_C = V \times \frac{R}{Z} \times \frac{X_C}{R}$$

$$V_C = V \times \frac{X_C}{Z} \qquad (44–4)$$

Formulas (44–3) and (44–4) may be used to calculate the voltages V_R and V_C in a series RC circuit when the applied voltage V, the resistance R, and the capacitive reactance X_C are known.

Problem. If 12 V is applied across a circuit consisting of a 47-Ω resistor in series with a capacitor whose reactance is 100 Ω, what are the values of the phase angle θ between V and I, the voltage V_R across R, and the voltage V_C across C?

Solution. Using formulas (44–3) and (44–4), we obtain the solutions using a scientific calculator and the following keystrokes:

$$\theta = \boxed{1}\ \boxed{0}\ \boxed{0}\ \boxed{\div}\ \boxed{4}\ \boxed{7}\ \boxed{=}\ \boxed{f}\ \boxed{\tan^{-1}}$$

$$= 64.826°$$

$$Z = \boxed{\cos}\ \boxed{\div}\ \boxed{4}\ \boxed{7}\ \boxed{=}\ \boxed{1/x}\ = 110.494\ \Omega$$

(Answers have been rounded off to three decimal places.)

$$V_R = V \times \frac{R}{Z} = 12 \times \frac{47}{110.494}$$

$$= 5.104\ \text{V}$$

$$V_C = V \times \frac{X_C}{Z} = 12 \times \frac{100}{110.494}$$

$$= 10.860\ \text{V}$$

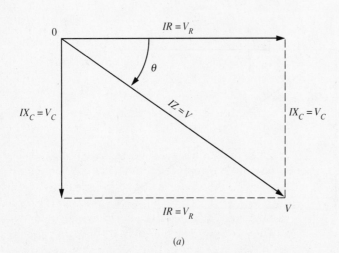

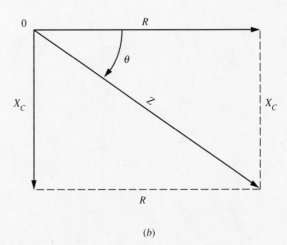

(a)   (b)

Figure 44–2. The phase angle of the voltage phasor diagram is the same as the phase angle of the impedance phasor diagram.

We can check the solutions using formula (44–1). By substituting the calculated values of V_R and V_C into formula (44–1), we find that

$$V = \sqrt{5.104^2 + 10.860^2}$$
$$= 12.000 \text{ V}$$

The calculated value of V is the same as the given applied voltage V and our solutions are therefore verified.

SUMMARY

1. In a series RC (capacitive) circuit, the current I leads the applied voltage V by an angle θ, called the phase angle.
2. The phase angle θ between V and I is the same as the angle θ between Z and R in the impedance phasor diagram of the RC circuit. The angle θ is also the same as the angle between V and V_R [Figure 44–1(b)].
3. The value of θ depends on the relative values of X_C, R, and Z and may be calculated from the formula

$$\theta = \tan^{-1}\left(\frac{X_C}{R}\right)$$

4. In a series RC circuit the voltage drop V_C across the capacitor lags the voltage drop V_R across the resistor by 90°.
5. Voltages V_C and V_R are phasors and are added using their phasors to obtain the applied voltage V in the circuit. That is, V is the phasor sum of V_R and V_C.
6. The relationship between the applied voltage V and the voltage drop V_R across R and the voltage drop V_C across C is given by the formula

$$V = \sqrt{V_R^2 + V_C^2}$$

7. If the applied voltage V and R, X_C, and Z are known in a series RC circuit, then V_R and V_C may be calculated from the formulas

$$V_R = V \times \frac{R}{Z}$$

$$V_C = V \times \frac{X_C}{Z}$$

SELF-TEST

Check your understanding by answering the following questions:

1. In the circuit of Figure 44–1(a), the measured values of V_R and V_C are, respectively, 5 V and 12 V. The applied voltage V must then equal _____ V.
2. The phase angle between voltages V and _____ is the same as the phase angle between V and I in the circuit of Figure 44–1(a).
3. The phase angle θ between V and I in question 1 is _____ °.
4. In the circuit of Figure 44–1(a), $X_C = 200 \ \Omega$, $R = 300 \ \Omega$, and $V = 6$ V. In this circuit,
   (a) $Z =$ _____ Ω;
   (b) $\theta =$ _____ °;
   (c) $V_R =$ _____ V;
   (d) $V_C =$ _____ V.
5. (True/False) In a series RC circuit, $\dfrac{X_C}{R} = \dfrac{R}{Z}$. _____

MATERIALS REQUIRED

Power Supplies:
- Isolation transformer
- Variable-voltage autotransformer (Variac or equivalent)

Instruments:
- Dual-trace oscilloscope
- EVM or DMM

Resistors ($\frac{1}{2}$-W, 5%):
- 1   5600-Ω
- 1   12,000-Ω

Capacitor:
- 1   0.5-μF or 0.47-μF

Miscellaneous:
- SPST switch
- Polarized line cord with on-off switch and fuse

PROCEDURE

1. Using an ohmmeter, measure the resistance of the 5600-Ω and 12,000-Ω resistors. Record the values in Table 44–1 (p. 317).
2. With the line cord unplugged, line switch **off**, and S_1 **open,** connect the circuit of Figure 44–3 (p. 314). The oscilloscope should be turned off and the autotransformer should be set at its lowest output voltage. Channel 1 and Channel 2 input jacks should be connected as shown. The EXT TRIG jack should be connected to the

capacitor as shown. Set the trigger switch to EXT.
3. Plug in the line cord, turn line switch **on,** and **close** S_1. Increase the output of the autotransformer to 10 V rms.
4. Channel 1 is the voltage reference channel. Turn **on** the oscilloscope. Adjust the controls of the scope so that a single sine wave, about 6 divisions peak-to-peak, fills the width of the screen. Use the horizontal and vertical controls to center the waveform on the screen.

oscilloscope

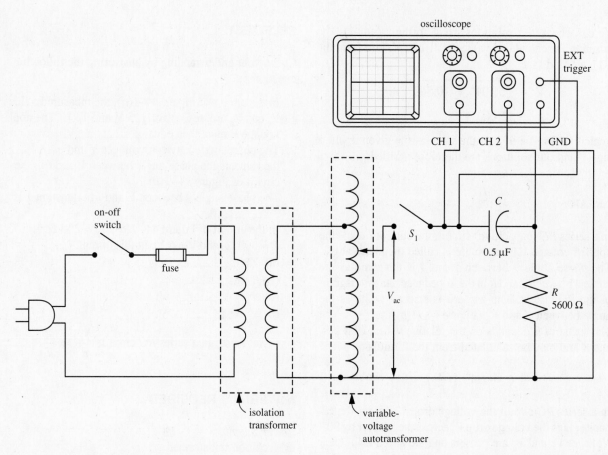

Figure 44–3. Circuit for procedure step 2.

5. Switch to Channel 2. This is the current channel. Adjust the controls so that a single sine wave, about 4 divisions peak-to-peak, fills the width of the screen. Use the vertical control to center the waveform vertically. *Do not use the horizontal control.*

6. Switch the oscilloscope to dual-channel mode. The Channel 1 and Channel 2 signals should appear together. Note where the curves cross the horizontal (x) axis. These are the zero points of the two sine waves. With a centimeter scale, accurately measure the horizontal distance d between the two positive or two negative peaks of the sine waves. Verify your measurement by measuring the distance between the corresponding zero points of the two waves (see Figure 44–4). Record the measurement in Table 44–1, in the 5600-Ω row. Also measure the distance D from 0 to 360° for the voltage sine wave. Record the value in Table 44–1 for the 5600-Ω resistor. Turn **off** the scope; **open** S_1; disconnect the 5600-Ω resistor.

7. Using the formula in Figure 44–4, calculate the phase angle θ between the voltage and current in the circuit of Figure 44–3. Record your answer in Table 44–1.

8. Replace the 5600-Ω resistor with a 12,000-Ω resistor in the circuit of Figure 44–3.

9. **Close** S_1. Repeat steps 4, 5, and 6 for the 12,000-Ω resistor. After taking the final measurement, turn **off** the scope; **open** S_1; and disconnect the scope from the circuit.

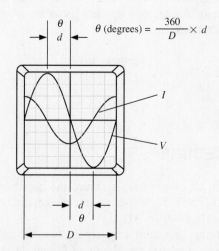

$$\theta \text{ (degrees)} = \frac{360}{D} \times d$$

Figure 44–4. Determining the phase angle between the voltage and current in an *RC* circuit using oscilloscope waveforms.

10. Repeat step 7 for the 12,000-Ω resistor. The circuit of Figure 44–3 with the capacitor and 12,000-Ω resistor connected in series will be used to verify the formula

$$\theta = \tan^{-1}\left(\frac{X_C}{R}\right)$$

where θ is the phase angle whose tangent is X_C/R.

11. **Close** S_1. Increase V_{ac} to 30 V. Measure the voltage across the resistor V_R and across the capacitor V_C. Record the value in Table 44–2 (p. 317), in the 12,000-Ω row.

12. Increase V_{ac} to 40 V; measure V_R and V_C. Record the values in Table 44–2 for the 12,000-Ω resistor.

13. Increase V_{ac} to 50 V; measure V_R and V_C. Record the values in Table 44–2 for the 12,000-Ω resistor. **Open** S_1; disconnect the 12,000-Ω resistor.

14. Calculate the current in the circuit for each value of V_{ac} using Ohm's law with the measured values of V_R and R. Record your answers in Table 44–2 for the 12,000-Ω resistor.

15. Calculate the capacitive reactance X_C of the capacitor for each value of V_{ac} using Ohm's law for capacitors with the measured value of V_C and the calculated value of I. Record your answers in Table 44–2 for the 12,000-Ω resistor.

16. Using the calculated values of X_C from step 15 and the measured value of R, calculate the phase angle θ for each value of V_{ac}.

$$\theta = \tan^{-1}\left(\frac{X_C}{R}\right)$$

Record your answers in Table 44–2 for the 12,000-Ω resistor.

17. Repeat steps 11 through 16 for the 5600-Ω resistor and the three values of V_{ac}.

18. Using the measured values of V_R and V_C for the 12,000-Ω resistor, calculate V_{ac} with the square-root formula $V = \sqrt{V_R^2 + V_C^2}$. Record your answers in the "Applied Voltage (Calculated)" column in Table 44–2. Repeat the calculation for V_R and V_C with the 5600-Ω resistor. Record your answers in Table 44–2.

ANSWERS TO SELF-TEST

1. 13
2. V_R
3. 67.4
4. (a) 361; (b) 33.7; (c) 5.0; (d) 3.32
5. false

Experiment 44   Name _____ Date _____

TABLE 44–1. Using the Oscilloscope to Find the Phase Angle θ in a Series *RC* Circuit

Resistance R, Ω		Capacitance C, μF	Width of Sine Wave D, cm	Distance Between Zero Points d, cm	Phase Angle θ, degrees
Rated Value	Measured Value				
5600		0.5			
12,000		0.5			

TABLE 44–2. Phase Angle θ and Voltage Relationships in a Series *RC* Circuit

Resistance Rated Value, Ω	Capacitance (Rated Value) C, μF	Applied Voltage V_{ac}, V	Voltage across Resistor V_R, V	Voltage across Capacitor V_C, V	Current (Calculated) I, A	Capacitive Reactance (Calculated) X_C, Ω	Phase Angle θ (Calculated from X_C and R), degrees	Applied Voltage (Calculated) V_{ac}, V
5600	0.5	30						
		40						
		50						
12,000	0.5	30						
		40						
		50						

QUESTIONS

1. Explain, in your own words, the relationship between the voltage across a resistance, the voltage across a capacitance, and the applied voltage in a series *RC* circuit.

_____

_____

_____

_____

_____

2. Refer to your data in Tables 44–1 and 44–2. Is the phase angle, θ, found using the oscilloscope the same (for comparable resistors) as the phase angle found using resistance/capacitive reactance formulas? Should they be? Explain your answer.

_____

_____

_____

_____

_____

Voltage Relationships in a Series RC Circuit **317**

3. Refer to your data in Table 44–2. Do the results of this experiment verify formulas (44–2), (44–3), and (44–4)? Support your answer with reference to specific data.

_____

_____

_____

_____

_____

_____

_____

_____

4. Do the results of this experiment verify formula (44–1)? Refer to specific data to support your answer.

_____

_____

_____

_____

_____

_____

_____

_____

5. What factors might introduce errors in using the oscilloscope to determine the phase angle θ?

_____

_____

_____

_____

_____

_____

_____

_____

_____

45

POWER IN AC CIRCUITS

OBJECTIVES

1. To differentiate between true power and apparent power in ac circuits
2. To measure power in an ac circuit

BASIC INFORMATION

Consumption of AC Power

Power dissipation in dc resistive circuits is defined as the product of the voltage and the current. That is,

$$P = V \times I \qquad (45\text{--}1)$$

where

P is power in watts (W),
V is voltage in volts (V),
I is current in amperes (A).

Power can also be calculated using the formulas

$$P = I^2 R$$

$$P = \frac{V^2}{R} \qquad (45\text{--}2)$$

In dc circuits V and I are constant, not varying, values. In ac circuits V and I are continuously varying and may be in phase or out of phase with one another. An ac circuit may also contain reactive components in addition to resistive components. Resistive components dissipate power in ac circuits just as they do in dc circuits. However, pure reactive components do not dissipate net power. In one part of the cycle they take power from the circuit; in the next part of the cycle they return power to the circuit. The result is that there is no total power consumed by the reactive component.

Apparent Power

Apparent power P_A is the input power to the ac circuit. It is defined as the product of the voltage V and the current I in the ac circuit. Thus,

$$P_A = V \times I$$

For lack of a better unit and to avoid confusion with the unit watts, apparent power is usually measured in voltamperes (VA).

True Power and Power Factor

The power in watts consumed by an electrical device with both resistive and reactive components is defined as true power. In a circuit containing both resistive and reactive components, true power is equal to or less than apparent power. The amount by which apparent power must be multiplied to obtain true power is called the *power factor* (PF) of the circuit. It is really the ratio of true power to apparent power. Thus,

$$\text{PF} = \frac{\text{true power}}{\text{apparent power}} = \frac{P(\text{W})}{P_A(\text{VA})}$$

and

$$P = P_A \times \text{PF} = V \times I \times \text{PF} \qquad (45\text{--}3)$$

The voltage phasor diagram in Figure 45–1(*a*) (p. 320) can be converted to a power diagram by multiplying each side of the voltage triangle by I, as in Figure 45–1(*b*). Thus,

$$\cos \theta = \frac{P(\text{W})}{P_A(\text{VA})} \qquad (45\text{--}4)$$

But from Figure 45–1(*a*) and Experiment 42,

$$\cos \theta = \frac{V_R}{V} = \frac{R}{Z} \qquad (45\text{--}5)$$

Therefore,

$$\text{PF} = \frac{R}{Z} \qquad (45\text{--}6)$$

If PF = 1, it means that the total impedance of the circuit equals the resistance. If PF = 0, it means that $R = 0$, and no true power is consumed in the circuit. Since R cannot be greater than Z, PF can never be greater than 1. And since R and Z are always positive, PF can never be a negative number.

True power in watts, represented by P, can be found using the same formulas as those used for dc circuits.

$$P = I^2 R$$

$$P = V_R I \qquad (45\text{--}7)$$

$$P = \frac{V_R^2}{R}$$

Formulas (45–7) can be used in ac circuits because the voltage across a resistor is always in phase with the current through the resistor.

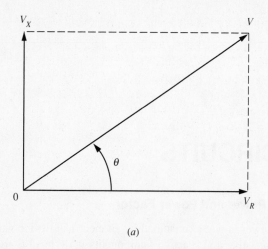

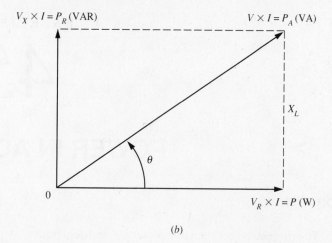

Figure 45–1. The voltage phasor diagram can be converted to a power phasor diagram.

Two examples show how power can be calculated in an ac series circuit.

Problem 1. In the circuit of Figure 45–2, a coil has an inductive reactance X_L of 1000 Ω and resistance R of 250 Ω. The voltage source is 20 V ac. We wish to find the apparent power and true power delivered by the supply, the PF of the circuit, and the angle by which the current I leads or lags the voltage V.

Solution. The impedance of the circuit is

$$Z = \sqrt{R^2 + X_L^2} = \sqrt{250^2 + 1000^2}$$

$$= 1031 \ \Omega$$

The current in the circuit is

$$I = \frac{V}{Z} = \frac{20}{1031}$$

$$= 0.0194 \text{ A, or } 19.4 \text{ mA}$$

Phase angle $\theta = \tan^{-1}\left(\dfrac{X_L}{R}\right) = \tan^{-1}\left(\dfrac{1000}{250}\right)$

On the calculator the keystrokes are

$$\theta = \boxed{1}\boxed{0}\boxed{0}\boxed{0}\boxed{\div}\boxed{2}\boxed{5}\boxed{0}\boxed{=}\boxed{f}\boxed{\tan^{-1}}$$

$$= 75.96°$$

Because the circuit is inductive, the current lags the voltage by 75.96°.

The apparent power P_A delivered by the supply is

$$P_A = VI = 20 \times 0.0194$$

$$= 0.388 \text{ VA}$$

The true power P is

$$P = I^2R = (0.0194)^2 \times 250$$

$$= 0.0941 \text{ W, or } 94.1 \text{ mW}$$

The power factor PF is

$$\text{PF} = \frac{R}{Z} = \frac{250}{1031}$$

$$= 0.242$$

Power factor is often expressed as a percentage. To find percent PF, multiply by 100:

$$\% \text{PF} = \text{PF} \times 100$$

Thus, in the preceding answer the percent power factor is

$$\% \text{PF} = \text{PF} \times 100$$

$$= 0.242 \times 100$$

$$= 24.2 \%$$

Expressed as a percentage, the power factor can never be greater than 100%.

Problem 2. The series circuit of Figure 45–3 contains a capacitor having a capacitive reactance $X_C = 300$ Ω and $R = 500$ Ω. The supply voltage is 12 V ac. We wish to find P_A, P, PF, and θ.

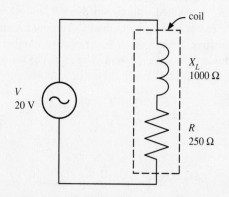

Figure 45–2. Circuit for problem 1.

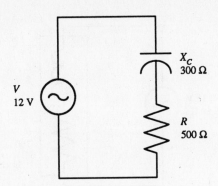

Figure 45–3. Circuit for problem 2.

Solution. We will find Z using the formulas

$$\theta = \tan^{-1}\left(\frac{X_C}{R}\right)$$

and

$$Z = \frac{R}{\cos \theta}$$

The calculator keystrokes are

$$\theta = \boxed{3}\,\boxed{0}\,\boxed{0}\,\boxed{\div}\,\boxed{5}\,\boxed{0}\,\boxed{0}\,\boxed{=}\,\boxed{f}\,\boxed{\tan^{-1}}$$

$$= 30.96°$$

Continuing the process to find Z,

$$Z = \boxed{\cos}\,\boxed{\div}\,\boxed{5}\,\boxed{0}\,\boxed{0}\,\boxed{=}\,\boxed{f}\,\boxed{1/x} = 583.10\ \Omega$$

$$I = \frac{V}{Z} = \frac{12}{583.1}$$

$$= 0.0206\,\text{A, or } 20.6\ \text{mA}$$

$$P_A = VI = 12 \times 0.0206 = 0.247\ \text{VA}$$

$$P = P_A \times \frac{R}{Z} = 0.247 \times \frac{500}{583.1}$$

$$P = 0.212\ \text{W}$$

This can be verified using the formula

$$P = I^2 R = 0.0206^2 \times 500$$

$$= 0.212\ \text{W}$$

The power factor is

$$\text{PF} = \frac{R}{Z} = \frac{500}{583.1} = 0.857,\ \text{or } 85.7\%$$

or

$$\text{PF} = \frac{P}{P_A} = \frac{0.212}{0.247}$$

$$= 85.8,\ \text{or } 85.8\%$$

(The difference of 0.1 is a result of the rounding process.)

Measuring AC Power—Voltage-Current Method

Power in an ac circuit may be determined by a series of measurements using familiar instruments, namely, the voltmeter and the oscilloscope. With the voltmeter we can measure the voltage V_R across R and the applied voltage V. We can then determine the power factor by substituting the measured values in the formula

$$\text{PF} = \frac{V_R}{V}$$

We can measure the resistance of R and calculate the current I using the formula

$$I = \frac{V_R}{R}$$

Then, using the formula

$$P = V \times I \times \text{PF}$$

we can calculate the true power P. We can also measure V_R and R and calculate P using the formula

$$P = \frac{V_R^2}{R}$$

The measurements are simple to make in a circuit such as the one in Figure 45–1. The calculations are straightforward.

If we wish to measure the phase angle in the circuit directly, we can use an oscilloscope. By the method discussed in Experiment 42, we can determine θ by comparing the phase angle of the applied voltage with the phase angle of the voltage across R (which is the same as the phase angle of I).

Wattmeter

Power can be measured directly using a wattmeter. For low-frequency measurements, such as 50–60 Hz, an analog meter having two coils is usually used. Figure 45–4 (p. 322) is a simple schematic representation of such a meter. Wattmeters of this type can have three or four terminals. Terminals A and B in Figure 45–4 are connected across the load; terminals C and D are connected in series with the load. In effect, A and B are for the voltage connections, and C and D are for the current connections. The meter pointer will indicate true power directly; that is, $P = V \times I \times \text{PF}$.

SUMMARY

1. Power in ac circuits is consumed only by the resistive components.
2. Apparent power P_A in an ac circuit is the product of the source voltage and the line current. $P = V \times I$, where V is the applied voltage and I is the current drawn by the circuit.

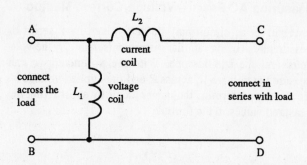

Figure 45–4. Simplified circuit diagram of a low-frequency analog wattmeter.

3. The true power dissipated by the circuit is the product of V and I and the power factor PF. The power factor is equal to the cosine of the angle between the voltage and current in the circuit, that is,

$$P = V \times I \times \cos \theta$$

4. Other formulas for true power are

$$P = I^2 R = \frac{V_R^2}{R}$$

where

I = current in the circuit in amperes
R = total resistance of the circuit in ohms
V_R = voltage measured across the total resistance of the circuit

5. The power factor ($\cos \theta$) of an ac circuit may be determined by measuring θ, the phase angle between the applied circuit voltage V and the current I, and calculating $\cos \theta$.

6. Other power factor formulas are:

$$\text{PF} = \cos \theta = \frac{R}{Z} = \frac{V_R}{V} = \frac{P}{P_A}$$

7. Power in an ac circuit may be determined by measuring the applied voltage V, the current I, and the phase angle θ and substituting the measured values in the formula

$$P = V \times I \times \cos \theta$$

8. True power may be measured directly, using a wattmeter.

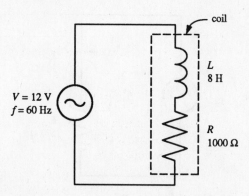

Figure 45–5. Circuit for Self-test.

SELF-TEST

Check your understanding by answering the following questions.

In the circuit of Figure 45–5, the coil has $L = 8$ H and $R = 1000\ \Omega$. The power source is $V = 12$ V at 60 Hz.

1. The inductive reactance of $L =$ _____ Ω.
2. The impedance of the circuit is _____ Ω.
3. The current I drawn by the circuit is _____ A.
4. $P_A =$ _____ VA.
5. The power factor of the circuit is _____ %.
6. The phase angle of the circuit is _____ °.
7. $I^2R =$ _____ W.
8. The ratio of true power to apparent power is called

_____ .

MATERIALS REQUIRED

Power Supplies:
■ Isolation transformer
■ Variable-voltage autotransformer (Variac or equivalent)

Instruments:
■ Dual-trace oscilloscope
■ EVM or DMM
■ Wattmeter
■ 0–25-mA ac ammeter or second DMM with ac ammeter scales

Resistor:
■ 1 100-Ω, 5-W

Capacitors:
■ 1 5-μF or 4.7-μF, 100-V
■ 1 10-μF, 100-V

Miscellaneous:
■ SPST switch
■ Polarized line cord with on-off switch and fuse

322 *Experiment 45*

PROCEDURE

A. Determining Power Using the Voltage-Current Method

A1. Using an ohmmeter, measure the resistance of the 100-Ω resistor and record the value in Table 45–1 (p. 327).

A2. With the line cord unplugged, line switch **off**, and S_1 **open**, connect the circuit of Figure 45–6. Set the auto-transformer to its lowest output voltage, and the ac ammeter to the 25-mA range.

A3. **Close S_1.** Increase the output voltage of the autotransformer until $V_{AB} = 50$ V. Measure the voltage across the resistor V_R and the current I. Record the values in Table 45–1 in the 5-μF row. **Open S_1;** disconnect the 5-μF capacitor.

A4. Calculate the apparent power P_A, the true power P, the power factor, and the phase angle of the circuit. Use the measured values of V_{AB}, V_R, and I, as appropriate, in your calculations. Record your answers in Table 45–1 in the 5-μF row.

A5. With S_1 **open** and the autotransformer set at its lowest output voltage, connect a 10-μF capacitor in series with the 100-Ω resistor.

A6. **Close S_1.** Increase the output of the autotransformer until $V_{AB} = 25$ V. Measure V_R and I and record the values in Table 45–1 in the 10-μF row. After your last measurement, **open S_1.**

A7. Repeat step A4 for the 100-Ω/10-μF series circuit. Record your answers in Table 45–1 in the 10-μF row.

B. Measuring Power with a Wattmeter

B1. With S_1 **open** and the autotransformer set to its lowest output voltage, connect a wattmeter in the series circuit of Part A. The wattmeter connections are as shown in Figure 45–7 (p. 324). Although the connections to the meter are typical, check the instruction manual of your meter to verify the connections.

B2. **Close S_1.** Increase the output voltage until $V_{AB} = 25$ V. Measure the true power P, using the wattmeter, and the voltage across the resistor V_R. Record the values in Table 45–2 (p. 327) in the 10-μF row. **Open S_1;** disconnect the 10-μF capacitor.

B3. Calculate the true power using V_R and the measured value of R (from Table 45–1). Calculate the power factor using V_R and V_{AB}. Record your answers in the 10-μF row of Table 45–2.

B4. With S_1 **open**, connect the 5-μF capacitor in series with the 100-Ω resistor.

B5. **Close S_1.** Increase the autotransformer output until $V_{AB} = 50$ V. Measure the power of the circuit using the wattmeter, and the voltage across the resistor V_R. Record the values in Table 45–2 in the 5-μF row. **Open S_1.**

B6. Calculate the true power using V_R and the measured value of R (from Table 45–1). Calculate the PF using V_R and V_{AB}. Record your answers in Table 45–2 in the 5-μF row.

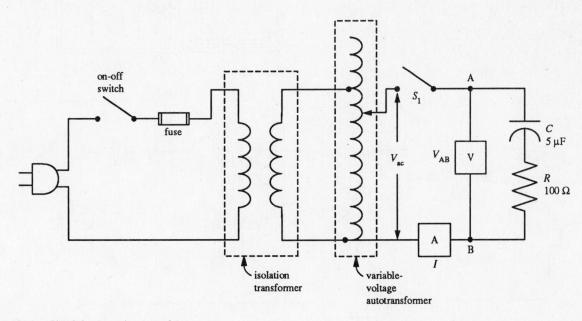

Figure 45–6. Circuit for procedure step A2.

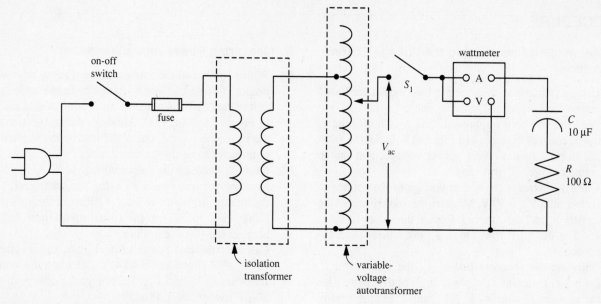

Figure 45–7. Circuit for procedure step B1.

C. Determining Power Factor with an Oscilloscope

C1. Connect the dual-trace oscilloscope to the series *RC* circuit, as shown in Figure 45–8. The autotransformer should be set at its lowest output voltage. The trigger switch should be set to EXT.

C2. **Close** S_1. Increase the output of the autotransformer to 10 V rms. Channel 1 is the voltage reference channel. Turn on the oscilloscope. Adjust the controls of the scope so that a single sine wave, about 6 div. peak-to-peak, fills the width of the screen. Use the horizontal and vertical controls to center the waveform on the screen.

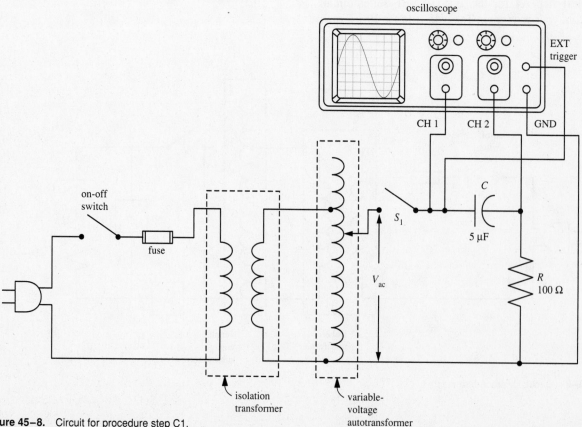

Figure 45–8. Circuit for procedure step C1.

C3. Switch to Channel 2. This is the current channel. Adjust the controls so that a single sine wave, about 4 div. peak-to-peak, fills the width of the screen. Use the vertical control to center the waveform vertically. *Do not use the horizontal control.*

C4. Switch the oscilloscope to dual-channel mode. The Channel 1 and Channel 2 signals should appear together. Note where the curves cross the horizontal (x) axis. These are the zero points of the two sine waves. With a centimeter scale, accurately measure the horizontal distance d between the two positive or two negative peaks of the sine waves. Verify your measurement by measuring the distance between the corresponding zero points of the two waves (see Figure 45–9). Record the measurement in Table 45–3 (p. 327) in the 5-μF row. Also measure the distance D from 0 to 360° for the voltage sine wave. Record the value in Table 45–3 for the 100-Ω resistor. Turn **off** the scope; **open** S_1; disconnect the 5-μF capacitor.

C5. Using the formula in Figure 45–9, calculate the phase angle θ between the voltage and current in the circuit of Figure 45–8. Using the calculated value of θ, calculate the power factor PF of the circuit. Record your answers in Table 45–3.

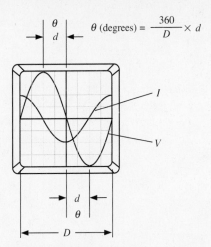

$$\theta \text{ (degrees)} = \frac{360}{D} \times d$$

Figure 45–9. Determining the power factor of an *RC* circuit using oscilloscope waveforms.

C6. Replace the 5-μF capacitor with a 10-μF capacitor in the circuit of Figure 45–8.

C7. Close S_1. Repeat steps C3 through C5 for the 10-μF capacitor. After taking the final measurement, turn **off** the scope; **open** S_1; and disconnect the scope from the circuit.

C8. Repeat step C5 for the 10-μF, 100-Ω series circuit.

ANSWERS TO SELF-TEST

1. 3016
2. 3177
3. 0.00378
4. 0.0454
5. 31.5
6. 71.65
7. 0.0143
8. PF

Experiment 45   Name _____ Date _____

TABLE 45–1. Power Measurement—Voltage-Current Method

Resistance R, Ω		Capacitance (Rated Value) C, μF	Applied Voltage V_{AB}, V	Voltage across Resistor V_R, V	Current (Measured) I, mA	Apparent Power P_A, VA	True Power P, W	Power Factor PF	Phase Angle θ, degrees
Rated Value	Measured Value								
100		5	50						
100		10	25						

TABLE 45–2. Power Measurement—Wattmeter Method

Resistance R (Rated Value), Ω	Capacitance (Rated Value) C, μF	Applied Voltage V_{AB}, V	Voltage Across Resistor V_R, V	Power (Measured) P, W	Power (Calculated) P, W	Power Factor PF, %
100	5	50				
100	10	25				

TABLE 45–3. Determining Power Factor with an Oscilloscope

Resistance (Rated Value) R, Ω	Capacitance (Rated Value) C, μF	Distance Between Zero Points, d, cm	Width of Sine Wave D, cm	Phase Angle (Calculated) θ, degrees	Power Factor (Calculated) PF, %
100	5				
100	10				

QUESTIONS

1. Explain, in your own words, the difference between true power and apparent power in an ac circuit.

_____

_____

_____

_____

_____

2. Refer to your data in Tables 45–1 and 45–2. Compare the values of true power obtained by the wattmeter method with those obtained using the voltage-current method. Should they be equal? Explain.

_____

_____

_____

_____

_____

Power in AC Circuits   **327**

3. A circuit in which the current leads the voltage has a leading power factor. If the current lags the voltage, the circuit has a lagging power factor. Refer to Tables 45–1 and 45–2. Is the power factor in this experiment leading or lagging? Explain why and support your answer with specific data.

4. Why is the power factor of an ac circuit significant? Can it be changed? If so, explain how.

5. Explain, in your own words, how phase angle and power factor are related.

46

FREQUENCY RESPONSE OF A REACTIVE CIRCUIT

OBJECTIVES

1. To study the effect on impedance and current of a change in frequency in a series *RL* circuit
2. To study the effect on impedance and current of a change in frequency in a series *RC* circuit

BASIC INFORMATION

Impedance of a Series *RL* Circuit

The impedance of a series *RL* circuit is given by the formula

$$Z = \sqrt{R^2 + X_L^2} \qquad (46\text{--}1)$$

If *R* remains constant, a change in X_L will affect *Z*. Thus, as X_L increases, *Z* also increases. As X_L decreases, *Z* decreases. Since

$$X_L = 2\pi f L \qquad (46\text{--}2)$$

we can change X_L by either increasing or decreasing the size of *L*, with *f* remaining constant. Or we can change X_L by increasing or decreasing *f*, with *L* remaining constant.

Let us consider the effect on X_L of varying *f*, with *L* constant. From formula (46–2) we see that as *f* increases, X_L increases; as *f* decreases, X_L decreases. The impedance *Z* will

therefore also increase or decrease, respectively, as *f* increases or decreases. An example will illustrate this variation.

Problem. In the circuit of Figure 46–1 (p. 330), $R = 30\ \Omega$ and $L = 63.7$ mH. Show how the impedance will vary as the frequency is increased from 0 to 500 Hz.

Solution. Using formulas (46–1) and (46–2), we calculate X_L and *Z* for a range of frequencies and record the results in Table 46–1. Note that as X_L increases, the value of *R* has a decreasing effect on the value of *Z*. Figure 46–2 (p. 330) is a graph of the data in Table 46–1 showing how the impedance increases as the frequency increases.

Current Versus Frequency in an *RL* Circuit

The current in an ac circuit is given by the formula

$$I = \frac{V}{Z} \qquad (46\text{--}3)$$

Current varies inversely with *Z*. Since *Z* increases with *f* in a series *RL* circuit, current will decrease as *f* increases. If the voltage in the problem were 100 V, the circuit current would be as given in Table 46–1.

In Figure 46–2 the graph of current *I* on the vertical scale versus the frequency *f* on the horizontal scale is known as the *frequency response curve* for the circuit.

TABLE 46–1. Variation of X_L and *I* with Frequency

f, Hz	X_L, Ω	$Z = \sqrt{R^2 + X_L^2}, \Omega$	$I = \dfrac{V}{Z}$, A
0	0	30	3.33
100	40	50	2.00
150	60	67.1	1.49
200	80	85.4	1.171
250	100	104.4	0.958
300	120	123.7	0.808
400	160	162.8	0.614
500	200	202.2	0.495

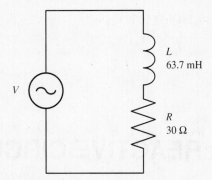

Figure 46–1. Circuit for illustrative problem.

Impedance of a Series *RC* Circuit

The impedance of a series *RC* circuit is given by the formula

$$Z = \sqrt{R^2 + X_C^2} \qquad \textbf{(46-4)}$$

Although formulas (46–4) and (46–1) are the same except for the substitution of X_C for X_L, the variation of Z with frequency is very different for the *RC* circuit. The reason is that X_C varies inversely with frequency. The formula for X_C is

$$X_C = \frac{1}{2\pi f C} \qquad \textbf{(46-5)}$$

As f increases, X_C decreases. As f decreases, X_C increases. Therefore, the impedance of a series *RC* circuit increases with a decrease in frequency but decreases with an increase in frequency.

Current Versus Frequency in an *RC* Circuit

In a series *RC* circuit, as f decreases, X_C increases, Z increases, and I decreases. As f increases, X_C decreases, Z decreases, and I increases. Again, this relationship is opposite that of a series *RL* circuit.

The foregoing discussion shows that the effects of a capacitor and an inductor on current in series *RC* and *RL* circuits are opposite. This fact is related to the phase relationship between voltage and current in a capacitor and an inductor. In an inductor, current lags voltage, whereas in a capacitor, current leads the voltage.

SUMMARY

1. In a series *RL* circuit, with R and L constant, as X_L increases, Z increases.
2. As the frequency f increases or decreases, X_L increases or decreases.
3. As f increases in a series *RL* circuit, Z increases. As f decreases in a series *RL* circuit, Z decreases.
4. As f increases in a series *RL* circuit, I in the circuit decreases. As f decreases, I increases.

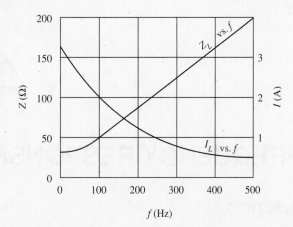

Figure 46–2. Graphs of impedance versus frequency and current versus frequency for a series *RL* circuit.

5. In a series *RC* circuit, with R and C constant, Z increases as f decreases.
6. In a series *RC* circuit, X_C decreases as the frequency f increases. The effect of f on X_C is opposite the effect of frequency on X_L.
7. In a series *RC* circuit I decreases as f decreases, and I increases when f increases.
8. The effects of a capacitor on impedance and current in a series-connected circuit are opposite the effects of an inductor in a series circuit.
9. The reactance of an inductor is

$$X_L = 2\pi f L$$

The impedance of a series *RL* circuit is

$$Z = \sqrt{R^2 + X_L^2}$$

10. The reactance of a capacitor is

$$X_C = \frac{1}{2\pi f C}$$

The impedance of a series *RC* circuit is

$$Z = \sqrt{R^2 + X_C^2}$$

11. The frequency response curve of a circuit is the graph of I on a vertical axis versus f on a horizontal axis.

SELF-TEST

Check your understanding by answering the following questions:

1. The X_L of a 3.19-H choke at 100 Hz is _____ Ω.
2. The Z of a series *RL* circuit at 1 kHz where $R = 1000$ Ω and $L = 3.19$ H, is _____ Ω.
3. The Z of the circuit of question 2, at 2 kHz, is _____ Ω.

330 *Experiment 46*

4. In a series *RL* circuit, *Z* _____ (increases/ decreases) with an increase in frequency.
5. In the circuit of question 2, if the applied voltage *V* = 15 V, the current *I* in the circuit is _____ A.
6. In the circuit of question 3, if the applied voltage *V* = 12 V, the current *I* in the circuit is _____ A.
7. In a series *RL* circuit, *I* _____ (increases/ decreases) with an increase in frequency.
8. In a series *RC* circuit, the X_C of the capacitor is 1000 Ω at 100 Hz. The X_C of the same capacitor at 200 Hz is _____ Ω, whereas at 50 Hz the X_C of the capacitor is _____ Ω .
9. The *Z* of a series *RC* circuit, with *R* and *C* constant, _____ (increases/decreases) with an increase in frequency.
10. If the *Z* of an *RC* circuit at 1 kHz = 1410 Ω, the *Z* of the same circuit at 2 kHz is _____ (greater/less) than 1410 Ω.
11. The current *I* in the *RC* circuit of question 10 _____ (increases/decreases) when *f* increases from 50 Hz to 100 Hz.

MATERIALS REQUIRED

Power Supply:
■ AF signal generator

Instrument:
■ EVM or DMM

Resistor (½-W, 5%):
■ 1   10,000-Ω

Capacitor:
■ 1   0.1-μF

Inductor:
■ Large iron-core choke, 8 or 9 H (see Experiment 36 for type and specifications)

PROCEDURE

A. Frequency Response of an *RL* Circuit

A1. Using an ohmmeter, measure the resistance of the 10-kΩ resistor and record the value in Table 46–2 (p. 333).
A2. With the AF signal generator **off,** connect the circuit of Figure 46–3. Set the signal generator to its lowest output voltage and frequency.
A3. Turn on the signal generator and set the output frequency to 50 Hz. Increase the output voltage until the voltage across the *RL* series circuit is *V* = 10 V rms. Maintain this voltage throughout the experiment. Measure the voltage across the resistor V_R and record the value in the 50-Hz row of Table 46–2.
A4. Increase the frequency to 100 Hz. Check to verify that *V* = 10 V; adjust output voltage if necessary. Measure V_R and record the value in Table 46–2 in the 100-Hz row.

A5. Repeat step A4 by increasing the frequency in 50-Hz steps: 150, 200, 250, 300, 350, 400, 450, 500 Hz. At each frequency measure V_R and record the value in Table 46–2. At each frequency verify that *V* = 10 V; adjust voltage if necessary. After all measurements have been taken, turn **off** the signal generator.
A6. Using the measured values of V_R and *R*, calculate the current in the circuit for each frequency. Record your answers in Table 46–2.
A7. Using the calculated value of current *I* and voltage *V*, calculate the circuit impedance *Z* for each frequency. Record your answers in Table 46–2.

B. Frequency Response of an *RC* Circuit

B1. With the signal generator **off,** connect the circuit of Figure 46–4 (p. 332). Set the signal generator to its lowest voltage output and frequency.
B2. Turn **on** the signal generator and set the output frequency to 500 Hz. Increase the output voltage of the signal generator until the voltage across the *RC* series circuit *V* = 10 V rms. Maintain this voltage throughout the experiment, checking periodically and adjusting the voltage if necessary.
B3. Measure the voltage across the resistor V_R, and record the value in Table 46–3 in the 500-Hz row.
B4. Decrease the frequency to 450 Hz. Check to see that *V* = 10 V rms; adjust if necessary. Measure V_R and record the value in the 450-Hz row of Table 46–3.

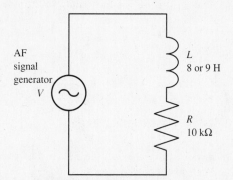

Figure 46–3. Circuit for procedure step A2.

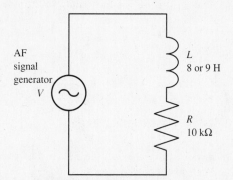 AF signal generator *V* — *L* 8 or 9 H — *R* 10 kΩ

Frequency Response of a Reactive Circuit

B5. Repeat step B4 in decreasing 50-Hz steps: 400, 350, 300, 250, 200, 150, 100, 50 Hz. At each frequency, measure V_R and check $V = 10$ V. Record the values at each frequency in Table 46–3. After all measurements have been taken, turn **off** the signal generator.

B6. Using the measured values of V_R (from Table 46–3) and R (from Table 46–2), calculate the current I in the circuit for each frequency. Record your answers in Table 46–3.

B7. Using the calculated values of current I and voltage V, calculate the circuit impedance for each value of frequency. Record your answers in Table 46–3.

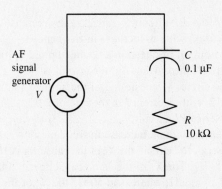

Figure 46–4. Circuit for procedure step B1.

ANSWERS TO SELF-TEST

1. 2004
2. 20.06 k
3. 40.08 k
4. increases
5. 0.000748
6. 0.0003
7. decreases
8. 500; 2000
9. decreases
10. less
11. increases

TABLE 46–2. Frequency Response of an *RL* Series Circuit

Frequency f, Hz	Applied Voltage V, V	Voltage Across R V_R, V	Circuit Current (Calc.) I, mA	Circuit Imped. (Calc.) Z, Ω
50	10			
100	10			
150	10			
200	10			
250	10			
300	10			
350	10			
400	10			
450	10			
500	10			

R (rated) = 10 kΩ: R (measured) _____

TABLE 46–3. Frequency Response of an *RC* Series Circuit

Frequency f, Hz	Applied Voltage V, V	Voltage Across R V_R, V	Circuit Current (Calc.) I, mA	Circuit Imped. (Calc.) Z, Ω
500	10			
450	10			
400	10			
350	10			
300	10			
250	10			
200	10			
150	10			
100	10			
50	10			

QUESTIONS

1. Explain, in your own words, how changes in frequency affect impedance and current in a series *RL* circuit.

_____

_____

_____

_____

_____

_____

_____

2. Explain, in your own words, how changes in frequency affect impedance and current in a series *RC* circuit.

_____

_____

_____

_____

_____

_____

_____

Frequency Response of a Reactive Circuit   **333**

3. On a separate sheet of 8½ × 11 graph paper plot a graph of impedance versus frequency using the data from Table 46–2. The horizontal (x) axis should be frequency, and the vertical (y) axis should be impedance. Label the axes as in Figure 46–2. On the same set of axes, plot current versus frequency using the data from Table 46–2. (See Figure 46–2 for the arrangement of axes.)

4. On a separate sheet of 8½ × 11 graph paper (not the sheet used in Question 3), plot a graph of impedance versus frequency using the data from Table 46–3. The horizontal (x) axis should be frequency, and the vertical (y) axis should be impedance. Label the axes as in Figure 46–2. On the same set of axes, plot current versus frequency using the data from Table 46–3. (See Figure 46–2 for the arrangement of axes.)

5. Refer to your graph of Z vs. f of Question 3. Is any portion of the graph linear? If so, what portion is linear? Should it be linear? Explain.

_____

_____

_____

_____

_____

6. Refer to your graph of Questions 3 and 4. How does the impedance-versus-frequency graph for the series RL circuit differ from the impedance-versus-frequency graph for the series RC circuit? Explain.

_____

_____

_____

_____

_____

_____

_____

7. Refer to your graphs of Questions 3 and 4. How does the current-versus-frequency graph for the series RL circuit differ from the current-versus-frequency graph for the series RC circuit? Explain.

_____

_____

_____

_____

_____

_____

_____

47

IMPEDANCE OF A SERIES *RLC* CIRCUIT

OBJECTIVE

To verify experimentally that the impedance Z of a series *RLC* circuit is

$$Z = \sqrt{R^2 + (X_L - X_C)^2}$$

BASIC INFORMATION

Impedance of an *RLC* Circuit

To understand the effects on alternating current of series-connected *RLC* circuits, it is necessary to recall the individual effect of each of these components.

The effects of a resistor in an ac circuit are the same as those in a dc circuit, because alternating current and voltage are in phase in a resistor. In an ac voltage divider made up of resistors only, the voltage drop across each individual resistor may be added arithmetically to find the voltage across the entire combination of resistors.

Reactances in an ac circuit depend on the frequency of the source. The value of reactance varies with changes in frequency. In addition, the current through a reactance and the voltage across the reactance are not in phase. For a pure inductance (that is, $R = 0$) the current through the inductance lags the voltage across the inductance by 90°. For a pure capacitance, the current through the capacitance leads the voltage across the capacitance by 90°.

The effects of an inductor connected in series with a resistor in an ac circuit also depend on the frequency and the size of the inductor. In a series *RL* circuit, the current lags the voltage by an angle less than 90°.

When a capacitor is connected in series with a resistor, the reactance of the capacitor together with the resistance of the resistor determines the effect on alternating current. The effects of a capacitor are determined both by its size (capacitance) and by the frequency. In a series *RC* circuit, the alternating current leads the voltage by a phase angle less than 90°.

It may be seen from the characteristics of an inductance and a capacitance that they have opposite effects on current and voltage in an ac circuit. This fact is demonstrated in the phasor diagram of X_L and X_C in the series *RLC* circuit of Figure 47–1.

The phasor sum of X_L and X_C is the arithmetic difference of their numerical values. The resultant phasor, usually labeled X, is in the direction of the larger reactance. If X_L is larger than X_C, the resultant phasor will be a vertical line in the *upward* direction from the origin. In that case the circuit will be inductive and current will *lag* voltage. If X_C is larger than X_L, the resultant phasor will be a vertical line in the *downward* direction from the origin. In that case the circuit will be capacitive, and current will *lead* voltage, as shown in Figure 47–2(*b*) (p. 336).

Figure 47–2 indicates that the impedance Z of the *RLC* circuit is given by the formula

$$Z = \sqrt{R^2 + (X_L - X_C)^2}$$

or

$$Z = \sqrt{R^2 + X^2} \qquad \textbf{(47–1)}$$

where X is the difference between X_L and X_C.

If $X = 0$ in formula (47–1), the impedance of the circuit will be completely resistive. That is, $Z = R$. In an *RLC* series circuit, $X = 0$, only if $X_L = X_C$. In other words, if the inductive reactance X_L in an *RLC* circuit cancels the capacitive reactance X_C, the circuit will behave as if it contained only resistance. At this point the impedance of the circuit has its lowest value.

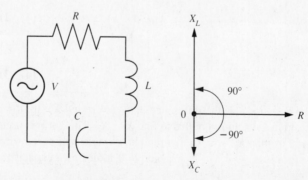

Figure 47–1. A series *RLC* circuit and its impedance phasor diagram. In this circuit the phasor diagram shows X_L is greater than X_C.

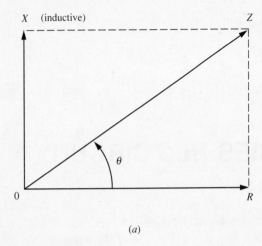

(a)

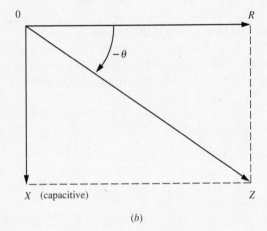

(b)

Figure 47–2. Impedance phasor diagram for an *RLC* series circuit.

Alternate Method for Finding the Impedance of an *RLC* Circuit

In previous experiments the impedance of a reactive circuit was found using the phase angle θ. The solution involved the use of a scientific calculator and the $\tan^{-1}$ function. A similar process can be used to find *Z* in an *RLC* circuit. From the phasor diagram in Figure 47–2 the phase angle can be found using

$$\theta = \tan^{-1}\left(\frac{X}{R}\right) \qquad (47\text{--}2)$$

and

$$Z = \frac{R}{\cos\theta} \qquad (47\text{--}3)$$

Because *X* is the difference between X_L and X_C, the first step in the solution of θ is to find $X_L - X_C$. This form is used rather than $X_C - X_L$ so that the phase angle will agree with those in Figure 47–2. A negative θ indicates a capacitive

circuit; a positive θ indicates an inductive circuit. The calculator keystrokes are as follows:

Input   Input   Input
Numerical   Numerical   Numerical
value   value   value

$$\theta = \boxed{X_L} \boxed{-} \boxed{X_C} \boxed{\div} \boxed{R} \boxed{=} \boxed{f} \boxed{\tan^{-1}}$$

and using this value of θ for the next calculation, we have

Input
Numerical
value

$$Z = \boxed{\cos} \boxed{\div} \boxed{R} \boxed{=} \boxed{1/x}$$

Illustrative Examples

As an illustrative example, we solve an *RLC* circuit using formulas (47–1), (47–2), and (47–3).

Problem 1. In the *RLC* circuit of Figure 47–3, $R = 82\ \Omega$, $X_L = 70\ \Omega$, and $X_C = 45\ \Omega$. The applied voltage is $V = 12$ V. Find the impedance of the circuit, whether the circuit is inductive or capacitve, the angle θ by which the current leads or lags the voltage, the power factor PF of the circuit, and power consumed in watts.

Solution. From formula (47–1),

$$Z = \sqrt{R^2 + X^2}$$

The difference *X* between X_L and X_C is

$$X = X_L - X_C = 70 - 45 = 25\ \Omega$$

Substituting this value in formula (47–1) gives

$$Z = \sqrt{82^2 + 25^2}$$
$$= 85.73\ \Omega$$

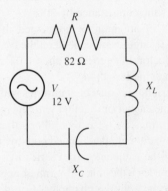

Figure 47–3. Circuit diagram for problems 1 and 2.

$$PF = \frac{R}{Z} = \frac{82}{85.73} = 0.9565 = 95.65\%$$

$$P = \frac{V^2}{Z} \times PF = \frac{12^2}{85.73} \times 0.9565$$

$$= 1.680 \times 0.9565$$

$$P = 1.607 \text{ W}$$

The phase angle θ can be found by many methods, but since $PF = \cos\theta$, we can apply this formula and—using a calculator—solve for θ directly. The keystrokes follow:

Input
Numerical
Value

$$\theta = \boxed{PF} \ \boxed{f} \ \boxed{\cos^{-1}} = \boxed{.} \boxed{9} \boxed{5} \boxed{6} \boxed{5} \ \boxed{f} \ \boxed{\cos^{-1}}$$

$$= 16.96°$$

The positive angle indicates the direction of X is upward—that is, in the inductive area. The current therefore lags the voltage by 16.96°.

Problem 2. We will use the circuit of Figure 47–3 with the values of X_L and X_C reversed. That is, $X_L = 45 \ \Omega$ and $X_C = 70 \ \Omega$.

Solution. We solve this circuit using formulas (47–2) and (47–3).

$$\theta = \tan^{-1}\left(\frac{X}{R}\right)$$

where $X = X_L - X_C = 45 - 70 = -25 \ \Omega$

Therefore

$$\theta = \tan^{-1}\left(\frac{-25}{82}\right)$$

The calculator keystroke sequence is as follows (the $+/-$ key changes 25 to -25):

$$\theta = \boxed{2} \boxed{5} \boxed{+/-} \boxed{\div} \boxed{8} \boxed{2} \boxed{=} \boxed{f} \boxed{\tan^{-1}}$$

$$= -16.96°$$

The negative sign indicates that the reactive phasor X is in the downward direction and the circuit is capacitive. This means the current will *lead* the voltage by 16.96°.

We continue the calculator keystrokes to find Z, where

$$Z = \frac{R}{\cos\theta} = \frac{82}{\cos(-16.96)}$$

$$Z = \boxed{8} \boxed{2} \boxed{\div} \boxed{1} \boxed{6} \boxed{.} \boxed{9} \boxed{6} \boxed{+/-} \boxed{\cos\theta}$$

$$= 85.73 \ \Omega$$

which is exactly the same value found in problem 1. This was to be expected since X and R are the same. However, in problem 1 the reactance X was positive, indicating an inductive circuit, whereas in problem 2 the reactance X was negative, indicating a capacitive circuit. In both cases, of course, Z was a positive number.

Because the numerical values of R, X, and Z are the same in problems 2 and 1, the power factor and power consumption of the circuit will be exactly the same. In problem 1 we say that the circuit has a *lagging* power factor; in problem 2 the circuit has a *leading* power factor.

SUMMARY

1. An inductor L and a capacitor C have opposite effects on current in an ac circuit. Thus, in an inductor, current lags, whereas in a capacitor, current leads the applied voltage.
2. The total reactance X of a series RLC circuit is the arithmetic difference of X_L and X_C.
3. If X_L is greater than X_C in a series RLC circuit, then the circuit is inductive and the current lags the source voltage by the phase angle θ.
4. If X_C is greater than X_L in a series RLC circuit, then the circuit is capacitive, and the current leads the source voltage by the phase angle θ.
5. If X_L is greater than X_C, the phase angle θ is positive; if X_C is greater than X_L, the phase angle θ is negative.
6. The impedance of an RLC series circuit can be found using the formula

$$Z = \sqrt{R^2 + X^2}$$

7. Impedance can also be found using the phase angle in the following formulas:

$$\theta = \tan^{-1}\left(\frac{X}{R}\right)$$

$$Z = \frac{R}{\cos\theta}$$

SELF-TEST

Check your understanding by answering the following questions:

1. In a series RLC circuit $X_L = 40\ \Omega$, $X_C = 70\ \Omega$, and $R = 40\ \Omega$. The net reactance X of this circuit is _____ Ω. This reactance is _____ (inductive/capacitive).

2. The impedance Z of the circuit in question 1 is _____ Ω.

3. The circuit in question 1 acts like an _____ (RC/RL) circuit.

4. In a series RLC circuit $X_L = 10\ \Omega$, $X_C = 15\ \Omega$, and $R = 12\ \Omega$. In this circuit the net _____ (inductive/capacitive) reactance is _____ , and the impedance phasor _____ (leads/lags) the resistance R.

5. In the circuit of question 4, the phase angle $\theta =$ _____ ° and is _____ (negative/positive).

6. In the circuit of question 4, $Z =$ _____ Ω.

7. The reason θ is negative is that $X_L - X_C$ is _____ (positive/negative).

MATERIALS REQUIRED

Power Supplies:
- Isolation transformer
- Variable-voltage autotransformer (Variac or equivalent)

Instrument:
- EVM or DMM

Resistor:
- 1   4700-Ω, ½-W, 5%

Capacitor:
- 1   0.5-μF or 0.47-μF

Inductor:
- Large iron-core choke, 8 or 9 H (see Experiment 36 for type and specifications)

Miscellaneous:
- SPST switch
- Polarized line cord with on-off switch and fuse

PROCEDURE

1. With the line cord unplugged, line switch **off**, and S_1 **open**, connect the circuit of Figure 47–4(a) (p. 339). Set the autotransformer to its lowest output voltage.

2. Plug in the line cord; line switch **on**; **close** S_1. Increase the output voltage of the autotransformer until $V_{AB} = 10$ V rms. Maintain this voltage throughout the experiment. Check it from time to time and adjust it if necessary.

3. Measure the voltage across the resistor V_R and the inductor, V_L. Record the values in Table 47–1 (p. 341) for the RL circuit. **Open** S_1.

4. Calculate the current in the circuit using the measured value of V_R and the rated value of R. Record your answer in Table 47–1 for the RL circuit.

5. Using the calculated value of I and the measured value of V_L, calculate X_L. Record your answer in the "RL" row of Table 47–1.

6. Calculate the total impedance of the circuit by two methods: Ohm's law (using the calculated value of I and the applied voltage V_{AB}) and the square-root formula (using R and X_L). Record your answers in the "RL" row of Table 47–1.

7. With S_1 **open**, add a 0.5-μF capacitor in series with the resistor and choke as in the circuit of Figure 47–4(b).

8. **Close** S_1. Check to see that $V_{AB} = 10$ V. Measure the voltage across the resistor V_R, the choke V_L, and the capacitor V_C. Record the values in the "RLC" row of Table 47–1. After taking all measurements, **open** S_1.

9. Calculate I and X_L as in steps 4 and 5. Similarly, using the measured value of V_C and the calculated value of I, calculate the capacitive reactance of the circuit. Record your answer in the "RLC" row of Table 47–1.

10. Calculate the impedance Z of the circuit by two methods: Ohm's law (using V_{AB} and I) and the square-root formula (using R, X_C, and X_L). Record your answers in the "RLC" row of Table 47–1.

11. With S_1 open, remove the choke from the circuit leaving only the resistor in series with the capacitor as in Figure 47–4(c).

12. **Close** S_1. Check V_{AB}; adjust if necessary. Measure V_R and V_C. Record the values in the "RC" row of Table 47–1. After taking all measurements, **open** S_1; turn line switch **off**; and unplug the line cord.

13. Using measured values of V_R and V_C and the rated value of R, calculate the current I in the circuit. Then using the calculated value of I, calculate X_C. Record your answers in Table 47–1 in the "RC" row.

14. Calculate the total impedance of the circuit by two methods: Ohm's law (using V_{AB} and I) and the square-root formula (using R and X_C). Record your answers in the "RC" row of Table 47–1.

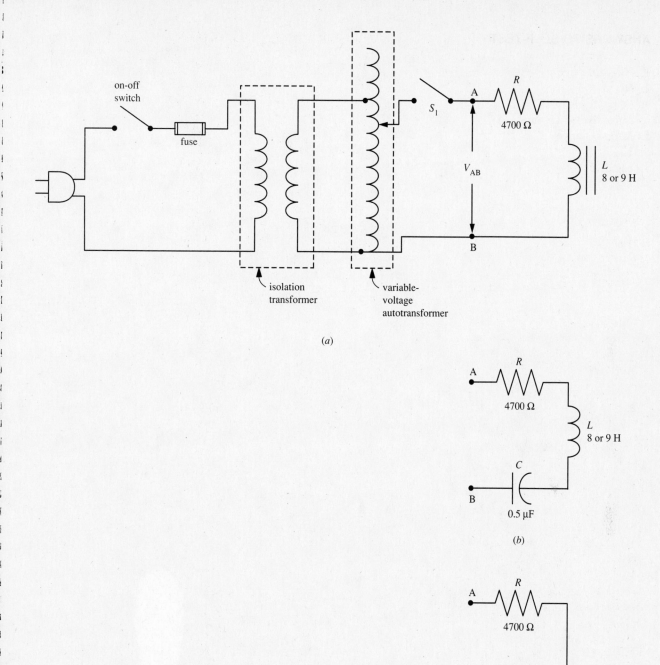

Figure 47–4. (*a*) Circuit for procedure step 1. (*b*) Circuit for procedure step 7.
(*c*) Circuit for procedure step 11.

Impedance of a Series RLC Circuit

ANSWERS TO SELF-TEST

1. 30; capacitive
2. 50
3. *RC*
4. capacitive; 5 Ω; lags
5. 22.6; negative
6. 13
7. negative

TABLE 47–1. Determining Impedance of an *RLC* Circuit

	Component			Applied Voltage V_{AB}, V	Voltage Across Resistor V_R, V	Voltage Across Inductor V_L, V	Voltage Across Capacitor V_C V	Current I, A	Reactance, Ω		Impedance Z, Ω	
Circuit	R, Ω	L, H	C, μF						Ind. X_L	Cap. X_C	Ohm's Law	Square-Root Formula
RL	4700	8 or 9	✕	10		✕				✕		
RLC	4700	8 or 9	0.5	10								
RC	4700	✕	0.5	10		✕			✕			

QUESTIONS

1. Explain, in your own words, the relationship between resistance, capacitance, inductance, and impedance in a series *RLC* circuit.

2. Refer to your data in Table 47–1. How was the impedance of the series *RL* circuit affected when the series capacitor was added?

3. Refer to Figure 47–4(b). Under what conditions will the impedance have its lowest value? What is the lowest possible value of impedance for this circuit? (The values of *R*, *L*, and *C* are fixed.)

4. Refer to Figure 47–4(*b*). Under what conditions will the series current have its maximum value? What is the maximum current possible in this circuit? (The values of *R*, *L*, and *C* are fixed.)

_____

_____

_____

_____

_____

_____

_____

_____

5. Refer to your data in Table 47–1. How does the Ohm's law value of impedance of the *RLC* circuit compare with the value calculated using the square-root formula? Explain any difference.

_____

_____

_____

_____

_____

_____

_____

_____

_____

_____

6. Refer to your data in Table 47–1. Is the *RLC* circuit inductive, capacitive, or resistive? Cite specific data to support your answer.

_____

_____

_____

_____

_____

_____

_____

_____

342   *Experiment 47*

48

EFFECTS OF CHANGES IN FREQUENCY ON IMPEDANCE AND CURRENT IN A SERIES *RLC* CIRCUIT

OBJECTIVE

To study experimentally the effect on impedance and current in a series *RLC* circuit of changes in frequency.

BASIC INFORMATION

In Experiment 47 we verified that the impedance Z of a series *RLC* circuit (Figure 48–1) is given by the formula

$$Z = \sqrt{R^2 + X^2} \qquad \textbf{(48–1)}$$

where X is the difference between X_L and X_C. In this experiment we will observe how a change in frequency of the voltage source affects circuit impedance and current.

Effect of Frequency on the Impedance of an *RLC* Circuit

Formula (48–1) shows that the minimum value of impedance in an *RLC* circuit, for a given value of R, occurs when $X_L = X_C$, or when $X_L - X_C = 0$. At that point $Z = R$, and the circuit acts like a pure resistance. The line current I is limited only by R. Therefore, I is maximum when $X_L = X_C$.

Because inductive reactance X_L and capacitive reactance X_C are dependent on frequency, there must be a frequency f_R

at which X_L and X_C are equal. In this experiment we study the effect on impedance of frequencies greater and less than f_R. In a later experiment we study the effects on impedance when $f = f_R$.

In the series *RLC* circuit we have been studying, as the frequency f of the sinusoidal voltage source increases beyond f_R, X_L increases and X_C decreases. Hence, the circuit acts like an inductance whose reactance X increases as f increases. As the frequency f decreases below f_R, X_C increases and X_L decreases. The circuit now acts like a capacitance whose reactance X increases as f decreases.

A graph of impedance versus frequency for a series *RLC* circuit is shown in Figure 48–2. This curve shows minimum Z at f_R. It shows also that Z is capacitive (Z_C) for frequencies less than f_R and inductive (Z_L) for frequencies greater than f_R.

The variation of current versus frequency is also shown graphically in Figure 48–2.

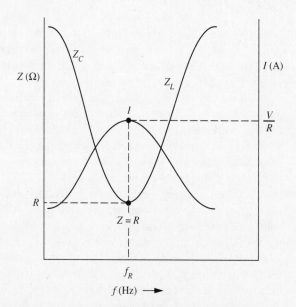

Figure 48–2. Variation of impedance (Z) and current (I) with changes in frequency (f) in an *RLC* circuit.

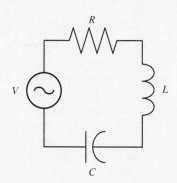

Figure 48–1. An *RLC* circuit.

SUMMARY

1. In a series *RLC* circuit there is a frequency f_R of the voltage source *V* at which $X_L = X_C$. At this frequency the impedance *Z* of the *RLC* circuit is minimum, and $Z = R$, the resistance in the circuit.

2. For frequencies higher than f_R, the X_L of the inductor increases while X_C decreases, and the circuit acts like an *RL* circuit, with

$$Z = \sqrt{R^2 + X^2}$$

As *f* increases beyond f_R, *Z* increases beyond *R*.

3. For frequencies lower than f_R, the X_C of the capacitor increases and X_L decreases. The circuit acts like an *RC* circuit. As *f* decreases below f_R, *Z* increases beyond *R*.

4. In a series *RLC* circuit the current can be calculated using Ohm's law,

$$I = \frac{V}{Z}$$

where *V* is the applied voltage and *Z* the circuit impedance.

SELF-TEST

Check your understanding by answering the following questions:

1. In the circuit of Figure 48–1 for a particular frequency f_R, $X_L = X_C = 1000 \ \Omega$ and $R = 4.7$ kΩ. The impedance *Z* of the circuit is _____ Ω.

2. The circuit in question 1 acts like a pure _____ .

3. For the frequency f_R + 100 Hz the impedance of the circuit in question 1 (Figure 48–1) is _____ (greater/less) than 4700 Ω, and the circuit is _____ (inductive/capacitive).

4. For the frequency f_R – 100 Hz the impedance of the circuit in question 1 is _____ (greater/less) than 4700 Ω, and the circuit is _____ (capacitive/inductive).

5. As *f* increases above f_R or decreases below f_R, the current *I* in the circuit _____ (increases/decreases/remains the same).

MATERIALS REQUIRED

Power Supply:
■ AF sine-wave generator

Instruments:
■ EVM or DMM
■ Oscilloscope

Resistor:
■ 1   4700-Ω, $\frac{1}{2}$-W, 5%

Capacitor:
■ 1   0.1-μF

Inductor:
■ Large iron-core choke, 8 or 9 H (see Experiment 36 for type and specifications)

Miscellaneous:
■ SPST switch

PROCEDURE

1. With the sine-wave generator **off** and set at its lowest output voltage and S_1 **open,** connect the circuit of Figure 48–3 (p. 345). The oscilloscope is connected across the inductor-capacitor combination (across AB).

2. Turn **on** the sine-wave generator and **close** S_1. Turn **on** the oscilloscope. Set the frequency of the generator to 150 Hz. Increase the voltage output of the generator to about half of its maximum voltage. Adjust the oscilloscope to display two cycles of a sine wave with an amplitude approximately 4 units peak-to-peak.

3. Increase the frequency output of the generator slowly while observing the waveform on the oscilloscope. If the amplitude of the waveform decreases, continue increasing the frequency until the amplitude begins to increase. Determine the frequency at which the amplitude is at a minimum. This is f_R. If the amplitude increased with an increase in frequency, decrease the frequency, again observing the amplitude of the sine wave on the scope.

Continue to decrease the frequency until you are able to determine the frequency f_R at which the amplitude of the wave is at a minimum. Measure the output voltage *V* of the generator at frequency f_R. Maintain this voltage throughout the experiment. Check it from time to time to verify its value; adjust the output voltage if necessary. Turn **off** the oscilloscope and disconnect it from the circuit.

4. With the generator output frequency set at f_R, measure the voltage across the resistor V_R, the capacitor V_C, the choke V_L, and across the capacitor-choke combination V_{LC}. Record the values in Table 48–1 in the "f_R" row.

5. Add 100 Hz to the value of f_R and set the sine-wave generator to this frequency. Record the value in Table 48–1. Check *V* (it must be the same as step 3; adjust if necessary). Measure V_R, V_C, V_L, and V_{LC}. Record the values in Table 48–1 in the "f_R + 100" row.

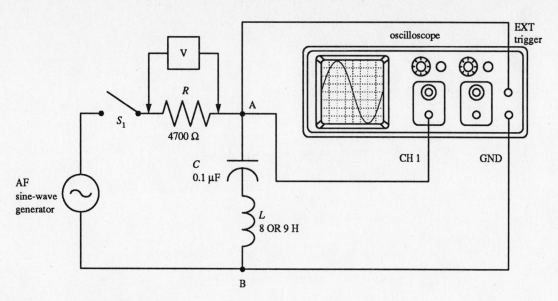

Figure 48–3. Circuit for procedure step 1.

6. Reduce the frequency of the generator by 20 Hz. Record this value in Table 48–1. Check V again, then measure V_R, V_C, V_L, and V_{LC}. Record the values in Table 48–1.
7. Continue to reduce the frequency in 20-Hz steps until the final setting is equal to $f_R - 100$ Hz. Do not repeat f_R. In each step verify and record V (and adjust if necessary to maintain the constant voltage of this experiment); also measure V_R, V_C, V_L, and V_{LC}. Record all values in Table 48–1. After all measurements have been taken, turn **off** the sine-wave generator and **open** S_1.
8. For each frequency in Table 48–1 calculate the difference between the V_L and V_C measurements. Record your answer as a positive number in Table 48–1.
9. For each frequency in Table 48–1 calculate the current in the circuit from the measured value of V_R and the rated value of R. Using the calculated value of I, calculate the impedance Z at each frequency by applying Ohm's law, $Z = V/I$.

10. Transfer the frequency steps from Table 48–1 to Table 48–2 (p. 347). Calculate X_C and X_L for each step, using the measured values of V_C and V_L from Table 48–1. Record your answers in Table 48–2. Calculate the circuit impedance at each step, using the square-root formula,

$$Z = \sqrt{R^2 + (X_L - X_C)^2}$$

and your calculated values of X_C and X_L and the rated value of R. Record your answers in Table 48–2. Also, transfer the calculated values of Z from Table 48–1 to Table 48–2.

ANSWERS TO SELF-TEST

1. 4.7 k
2. resistance
3. greater; inductive
4. greater; capacitive
5. decreases

Experiment 48   Name _____ Date _____

TABLE 48–1.   Effect of Frequency on Impedance of a Series *RLC* Circuit

Step	Frequency Hz	Voltage Across Res. V_R, V	Voltage Across Ind. V_L, V	Voltage across Cap. V_C, V	Voltage Across AB V_{LC}, V	Voltage Difference $V_L - V_C$, V	Current (Calculated) I, A	Impedance Z (Ohm's Law Cal.), Ω
f_R								
$f_R + 100$								
$f_R + 80$								
$f_R + 60$								
$f_R + 40$								
$f_R + 20$								
$f_R - 20$								
$f_R - 40$								
$f_R - 60$								
$f_R - 80$								
$f_R - 100$								

TABLE 48–2.   Comparison of Impedance Calculations for a Series *RLC* Circuit

Step	Frequency, Hz	Inductive Reactance (Calculated) X_L, Ω	Capacitive Reactance (Calculated) X_C, Ω	Impedance (Calculated — Square-Root Formula) Z, Ω
f_R				
$f_R + 100$				
$f_R + 80$				
$f_R + 60$				
$f_R + 40$				
$f_R + 20$				
$f_R - 20$				
$f_R - 40$				
$f_R - 60$				
$f_R - 80$				
$f_R - 100$				

Frequency Changes in a Series RLC Circuit   **347**

QUESTIONS

1. Explain, in your own words, the effect that changes in frequency have on the impedance and current in a series *RLC* circuit.

_____

_____

_____

_____

_____

_____

2. On a separate sheet of $8\frac{1}{2} \times 11$ graph paper, plot a graph of frequency versus impedance using the data from Table 48–1. The horizontal (x) axis should be frequency, and the vertical (y) axis should be impedance. On the same axes, plot a graph of frequency versus circuit current using the data from Table 48–1. Label the axes and the graph as shown in Figure 48–2.

3. Discuss how the graphs of Question 2 support your explanation of Question 1. Refer to specific parts of the graphs in your discussion.

_____

_____

_____

_____

_____

4. Refer to your data in Table 48–1. Compare the voltage across the inductance and capacitance V_{LC} and the voltage difference $V_L - V_C$ for each frequency in the table. Should the values for each comparable frequency be the same? Explain your answer.

_____

_____

_____

_____

_____

5. Compare the impedance calculated in Table 48–2 with the value in Table 48–1 for each frequency. Do the results confirm Formula (48–1)? Explain your answer with reference to specific data in the tables.

_____

_____

_____

_____

_____

EXPERIMENT

49

IMPEDANCE OF PARALLEL *RL* AND *RC* CIRCUITS

OBJECTIVES

1. To determine experimentally the impedance of a parallel *RL* circuit
2. To determine experimentally the impedance of a parallel *RC* circuit

BASIC INFORMATION

Impedance of a Parallel *RL* Circuit

The circuit of Figure 49–1 consists of two parallel branches, a branch consisting of a resistor *R* only, and a branch consisting of an inductor *L* only. Since *R* and *L* are connected in parallel, the same voltage *V* appears across each branch. By Ohm's law the current through each branch is

$$\text{Current in } R \text{ branch:} \quad I_R = \frac{V}{R}$$

$$\text{Current in } L \text{ branch:} \quad I_L = \frac{V}{X_L}$$

By Kirchhoff's current law, the total current I_T delivered by *V* is the sum of I_R and I_L. But I_R and I_L are not in phase; I_R is in phase with the voltage, and I_L is out of phase and lagging the voltage by 90°. The currents must therefore be added using phasors.

With *V* as the reference phasor, Figure 49–2 is the current phasor diagram for the *RL* parallel circuit of Figure 49–1.

From this diagram we can derive the formula for I_T using the right-triangle relationship

$$I_T = \sqrt{I_R^2 + I_L^2} \qquad (49\text{–}1)$$

The current phasor diagram (Figure 49–2) also shows that the current delivered by the source lags the voltage by the phase angle θ. We can solve for this angle using the formula

$$\theta = \tan^{-1}\left(\frac{I_L}{I_R}\right) \qquad (49\text{–}2)$$

Again the scientific calculator provides a convenient method for finding θ. The keystrokes are as follows:

$$\theta = \boxed{\begin{array}{c}\text{Input}\\\text{Numerical}\\\text{value}\\I_L\end{array}} \div \boxed{\begin{array}{c}\text{Input}\\\text{Numerical}\\\text{value}\\I_R\end{array}} = \boxed{f} \boxed{\tan^{-1}}$$

The total impedance of the parallel *RL* circuit can be found from the total current I_T and the applied voltage *V* using Ohm's law

$$Z = \frac{V}{I_T} \qquad (49\text{–}3)$$

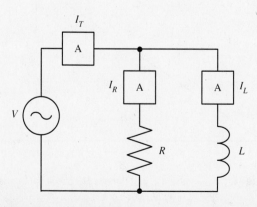

Figure 49–1. Parallel *RL* circuit.

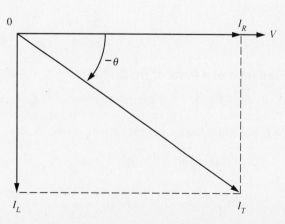

Figure 49–2. Total current I_T in a parallel *RL* circuit is the phasor sum of I_R and I_L.

Problem 1. In the circuit of Figure 49–1, $R = 20\ \Omega$, $X_L = 45\ \Omega$, and $V = 15$ V. Find the total current I_T, the impedance Z, and the angle θ by which the line current lags the voltage V.

Solution. The current in each branch is found using Ohm's law:

$$I_R = \frac{V}{R} = \frac{15}{20} = 0.75 \text{ A}$$

$$I_L = \frac{V}{X_L} = \frac{15}{45} = 0.333 \text{ A}$$

The total current circuit is found using formula (49–1):

$$I_T = \sqrt{I_R^2 + I_L^2} = \sqrt{0.75^2 + 0.333^2}$$

$$= 0.821 \text{ A}$$

From formula (49–3) the impedance of the circuit is

$$Z = \frac{V}{I_T} = \frac{15}{0.821}$$

$$= 18.270\ \Omega$$

The phase angle θ is found using formula (49–2) and the scientific calculator:

$$\theta = \boxed{.}\ \boxed{3}\ \boxed{3}\ \boxed{3}\ \boxed{\div}\ \boxed{.}\ \boxed{7}\ \boxed{5}\ \boxed{=}\ \boxed{f}\ \boxed{\tan^{-1}}$$

$$= 23.94°$$

Since the circuit is inductive, the current lags the voltage by 23.94°.

The power factor of a parallel circuit can be found using the phasor diagram of Figure 49–2. Since PF = cos θ, Figure 49–2 shows that

$$\cos\theta = \frac{I_R}{I_T}$$

The power factor PF of the circuit in problem 1 is

$$\text{PF} = \frac{I_R}{I_T} = \frac{0.75}{0.821}$$

$$= 0.914 = 91.4\% \text{ lagging}$$

Impedance of a Parallel *RC* Circuit

The circuit of Figure 49–3 consists of a capacitor C in parallel with a resistor R. The supply voltage V is applied across both C and R. By Ohm's law the current in each branch is

Current in R Branch: $I_R = \dfrac{V}{R}$

Current in C Branch: $I_C = \dfrac{V}{X_C}$

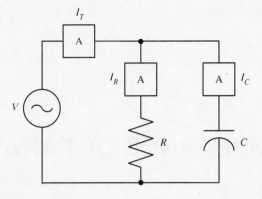

Figure 49–3. Parallel *RC* circuit.

By Kirchhoff's current law, the sum of I_R and I_C is equal to the total current I_T delivered by the supply. But I_R and I_C cannot be added arithmetically, because they are not in phase; I_R is in phase with the applied voltage, and I_C is out of phase and leading the applied voltage by 90°. The two branch currents I_R and I_C must be combined by phasor addition. With the applied voltage V as the reference phasor, Figure 49–4 is the current phasor diagram for the *RC* parallel circuit. From this diagram we can derive the formula for the total current I_T delivered by the voltage source based on the right-triangle relationship

$$I_T = \sqrt{I_R^2 + I_C^2} \qquad \textbf{(49–4)}$$

The phasor diagram (Figure 49–4) also shows that the current delivered by the supply leads the voltage by the phase angle θ. We can solve for this angle using the formula

$$\theta = \tan^{-1}\left(\frac{I_C}{I_R}\right) \qquad \textbf{(49–5)}$$

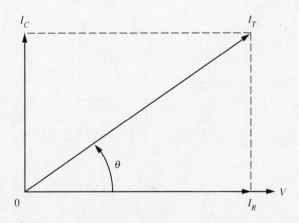

Figure 49–4. Total current I_T in a parallel *RC* circuit is the phasor sum of I_R and I_C.

The keystrokes on a scientific calculator are:

Input
Numerical
value

Input
Numerical
value

$\theta = \boxed{I_C} \; \boxed{\div} \; \boxed{I_R} \; \boxed{=} \; \boxed{f} \; \boxed{\tan^{-1}}$

The total impedance of the parallel RC circuit can be calculated using the total current I_T and the supply voltage V. By Ohm's law,

$$Z = \frac{V}{I_T} \qquad \qquad (49\text{-}6)$$

Problem 2. The circuit of Figure 49–3 consists of a resistor $R = 470 \; \Omega$ in parallel with a capacitive reactance $X_C = 200 \; \Omega$. The applied voltage is $V = 20$ V. Find the total current I_T, the total impedance Z, and the angle θ by which I_T leads V.

Solution. The current in each branch is found using Ohm's law:

$$I_R = \frac{V}{R} = \frac{20}{470}$$

$$= 0.0426 \text{ A}$$

$$I_C = \frac{V}{X_C} = \frac{20}{200}$$

$$= 0.100 \text{ A}$$

The total current is found using formula (49–4)

$$I_T = \sqrt{I_R^2 + I_C^2} = \sqrt{0.0426^2 + 0.100^2}$$

$$= 0.109 \text{ A}$$

From formula (49–6)

$$Z = \frac{V}{I_T} = \frac{20}{0.109}$$

$$= 183 \; \Omega$$

The phase angle θ is calculated with a scientific calculator using the following keystrokes:

$\theta = \boxed{.} \; \boxed{1} \; \boxed{\div} \; \boxed{.} \; \boxed{0} \; \boxed{4} \; \boxed{2} \; \boxed{6} \; \boxed{=} \; \boxed{f} \; \boxed{\tan^{-1}}$

$$= 67°$$

Since the circuit is capacitive, the current leads the voltage by 67°.

The power factor of the circuit is

$$PF = \cos \theta = \frac{I_R}{I_T} = \frac{0.0426}{0.109}$$

$$= 0.391 = 39.1\% \text{ leading}$$

SUMMARY

1. In a parallel RL or RC circuit, the voltage across each branch of the circuit is the same.
2. In a parallel RL circuit, branch currents I_R and I_L are phasors. The total (line) current I_T is a phasor equal to the phasor sum of I_R and I_L.
3. In a parallel RC circuit, branch currents I_R and I_C are phasors. The total (line) current I_T is a phasor equal to the phasor sum of I_R and I_C.
4. In a parallel RL or RC circuit the applied voltage V is used as the reference phasor.
5. In a parallel RL circuit the current in L lags the current in R by 90°. The value of I_T can be calculated from the formula

$$I_T = \sqrt{I_R^2 + I_L^2}$$

6. In a parallel RL circuit the total current lags the applied voltage V by some angle θ, where

$$\theta = \tan^{-1} \left(\frac{I_L}{I_R} \right)$$

7. In a parallel RC circuit, the current in C leads the current in R by 90°. The value of I_T can be calculated from the formula

$$I_T = \sqrt{I_R^2 + I_C^2}$$

8. In a parallel RC circuit, the total current leads the applied voltage V by some angle θ, where

$$\theta = \tan^{-1} \left(\frac{I_C}{I_R} \right)$$

9. The impedance Z of an RL or an RC circuit may be calculated from the formula

$$Z = \frac{V}{I_T}$$

where V is the applied voltage, and I_T is the total, or line, current.

SELF-TEST

Check your understanding by answering the following questions:

1. In a parallel RL circuit the applied voltage is $V = 10$ V; the measured values are $I_L = 0.2$ A and $I_R = 0.5$ A. The total current is $I_T = $ _____ A.
2. The angle θ by which total current I_T (in question 1) _____ (leads/lags) the applied voltage V is _____ °.

3. The impedance of the circuit in question 1 is _____ Ω.
4. In a parallel RC circuit the applied voltage is $V = 6$ V, and the measured values are $I_C = 0.1$ A and $I_R = 0.2$ A. The total current is $I_T =$ _____ A.
5. The angle θ by which the total current I_T (in question 4) _____ (leads/lags) the applied voltage V is _____ °.
6. The impedance of the circuit in question 4 is _____ Ω.

MATERIALS REQUIRED

Power Supplies:
- Isolation transformer
- Variable-voltage autotransformer (Variac or equivalent)

Instruments:
- EVM or DMM
- 0–25-mA ac ammeter or second DMM with ac ammeter ranges

Resistor:
- 1   5100-Ω, 1-W

Capacitor:
- 1   0.5-μF or 0.47-μF

Inductor:
- Large iron-core choke, 8 or 9 H (see Experiment 36 for type and specifications)

Miscellaneous:
- 3 SPST switches
- Polarized line cord with on-off switch and fuse

PROCEDURE

A. Impedance of a Parallel *RL* Circuit

A1. With the line cord unplugged, line switch **off,** S_1, S_2, and S_3 **open,** connect the circuit of Figure 49–5. Set the autotransformer to its lowest output voltage. The ammeter should be set to the 25-mA (or higher) range, and the voltmeter should be set at the 150-V (or higher) range.

A2. Plug in the line cord; line switch **on. Close** S_1. Slowly increase the output voltage of the autotransformer V until $V = 50$ V. Maintain this voltage throughout the experiment; check it from time to time and adjust if necessary.

A3. **Close** S_2. Check to see that $V = 50$ V. Measure the current in the circuit. With S_3 **open,** the ammeter will measure only I_R. Record this value in Table 49–1 (p. 355). **Open** S_2.

A4. **Close** S_3. Check that $V = 50$ V. Measure the current in the circuit. With S_2 **open,** the ammeter will measure only I_L. Record this value in Table 49–1.

A5. With S_3 still **closed, close** S_2. Measure the current in the circuit. With both S_2 and S_3 **closed,** the ammeter will measure the total current I_T in the parallel *RL* circuit. Record this value in Table 49–1. After completing the measurement, **open** S_1, S_2, and S_3. Disconnect the choke from the circuit.

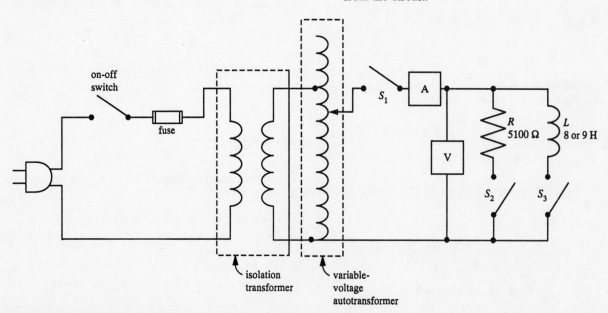

Figure 49–5. Circuit for procedure step A1.

A6. Using the measured values of I_R and I_L, calculate I_T by applying the square-root formula (49–1). Record your answer in Table 49–1. Calculate the circuit impedance with Ohm's law using V and the measured value of I_T. Record your answer in Table 49–1.

B. Impedance of a Parallel *RC* Circuit

B1. With S_1, S_2, and S_3 **open** and the autotransformer set at its lowest output voltage, connect the circuit of Figure 49–6.

B2. Close S_1. Increase the output voltage of the autotransformer V until $V = 50$ V. Maintain this voltage throughout the experiment. Check it from time to time and adjust the voltage if necessary.

B3. Close S_2. With $V = 50$ V, measure the current in the circuit. Since S_3 is **open,** the only current being measured is I_R. Record this value in Table 49–2 (p. 355). **Open** S_2.

B4. With S_2 **open, close** S_3. Measure the current in the circuit. The ammeter will measure only I_C. Record this value in Table 49–2.

B5. With S_3 **closed, close** S_2. Measure the current in the circuit. With both S_2 and S_3 **closed,** the ammeter will measure the total current I_T in the parallel *RC* circuit. Record this value in Table 49–2. After completing the measurement, **open** S_1, S_2, and S_3, and turn the line switch **off.** Unplug the line cord.

B6. Using the measured values of I_R and I_C, calculate the total line current I_T by applying the square-root formula (49–4). Record your answer in Table 49–2. Calculate the circuit impedance with Ohm's law using V and the measured value of I_T. Record your answer in Table 49–2.

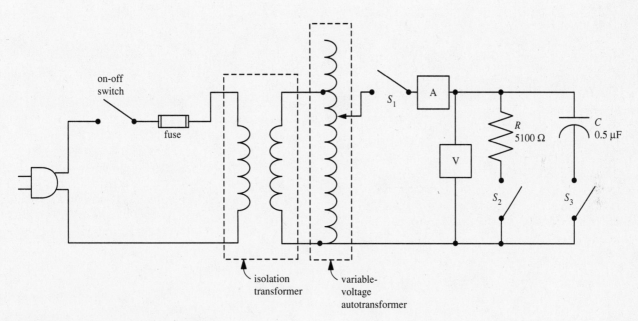

Figure 49–6. Circuit for procedure step B1.

ANSWERS TO SELF-TEST

1. 0.539
2. lags; 21.8
3. 18.6
4. 0.224
5. leads; 26.6
6. 26.8

Experiment 49 Name ______________________ Date ____________

TABLE 49–1. Impedance of a Parallel *RL* Circuit

Applied Voltage V, V	*Current in Resistor Branch* I_R, A	*Current in Inductor Branch* I_L, A	*Total Line Current (Measured)* I_T, A	*Total Line Current (Calculated using Square-Root Formula)* I_T, A	*Circuit Impedance (Calculated using Ohm's Law)* Z, Ω
50					

TABLE 49–2. Impedance of a Parallel *RC* Circuit

Applied Voltage V, V	*Current in Resistor Branch* I_R, A	*Current in Capacitor Branch* I_C, A	*Total Line Current (Measured)* I_T, A	*Total Line Current (Calculated using Square-Root Formula)* I_T, A	*Circuit Impedance (Calculated using Ohm's Law)* Z, Ω
50					

QUESTIONS

1. Refer to your data in Table 49–1. Do the measured values of I_R, I_L, and I_T verify formula (49–1)? Explain your answer with reference to specific data in the table.

2. Refer to your data in Table 49–2. Do the measured values of I_R, I_C, and I_T verify formula (49–4)? Explain your answer with reference to specific data in the table.

3. Calculate the phase angle of the *RL* circuit from your data in Table 49–1. State whether it is a leading or lagging angle.

_____

_____

4. Calculate the phase angle of the *RC* circuit from your data in Table 49–2. State whether it is a leading or lagging angle.

_____

_____

5. On a separate sheet of 8½ × 11 cross-section paper, draw a phasor diagram to scale showing the branch currents and total current in the *RL* circuit of Part A. Label all phasors and the phase angle. On the same sheet of cross-section paper, draw a phasor diagram to scale, showing the branch currents and total current in the *RC* circuit of Part B. Label all phasors and the phase angle.

50

IMPEDANCE OF A PARALLEL *RLC* CIRCUIT

OBJECTIVE

To determine experimentally the impedance of a circuit containing a resistance R in parallel with an inductance L in parallel with a capacitance C.

BASIC INFORMATION

The simplest type of parallel *RLC* circuit is illustrated in Figure 50–1. The ac voltage V is common to each leg in the circuit. The current I_R in R may be calculated using Ohm's law,

$$I_R = \frac{V}{R} \qquad \textbf{(50–1)}$$

where I is in amperes, V is in volts, and R is in ohms. The currents I_C in C and I_L in L may be calculated similarly.

$$I_C = \frac{V}{X_C} \qquad \textbf{(50–2)}$$

$$I_L = \frac{V}{X_L} \qquad \textbf{(50–3)}$$

Because R, L, and C are in parallel, the voltage across each leg is the applied voltage V.

The total line current I_T is equal to the phasor sum of I_R, I_C, and I_L. This is consistent with the results of Experiment 49, where parallel *RL* and *RC* circuits were studied.

Figure 50–2 is a phasor diagram showing the phase relationship between the applied voltage V and the currents in each leg. Thus, I_R is in phase with V, I_C leads V by 90°, and I_L lags V by 90°. Since I_C and I_L are 180° out of phase, they may be subtracted arithmetically to give the reactive current in the circuit, as in Figure 50–3 (p. 358). If I_L is greater than I_C, the circuit is inductive. If I_C is greater than I_L, the circuit is capacitive, as in Figure 50–3. A special case exists when $I_L = I_C$. This is discussed in a later experiment. The line current may be obtained by finding the phasor sum of I_R and the reactive current I_X, where I_X is the difference between I_L and I_C, as in Figure 50–3.

The line current may also be calculated by substituting the values of I_R and I_X in the formula

$$I_T = \sqrt{I_R^2 + I_X^2} \qquad \textbf{(50–4)}$$

This formula can be used as long as there is no resistance in series with L or C in the circuit of Figure 50–1.

The line current I_T is the phasor sum of the resistive and reactive currents and is not in phase with either current. It differs from I_R, the resistive current, by the angle θ, with the applied voltage V as the reference phasor. This angle may be calculated from the formula

$$\theta = \tan^{-1}\left(\frac{I_X}{I_R}\right) \qquad \textbf{(50–5)}$$

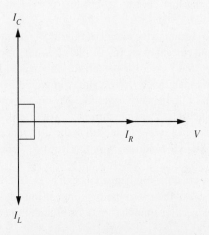

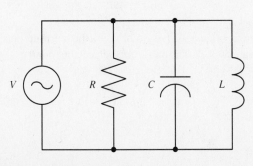

Figure 50–1. Parallel *RLC* circuit.

Figure 50–2. Current phasor diagram for a parallel *RLC* circuit. The applied voltage V is used as the reference phasor.

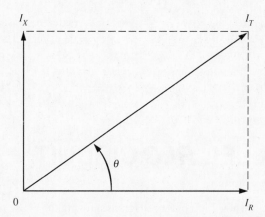

Figure 50–3. The reactive current I_X is the difference between I_C and I_L. In this case I_C is greater than I_L, so θ is positive and the *RLC* circuit is capacitive.

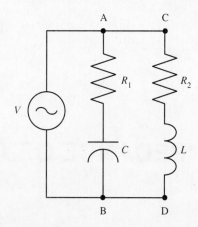

Figure 50–4. Parallel *RLC* circuit consisting of series *RC* and *RL* branches.

and using the following keystrokes on a scientific calculator:

Input Numerical value	Input Numerical value			

$$\theta = \boxed{I_X} \;\; \boxed{\div} \;\; \boxed{I_R} \;\; \boxed{=} \;\; \boxed{f} \;\; \boxed{\tan^{-1}}$$

Having found the line current I_T, we can now calculate the value of impedance Z in ohms using Ohm's law,

$$Z = \frac{V}{I_T} \qquad (50\text{–}6)$$

Differences Between AC and DC Parallel Circuits

There is an important difference between dc and ac parallel circuits. In dc parallel circuits the total line current I_T increases with every parallel branch added. In ac parallel circuits the line current does not always increase as more branches are added. For example, if a circuit initially consists of a resistive branch in parallel with a capacitive branch, the effect of adding an inductive branch *may* be to *reduce* the line current, a fact that is evident from formula (50–4).

The line current in the circuit of Figure 50–1 would be less than the line current of an *RC* parallel circuit if the difference between I_L and I_C were less than I_C. This would happen if I_L were less than twice I_C. The situation would be similar if the original circuit were a parallel *RL* circuit and *C* were added in parallel. The line current I_T would drop if the inductive current I_L were greater than the difference between I_L and I_C. Such a condition would arise if I_C were less than twice I_L.

It follows that the total impedance of ac parallel circuits is not always *less* than the smallest branch impedance but is sometimes *greater* than the smallest branch impedance.

Parallel Circuits with *RL* and *RC* Series Branches

A parallel *RLC* circuit may contain branches with R and L or R and C in series, as in Figure 50–4. Branch AB consists of a resistor R_1 in series with a capacitor C. Branch CD consists of a resistor R_2 in series with an inductor L.

The impedance of the series combinations in each branch can be found using formulas studied in previous experiments.

For branch AB

$$Z_{AB} = \sqrt{R_1^2 + X_C^2} \qquad (50\text{–}7)$$

The current I_{AB} in branch AB is

$$I_{AB} = \frac{V}{Z_{AB}} \qquad (50\text{–}8)$$

Because branch AB is capacitive, the current leads the voltage across AB. The phase angle θ can be found using the formula

$$\theta_1 = \tan^{-1}\left(\frac{X_C}{R_1}\right) \qquad (50\text{–}9)$$

For branch CD

$$Z_{CD} = \sqrt{R_2^2 + X_L^2} \qquad (50\text{–}10)$$

The current I_{CD} in branch CD is

$$I_{CD} = \frac{V}{Z_{CD}} \qquad (50\text{–}11)$$

Because branch CD is inductive, the current lags the voltage across CD. The phase angle θ_2 can be found using the formula

$$\theta_2 = \tan^{-1}\left(\frac{X_L}{R}\right) \qquad (50\text{–}12)$$

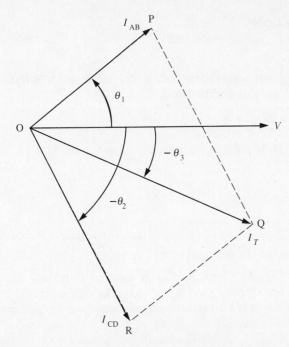

Figure 50–5. Current phasor diagram for an *RLC* circuit with *RL* and *RC* branches. The applied voltage *V* is used as the reference phasor.

The total, or line, current delivered by the voltage source is the phasor sum of I_{AB} and I_{CD}. The phasor diagram of Figure 50–5 shows the reference voltage V at 0°. Current I_{AB} leads the voltage at an angle θ_1. Current I_{CD} lags the voltage at an angle θ_2.

To find I_T, the parallelogram OPQR is completed by drawing PQ parallel to phasor OR and QR parallel to phasor OP. The phasor OQ is the total current I_T, and θ_3 is the angle by which the total current is out of phase with the applied voltage. In Figure 50–5 angle θ_3 is a negative angle, indicating that I_T lags the applied voltage, and the parallel circuit is behaving inductively. This is indicated by the fact that phasor I_{CD} is larger than phasor I_{AB}. If the reverse were true, θ_3 would be positive, and the total circuit would be capacitive.

SUMMARY

A method for finding the impedance of a parallel *RLC* circuit requires the following steps:

1. Calculate the impedance of each branch, or leg, of the parallel circuit by the methods already discussed in previous experiments. Calculate the value of *Z* using *V* and *I* and Ohm's law. Calculate the phase angle θ using tan⁻¹.
2. Find the current in each branch and the phase angle between the applied voltage *V* and the branch current. Branch current is computed using the formula

$$I = \frac{V}{Z}$$

Note that the phase angle (in degrees) is the same as the angle θ in statement 1, but it is opposite in sign. Thus, the phase angle of a capacitive current is positive (+) and that of an inductive current is negative (−).

3. Obtain the phasor sum of all branch currents. The resultant is the total, or line, current. The value of current and phase angle can be found from the phasor diagram.
4. If the parallel *RLC* circuit has no resistance in the capacitive or inductive branch—that is, if the circuit is exactly as in Figure 50–1—then the net reactive current I_X is the difference between I_L and I_C.
5. Find the total current I_T for the circuit in statement 4 by using the formula

$$I_T = \sqrt{I_R^2 + I_X^2}$$

and the phase angle θ between the applied voltage *V* and the total current is

$$\theta = \tan^{-1}\left(\frac{I_X}{I_R}\right)$$

6. Find the impedance *Z* of the parallel *RLC* circuit by using the formula

$$Z = \frac{V}{I_T}$$

where *V* is the voltage across the parallel circuit and I_T is the total, or line, current.

SELF-TEST

Check your understanding by answering the following questions:

1. In the circuit of Figure 50–1 $R = 12\ \Omega$, $X_C = 15\ \Omega$, and $X_L = 18\ \Omega$. The applied voltage $V = 6$ V.
   (a) The current I_R in $R =$ _____ A.
   (b) The phase angle between *V* and I_R is _____°.
2. For the circuit in question 1, (a) the current I_C in C = _____ A, and (b) the phase angle between *V* and $I_C =$ _____°.
3. For the circuit in question 1, (a) the current I_L in L = _____ A, and (b) the phase angle between *V* and I_L is _____°.
4. The net reactive current in the circuit in question 1 is _____ A, and it is _____ (capacitive/inductive).
5. The total current I_T in the circuit of question 1 is _____ A.
6. The phase angle between the applied voltage *V* and I_T in the circuit of question 1 is _____° and its sign is _____ (+/−).
7. The total impedance in the circuit of question 1 is _____ Ω.

MATERIALS REQUIRED

Power Supplies:
- Isolation transformer
- Variable-voltage autotransformer (Variac or equivalent)

Instruments:
- EVM or DMM
- 0–25-mA ac ammeter or second DMM with ac ammeter ranges

Resistor:
- 1   5100-Ω, 1-W

Capacitor:
- 1   0.5-μF or 0.47-μF

Inductor:
- Large iron-core choke, 8 or 9 H (see Experiment 36 for type and specifications)

Miscellaneous:
- 4 SPST switches
- Polarized line cord with on-off switch and fuse

PROCEDURE

1. With the line cord unplugged, line switch **off,** and switches S_1 through S_4 **open,** connect the circuit of Figure 50–6. Set the autotransformer to its lowest output voltage. Set the ammeter to the 25-mA (or higher) range. Set the voltmeter to the 150-V range.

2. Plug in the line cord; line switch **on. Close** S_1 and slowly increase the output voltage V to $V = 60$ V. Maintain this voltage throughout the experiment. Check the voltage from time to time and adjust it if necessary.

3. **Close** S_2. Check to see that $V = 60$ V; adjust if necessary. Measure the current. Since S_3 and S_4 are **open,** the only current in the circuit is the current in the resistor I_R. Record the value in Table 50–1. **Open** S_2.

4. **Close** S_3. Check that $V = 60$ V. Measure the current. Since S_2 and S_4 are **open,** the only current in the circuit is the current in the inductor (choke) I_L. Record the value in Table 50–1. **Open** S_3.

5. **Close** S_4. Check V; adjust if necessary. Measure the current. Since S_2 and S_3 are **open,** the only current in the circuit is the current in the capacitor branch I_C. Record the value in Table 50–1.

6. **Close** S_2 (S_4 is still **closed**). Verify that $V = 60$ V. Measure the current in the circuit. With S_2 and S_4 **closed** and S_3 **open,** the current in the circuit is the total of I_R and I_C, or I_{RC}. Record the value in Table 50–1. **Open** S_4.

7. **Close** S_3 (S_2 is still **closed**). $V = 60$ V. Measure the current in the circuit. With S_2 and S_3 **closed** and S_4 open, the current in the circuit is the total of I_R and I_L, or I_{RL}. Record the value in Table 50–1.

8. **Close** S_4. Now S_2, S_3, and S_4 are **closed.** Check V. Measure the current in the circuit. Since all branch circuit switches are **closed,** the ammeter will measure the total current I_T of the parallel *RLC* circuit. Record the value in Table 50–1. **Open** all switches; turn line switch **off** and unplug the line cord.

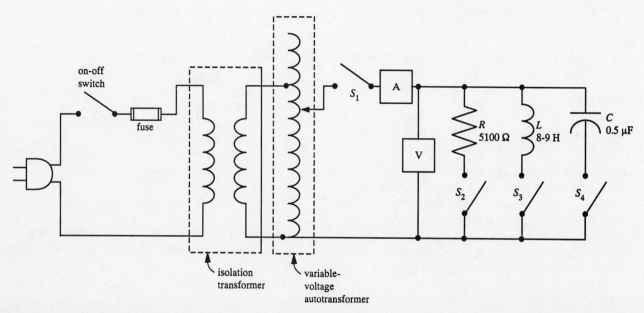

Figure 50–6. Circuit for procedure step 1.

9. Calculate the line current I_T using the measured values of I_R, I_L, and I_C and the square-root formula (50–4). Record your answer in Table 50–1.
10. Using the measured value of V (it should be 60 V) and the measured value of I_T, calculate the circuit impedance and indicate whether the circuit is inductive, capacitive, or resistive. Record your answers in Table 50–1.
11. Calculate the phase angle and power factor for the parallel RLC circuit and indicate whether the circuit has a leading or lagging power factor. Record your answers in Table 50–1.

ANSWERS TO SELF-TEST

1. (a) 0.5; (b) 0
2. (a) 0.4; (b) 90
3. (a) 0.333; (b) –90
4. 0.0667; capacitive
5. 0.504
6. 7.6; +
7. 11.9

Experiment 50 Name ________________________ Date ____________

TABLE 50–1. Determining the Impedance of a Parallel *RLC* Circuit

Applied Voltage V, V	*Resistor Current* I_R, A	*Inductor Current* I_L, A	*Capacitor Current* I_C, A	*Resistor and Capacitor Current* I_{RC}, A	*Resistor and Inductor Current* I_{RL}, A	*Total Current in RLC Circuit (Measured)* I_T, A	*Total Current (Calculated using Square Root Formula)* I_T, A	*Circuit Impedance Z (R, L, or C),* Ω
60								

Power factor ____________ % Leading/lagging? ____________ Phase Angle (degrees) ____________

QUESTIONS

1. Refer to your data in Table 50–1. Do the measured values of I_R, I_L, I_C, and I_T verify Formula (50–4)? Explain your answer with reference to specific data in the table.

2. Refer to your data in Table 50–1. Compare I_{RC} with I_T. Which is greater? Explain.

3. Using your data in Table 50–1, calculate the impedance of the *RL* section alone. Also, calculate the impedance of the *RC* section alone. Compare the two impedances with the total impedance of the *RLC* circuit. Are the results as expected? Explain.

4. On a separate sheet of $8^{1}/_{2} \times 11$ cross-section paper, draw a phasor diagram to scale of the currents in the parallel *RLC* circuit. Label all phasors and the phase angle of the circuit. On the same sheet of cross-section paper, draw the series circuit equivalent to the *RLC* circuit. Label the values of the series components. Show all calculations.

EXPERIMENT

51

RESONANT FREQUENCY AND FREQUENCY RESPONSE OF A SERIES *RLC* CIRCUIT

OBJECTIVES

1. To determine experimentally the resonant frequency f_R a series *LC* circuit
2. To verify that the resonant frequency of a series *LC* circuit is given by the formula

$$f_R = \frac{1}{2\pi\sqrt{LC}}$$

3. To develop experimentally the frequency-response curve of a series *LC* circuit

BASIC INFORMATION

Resonant Frequency of Series *RLC* Circuit

In the circuit of Figure 51–1 the voltage *V* is produced by an ac generator whose frequency and output voltage are manually adjustable. For a particular frequency *f* and output voltage *V* there will be a current *I* such that $I = V/Z$, where *Z* is the impedance of the circuit. The voltage drops across *R*, *L*, and *C* will be given by IR, IX_L, and IX_C, respectively.

If the generator frequency is changed but *V* remains constant, the current *I* and the voltage drops across *R*, *L*, and *C* will change.

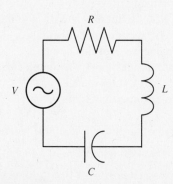

Figure 51–1. Series *RLC* circuit.

There is a frequency f_R, called the *resonant frequency*, at which

$$X_L = X_C \qquad \textbf{(51–1)}$$

The resonant frequency can be found as follows:

$$X_L = 2\pi f L$$

and $\qquad\qquad\qquad\qquad\qquad\qquad$ **(51–2)**

$$X_C = \frac{1}{2\pi f C}$$

At f_R, when $X_L = X_C$,

$$2\pi f_R L = \frac{1}{2\pi f_R C} \qquad \textbf{(51–3)}$$

Formula (51–3) can be solved for f_R:

$$f_R^2 = \frac{1}{(2\pi)^2 LC} \qquad \textbf{(51–4)}$$

or

$$f_R = \frac{1}{2\pi\sqrt{LC}} \qquad \textbf{(51–5)}$$

where f_R is given in hertz, *L* in henrys, and *C* in farads.

This is the formula for the resonant frequency f_R of a resonant series *RLC* circuit. If the value of *L* and *C* are known, the resonant frequency can be calculated directly using formula (51–5).

Problem. Find the resonant frequency of a series *RLC* circuit where $L = 8$ H, $C = 0.01$ μF, and $R = 47$ Ω.

Solution. Formula (51–5) will be used to find f_R.

$$f_R = \frac{1}{2\pi\sqrt{LC}} = \frac{1}{2\pi\sqrt{8 \times 0.01 \times 10^{-6}}}$$

$$f_R = \frac{10^4}{2\pi\sqrt{8}}$$

$$f_R = 563 \text{ Hz}$$

Note that the value of *R* did not enter into the calculation of f_R.

Characteristics of a Series Resonant Circuit

Some characteristics of a series RLC circuit were investigated in a previous experiment. The following facts were observed:

1. The voltage drop across a reactive component is equal to the product of the current I in the circuit and the reactance X of the component.
2. The total reactive effect on a circuit is the difference between the capacitive reactance X_C and the inductive reactance X_L.
3. The impedance Z of a series RLC circuit is

$$Z = \sqrt{R^2 + X^2}$$

4. The impedance Z of the circuit is minimum when $X_L = X_C$, and circuit current I is maximum at this point.

When $X_L = X_C$ the circuit is said to be resonant. *At resonance, the impedance of a series RLC circuit is minimum and the current I in the circuit is maximum.* This fact, which was established experimentally in a previous experiment, can also be found using the formula

$$Z = \sqrt{R^2 + X^2}$$

At f_R, $X = 0$,

$$Z = \sqrt{R^2 + (0)^2} = R$$

Therefore, a *minimum impedance $Z = R$* exists at f_R. Also, since

$$I = \frac{V}{Z} = \frac{V}{R}$$

and V is assumed to be fixed, there will be *maximum current I* in the circuit at resonance. Note that at f_R the applied voltage V appears across R, and resonant current I may be calculated from

$$I = \frac{V}{R} \qquad (51\text{-}6)$$

Because I is limited at resonance only by the value of R, the circuit is said to be *resistive*. For all frequencies higher than f_R, X_L is greater than X_C, and the circuit is *inductive*. For all frequencies lower than f_R, X_C is greater than X_L, and the circuit is *capacitive*.

At resonance the voltage V_L across L and the voltage V_C across C are *maximum* and equal. Theoretically, at the resonant frequency when R becomes zero, the current I becomes infinite and the voltages V_L and V_C become infinitely large. Practically, this condition is never realized because L has some resistance R_L that may be considered in series with it.

Frequency Response of a Series Resonant Circuit

The frequency-response characteristic of a series RLC circuit can be found experimentally by applying a constant-amplitude signal voltage to the circuit at the resonant frequency and at frequencies on both sides of resonance. The voltage across L or C is measured, and a graph of V_L or V_C versus f is plotted. This is one form of the frequency-response curve of the circuit.

The circuit current I can also be measured at f_R and at frequencies on both sides of f_R. A graph of I versus f is another form of the frequency-response curve of the circuit.

SUMMARY

1. In a series RLC circuit there is a frequency f_R, called the resonant frequency, which may be calculated from the formula

$$f_R = \frac{1}{2\pi\sqrt{LC}}$$

At resonance $X_L = X_C$.
2. At f_R the impedance of the circuit is minimum, and $Z = R$.
3. At f_R the current I in the circuit is maximum, and $I = V/R$ where V is the voltage applied across the circuit.
4. At resonance the voltages V_L across L and V_C across C are maximum and equal.
5. At resonance a series RLC circuit acts as a resistance; above resonance it is inductive, and below resonance it is capacitive.

SELF-TEST

Check your understanding by answering the following questions:

1. In a series resonant circuit, $X_L = 120\ \Omega$ at f_R. The value of X_C at resonance is _____ Ω.
2. When the frequency f of the source applied to a series RLC circuit is higher than the resonant frequency f_R, then X_L _____ (equals/is greater than/is less than) X_C and the circuit is _____ .
3. In a series RLC circuit $R = 1\ \text{k}\Omega$, $L = 100\ \mu\text{H}$, $C = 0.001\ \mu\text{F}$, and the applied voltage $V = 12\ \text{V}$. The resonant frequency f_R of this circuit is _____ Hz.
4. In the circuit of question 3, the circuit current $I =$ _____ A at resonance.
5. In the circuit of question 3, the voltage V_L across L is _____ V.
6. In the circuit of question 3, the impedance Z at resonance is _____ Ω.

MATERIALS REQUIRED

Power Supply:
■ AF sine-wave generator

Instruments:
■ EVM or DMM
■ Oscilloscope

Resistor:
■ 1   100-Ω, ½-W, 5%

Capacitors:
■ 1   0.001-μF
■ 1   0.01-μF
■ 1   0.05-μF or 0.047-μF

Inductor:
■ 30-mH choke

PROCEDURE

A. Determining the Resonant Frequency of a Series *RLC* Circuit

A1. With the sine-wave generator and oscilloscope **off**, connect the circuit of Figure 51–2.

A2. Turn **on** the sine-wave generator and set the frequency for 30 kHz. Turn **on** the oscilloscope and calibrate it for voltage measurements. Adjust the scope to view the sine-wave output of the generator. Increase the output of the generator until the scope indicates a 2-V p-p voltage. Maintain this voltage throughout the experiment.

A3. Observe the rms voltage across the capacitor V_C as the frequency of the generator is varied above and below 30 kHz. Note the frequency at which V_C is a maximum. In a series *RLC* circuit, V_C is a maximum at the resonant frequency, f_R. Record the value of f_R in Table 51–1 (p. 369) in the 0.001-μF row.

A4. Replace the 0.001-μF capacitor with a 0.05-μF capacitor. Check the generator output voltage to verify that it is 2 V p-p; adjust the voltage if necessary.

A5. Set the frequency of the generator to 13 kHz. Observe the voltage across the capacitor V_C as the frequency is varied above and below 13 kHz. At the point of maximum V_C the frequency is f_R. Record this value in Table 51–1 in the 0.05-μF row.

A6. Replace the 0.05-μF capacitor with a 0.01-μF capacitor. Check the generator output voltage and adjust to maintain 2 V p-p if necesary.

A7. Set the frequency of the generator to 9 kHz. Observe the voltage across the capacitor V_C as the frequency is varied above and below 9 kHz. At the resonant frequency f_R the voltage across the capacitor will be maximum. Record the value of f_R in Table 51–1 in the 0.01-μF row.

A8. Calculate the resonant frequencies for 30-mH–0.001-μF; 30-mH–0.05-μF; and 30-mH–0.01-μF series *LC* combinations. Use formula (51–5) and the rated values of L and C. Record your answers in Table 51–1.

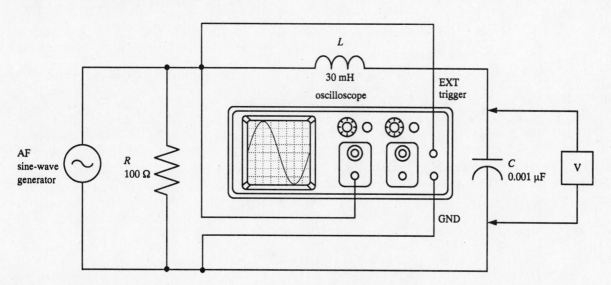

Figure 51–2.   Circuit for procedure step A1.

B. Plotting a Frequency-Response Curve

B1. With the circuit of Figure 51–2 still connected and the 0.01- μF capacitor in the circuit, check the oscilloscope to verify that the output voltage is still 2 V p-p. Also verify the value of f_R for the 30-mH and 0.01-μF circuit. (It should be the same as that obtained in step A7.)

B2. Examine Table 51–2. In this part of the experiment you will take a series of readings at frequencies above and below the resonant frequency. For each frequency setting you will measure and record the voltage across the 0.01-μF capacitor. Since f_R may not be a round number, you may not be able to set the exact frequencies on the generator. Therefore, choose values of frequency as close to the step values as possible. For example, if $f_R = 9227$, $f_R + 3000 = 12{,}227$; in that case, choose the closest frequency for which an exact setting can be made. It is important to continue monitoring the output voltage of the generator and adjusting it to 2 V p-p if necessary. Complete Table 51–2. After completing all measurements, turn **off** the oscilloscope and sine-wave generator.

ANSWERS TO SELF-TEST

1. 120
2. is greater than; inductive
3. 503,292
4. 0.012
5. 3.79
6. 1000

Name _____ Date _____

TABLE 51–1. Resonant Frequency of a Series *RLC* Circuit

Inductor L, mH	Capacitor C, μF	Resonant Frequency f_R, Hz	
		Measured	Calculated
30	0.001		
30	0.05		
30	0.01		

TABLE 51–2. Frequency Response of a Series *RLC* Circuit

Step	Frequency f, Hz	Voltage Across Capacitor V_C, V rms
$f_R - 6000$		
$f_R - 5000$		
$f_R - 4000$		
$f_R - 3000$		
$f_R - 2000$		
$f_R - 1000$		
$f_R - 500$		
f_R		
$f_R + 500$		
$f_R + 1000$		
$f_R + 2000$		
$f_R + 3000$		
$f_R + 4000$		
$f_R + 5000$		
$f_R + 6000$		

QUESTIONS

1. Explain, in your own words, what the resonant frequency of an *RLC* circuit is.

_____

_____

_____

_____

_____

2. Refer to your data in Table 51–2. In a series *RLC* circuit, under what conditions is it possible for the voltage across the inductance or capacitance to be greater than the applied voltage? Explain.

_____

_____

_____

_____

Resonant Frequency and Frequency Response of a Series RLC Circuit

3. What was the impedance of your series *RLC* circuit at resonance? Explain.

_____

_____

_____

_____

_____

_____

4. Refer to your data in Table 51–1. Compare the calculated value of resonant frequency with the measured value. Explain any unexpected results.

_____

_____

_____

_____

_____

5. Discuss the effect on the resonant frequency of changes in capacitance with fixed resistance and inductance in a series *RLC* circuit. Refer to your data in Table 51–1.

_____

_____

_____

_____

_____

6. On a separate sheet of 8½ × 11 graph paper, plot a graph of frequency versus voltage across the capacitor using your data in Table 51–2. The horizontal (*x*) axis should be frequency, and the vertical (*y*) axis should be voltage across the capacitor. Draw a broken vertical line at the resonant frequency extending from the frequency axis to the frequency response curve. Label all axes; label the resonant frequency line f_R.

7. Why was it important to constantly monitor the output voltage of the signal generator during the frequency response measurements?

_____

_____

_____

_____

_____

_____

EXPERIMENT

52

EFFECT OF *Q* ON FREQUENCY RESPONSE AND BANDWIDTH OF A SERIES RESONANT CIRCUIT

OBJECTIVES

1. To measure the effect of circuit Q on frequency response
2. To measure the effect of circuit Q on bandwidth at the half-power points

BASIC INFORMATION

Circuit *Q* and Frequency Response

In Experiment 51 we studied the frequency-response curve of a series resonant LC circuit. In that experiment the only resistance in the circuit was that associated with the coil. Theoretically, at resonance $X_L = X_C$ and the impedance $Z = R_L$, where R_L equals the resistance of the coil.

The amount of coil resistance R_L, therefore, determines the current flow through the circuit at resonance if there is no resistance other than the coil resistance. The R_L and X_L of the coil determine the *quality*, or Q, of the circuit, which is given by the formula

$$Q = \frac{X_L}{R_L} \qquad (52\text{--}1)$$

The Q of the circuit also determines the rise in voltage across L and C at the resonant frequency f_R. The voltage developed across L is given by the formulas

$$V_L = I X_L$$

$$= \frac{V}{R} \times X_L \qquad (52\text{--}2)$$

If the circuit resistance R is the coil resistance R_L, then

$$V_L = V \times \frac{X_L}{R_L}$$

$$V_L = VQ \qquad (52\text{--}3)$$

Also, since $X_L = X_C$ at resonance,

$$I X_L = I X_C$$

and

$$V_L = V_C$$

Therefore,

$$V_C = VQ \qquad (52\text{--}4)$$

Formulas (52–3) and (52–4) become significant for values of $Q > 1$. For such values, V_C and V_L are greater than the applied voltage V. Also, the higher the value of Q, the greater the voltage gain in the circuit. This is the *first* example of voltage gain.

Circuit Q is also significant when we consider the frequency response of a series resonant circuit. The frequency-response characteristic can be determined by applying a constant-amplitude signal voltage V into the circuit at the resonant frequency and at frequencies on either side of resonance. The voltage across L or C is measured, and a graph of V_L or V_C versus f is plotted. This is one form of the frequency-response curve of the circuit.

The circuit current I can also be determined. A graph of I versus f is another form of the frequency-response curve of the circuit.

Circuit *Q* and Bandwidth

Figure 52–1 (p. 372) is a graph of the response of a series resonant circuit. Three significant points have been marked on the curve. These are f_R, the resonant frequency, and f_1 and f_2. Points f_1 and f_2 are located at 70.7 percent of maximum (maximum is at f_R) on the curve. They are called the *halfpower points*, and the frequency separation between them is $f_2 - f_1$. This frequency separation is called the *bandwidth* (*BW*) of the circuit. Bandwidth, therefore, is given by the formula

$$BW = f_2 - f_1 \qquad (52\text{--}5)$$

Bandwidth is related to Q. It can be shown that *BW* is given by the formula

$$BW = \frac{f_R}{Q} \qquad (52\text{--}6)$$

As was noted in the preceding experiment, the resonant frequency f_R of a series RLC circuit can be calculated from the formula

$$f_R = \frac{1}{2\pi\sqrt{LC}} \qquad (52\text{--}7)$$

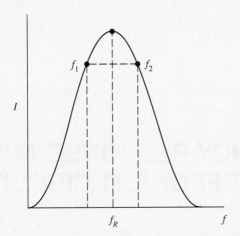

Figure 52-1. Frequency-responce curve of a series resonant circuit.

where f_R is in hertz, L is in henrys, and C is in farads. Since the formula does not include R, it is apparent that the resonant frequency of the series RLC circuit is not affected by the size of the resistance R. However, R does affect the bandwidth and amplitude of the response curve. The higher the value of R, the lower is the value of Q, as shown in formula (52-1). The higher the resistance R, the wider is the bandwidth, as shown by formulas (52-1) and (52-6). Moreover, the lower the value of Q, the lower is the value of current I in the circuit, and the lower are the voltages V_L across L and V_C across C.

Series resonant circuits are used in communication, video, and industrial electronics as frequency-selective circuits and as traps to eliminate unwanted signals. Normally such circuits require a highly peaked response with narrow bandwidth. To achieve such a response, the Q of the circuit must be high. Hence high-Q coils are employed. In such circuits the Q of the circuit is mainly determined by the Q of the coil.

However, there are some applications in electronics where wideband frequency-selective circuits are required. In such cases coil "loading" is achieved by the use of external resistors.

SUMMARY

1. The quality, or Q, of a coil is defined by the formula

$$Q = \frac{X_L}{R}$$

2. In a series resonant circuit, where the only resistance in the circuit is coil resistance R_L the Q of the circuit is determined by the Q of the coil.

3. In a series resonant circuit the voltages across L and C are equal and are affected by the Q of the circuit. These voltages may be calculated from the formula

$$V_L = V_C = VQ$$

where V is the applied voltage and V_L and V_C are the voltages across L and C, respectively.

4. If circuit Q is greater than one, then the voltages V_L and V_C are higher than the applied voltages. The voltages in a series resonant circuit provide the first example of amplification.

5. Bandwidth (BW) of a response curve is defined as the difference between the two frequencies f_2 and f_1 at the half-power points (Figure 52-1). The half-power points are those points that appear at 70.7 percent of maximum on the frequency-response curve. Maximum output occurs at the resonant frequency f_R.

6. Bandwidth is related to circuit Q and is given by the formula

$$BW = \frac{f_R}{Q}$$

7. The lower the Q of the circuit, the wider the bandwidth and the flatter the response curve.

8. The lower the Q of the circuit, the lower the amplitude of the response curve and the lower the circuit gain.

SELF-TEST

Check your understanding by answering the following questions.

In the circuit of Figure 52-2, $L = 50$ mH, $C = 0.01$ μF, and $R = 20$ Ω. Assume that coil resistance $R_L = 0$. The applied voltage $V = 5.0$ V. The output voltage is V_C.

1. The resonant frequency is $f_R =$ _____ kHz.
2. At resonance $X_L =$ _____ Ω.
3. The circuit Q is _____ at resonance.
4. The voltage V_L across L is _____ V at resonance.
5. The bandwidth of the circuit is _____ Hz.
6. The voltage V_C across C at the half-power points is _____ V.

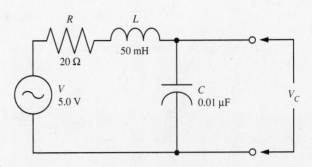

Figure 52-2. Circuit diagram for the Self-test.

MATERIALS REQUIRED

Power Supply:
■ AF sine-wave generator

Instruments:
■ EVM or DMM
■ Oscilloscope

Resistors (¹/₂-W, 5%):
■ 1 10-Ω
■ 1 47-Ω
■ 2 100-Ω

Capacitor:
■ 1 0.01-μF

Inductor:
■ 30-mH choke

PROCEDURE

A. Circuit *Q* and Frequency Response of a Series Resonant Circuit

A1. With the sine-wave generator and oscilloscope turned **off**, connect the circuit of Figure 52–3. The oscilloscope should be calibrated to measure the output voltage of the generator.

A2. Turn **on** the generator and scope. Adjust the output of the generator *V* to 1 V rms as measured by the oscilloscope. Maintain this voltage throughout the experiment, checking it whenever the frequency of the generator is changed; adjust to 1 V rms if necessary.

A3. Set the sine-wave generator to 10 kHz. Vary the frequency above and below 10 kHz until the maximum voltage across the capacitor V_C is determined. The maximum V_C is reached at the resonant frequency, f_R. Record f_R and V_C in Table 52–1 (p. 375).

A4. Examine Table 52–1. You will measure the voltage across the capacitor V_C as you vary the frequency from 6000 Hz below the resonant frequency to 6000 Hz above f_R in steps of 1000 Hz (except for 500 Hz above and below f_R). Choose the frequency on the generator as close as possible to the deviation shown. Record the

actual frequency in the column shown. Record each voltage in the column marked "10-Ω resistor." After completing all measurements, turn **off** the signal generator and disconnect the 10-Ω resistor from the circuit.

A5. Replace the 10-Ω resistor with a 47-Ω resistor. Turn **on** the generator and adjust the output voltage *V* to 1 V rms as measured by the scope. Maintain this voltage throughout the experiment.

A6. Measure the voltage across the capacitor for each of the frequencies in Table 52–1. Record the values in Table 52–1 in the "47-Ω Resistor" column. After completing all the measurements, turn **off** the generator and disconnect the 47-Ω resistor.

A7. Replace the 47-Ω resistor with a 100-Ω resistor. Turn **on** the generator and adjust the output *V* to 1 V rms as measured by the oscilloscope. Maintain this voltage throughout the experiment.

A8. Measure the voltage across the capacitor V_C for each frequency in Table 52–1. Record the values in Table 52–1 in the "100-Ω resistor" column. After completing all the measurements, turn **off** the generator and the oscilloscope; disconnect the 100-Ω resistor.

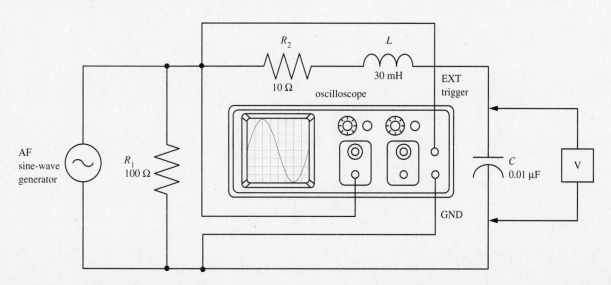

Figure 52–3.   Circuit for procedure step A1.

Effect of Q in Series Resonant Circuit   **373**

B. Effect of Resistance on Resonant Frequency; Determining the Phase Angle of a Resonant Circuit

B1. Reconnect the circuit of Figure 52–3 with the 10-Ω resistor and the voltmeter across the resistor.

B2. Turn **on** the generator and the oscilloscope. Set the generator output V at 1 V rms as measured by the scope. Maintain this voltage throughout the experiment; adjust if necessary.

B3. Vary the frequency until the voltage across the resistor V_R is a maximum. At maximum V_R the frequency is the resonant frequency of the circuit. Record f_R and V_R in Table 52–2 (p. 375) in the 10-Ω row. Measure the voltage across the capacitor-inductor combination, V_{LC}. Record the value in Table 52–2 in the 10-Ω row. Turn **off** the generator and disconnect the 10-Ω resistor.

B4. Connect the 47-Ω resistor and repeat step B3. Record the frequency in the 47-Ω rows. Measure the voltage across the capacitor-inductor combination, V_{LC}. Record the value in Table 52–2 in the 47-Ω row. Turn **off** the generator and disconnect the 47-Ω resistor.

B5. Replace the 47-Ω resistor with a 100-Ω resistor and repeat step B3. Record the frequency in the 100-Ω row. Measure the voltage across the capacitor-inductor combination, V_{LC}. Record the value in Table 52–2 in the 100-Ω row. Turn **off** the generator and the oscilloscope. Disconnect the circuit.

B6. Measure the resistance of the inductor and record the value in Table 52–2.

B7. For each value of resistor, calculate the current in the circuit, using the measured value of V_R and the rated value of R. Record your answers in Table 52–2.

B8. Calculate the phase angle of each circuit using rated values of L and C and the formula

$$\theta = \tan^{-1}\left(\frac{X}{R}\right)$$

Record your answers in Table 52–2.

ANSWERS TO SELF-TEST

1. 7.118
2. 2236
3. 111.8
4. 559
5. 63.7
6. 395

TABLE 52–1.   Circuit Q and Frequency Response of a Series Resonant Circuit

Frequency Deviation	Frequency f, Hz	10-Ω Resistor Voltage across Capacitor V_C, V	47-Ω Resistor Voltage across Capacitor V_C, V	100-Ω Resistor Voltage across Capacitor V_C, V
$f_R - 6000$				
$f_R - 5000$				
$f_R - 4000$				
$f_R - 3000$				
$f_R - 2000$				
$f_R - 1000$				
$f_R - 500$				
f_R				
$f_R + 500$				
$f_R + 1000$				
$f_R + 2000$				
$f_R + 3000$				
$f_R + 4000$				
$f_R + 5000$				
$f_R + 6000$				

TABLE 52–2.   Effect of Resistance on a Series Resonant Circuit

Resistor R, Ω	Resonant Frequency f_R, Hz	Voltage across Resistor V_R, V	Voltage across Inductor/ Capacitor Combination V_{LC}, V	Circuit Current (Calculated) I, A	Phase Angle (Calculated) θ, degrees
10					
47					
100					

R_L (resistance of 30-mH inductor) = _____ Ω

QUESTIONS

1. On a separate sheet of $8^{1}/_{2} \times 11$ graph paper, plot a graph of frequency versus voltage across the capacitor using your data in Table 52–1. Plot a separate frequency response curve for each value of resistance. The horizontal (x) axis should be frequency, and the vertical (y) axis should be voltage. Label the curves with the appropriate resistor values. Indicate the resonant frequency on the frequency axis. Also indicate the half-power points for each of the curves. Label each curve with its bandwidth and Q.

Effect of Q in Series Resonant Circuit   **375**

2. Explain, in your own words, the relationship between the Q of a circuit and the frequency response of the circuit.

_____

_____

_____

_____

_____

_____

_____

3. Explain, in your own words, the relationship between the Q of a circuit and the bandwidth at the half-power points.

_____

_____

_____

_____

_____

_____

_____

4. Refer to your data in Table 52–2. What effect, if any, did changes in resistance have on the resonant frequency of the circuit (Figure 52–3)? Explain any unexpected results.

_____

_____

_____

_____

_____

_____

5. Refer to your data in Table 52–2. At the resonant frequency, what is the phase angle between the applied voltage and the circuit current? Explain any unexpected results.

_____

_____

_____

_____

_____

53

CHARACTERISTICS OF PARALLEL RESONANT CIRCUITS

OBJECTIVES

1. To determine experimentally the resonant frequency of a parallel *RLC* circuit
2. To measure the line current and impedance of a parallel *RLC* circuit at the resonant frequency
3. To measure the effect of variations in frequency on the impedance of a parallel *RLC* circuit

BASIC INFORMATION

Characteristics of a Parallel Resonant Circuit

Resonant Frequency of a High-*Q* Circuit

The circuit of Figure 53–1 consists of C and L in parallel, with the coil resistance R_L shown in series with L. It is assumed here that the Q of this circuit is high (that is, that R_L is small compared with X_L) and that the resistance of C and the wiring resistance of the circuit are negligible and may be ignored.

There is a particular frequency at which $X_L = X_C$. This frequency may be defined as the condition for parallel resonance in a high-*Q* circuit and is similar to the condition for series resonance.

There are other definitions for parallel resonance. Thus, parallel resonance may be considered as the frequency at which the impedance of the parallel circuit is maximum. Also, parallel resonance may be considered as the frequency at which the parallel impedance of the circuit has unity power factor. These three definitions may lead to three different frequencies, each of which may be considered as the resonant frequency. In circuits whose Q is greater than 10, however, the three conditions lead to the same resonant frequency.

In a high-*Q* circuit, the formula for the resonant frequency f_R is the same as in the case of series resonance and is given by

$$f_R = \frac{1}{2\pi\sqrt{LC}} \qquad (53\text{–}1)$$

Line Current

If the resistance R_L of the inductance L in the circuit of Figure 53–1 is small, then at resonance the impedance X_C of the capacitive branch is practically equal to the impedance of the inductive branch. The current in each branch may therefore be considered equal. These currents, however, are practically 180° out of phase. Their phasor sum, which is the line, or total, current I_T, is very small. Because the impedance that this parallel resonant circuit presents to the circuit is equal to V/I_T, it is apparent that the impedance is very high (since I_T is low). Although the line current is low, the circulating current in the parallel resonant circuit is high at resonance.

Frequency Response

The characteristics of a parallel resonant circuit to frequencies on either side of the resonant frequency may now be studied. For a frequency f_a higher than f_R, the reactance X_C of the capacitive branch is lower than the reactance X_L of the inductive branch. There is, therefore, a greater capacitive than inductive current, and the circuit may be considered *capacitive*. Similarly, it may be shown that for a frequency f_b lower than f_R, the circuit is *inductive*. The graph of Figure 53–2 (p. 378) illustrates these relationships. Individual graphs for I_L, I_C, and I_T are shown. It is evident that at resonance, line current is minimum, that I_L equals I_C, and that I_L and I_C are each greater than the line, or total, current I_T.

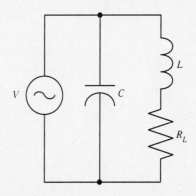

Figure 53–1. Parallel resonant circuit.

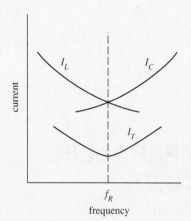

Figure 53-2. Current versus frequency in a parallel resonant circuit. The line current, I_T, is the phasor sum of I_C and I_L.

Impedance

Figure 53-3 is a graph of impedance versus frequency in a parallel resonant circuit. This graph resembles the frequency-response characteristic of a series resonant circuit. It shows that circuit impedance is maximum at resonance and falls off on either side of resonance.

The resonant frequency f_R of a parallel LC circuit such as in Figure 53-4 can be found experimentally using a variable-frequency source. By measuring the voltage across the known resistor R in the circuit I_T we can calculate the line, or total, current of the circuit using Ohm's law:

$$I_R = \frac{V_R}{R} = I_T \qquad (53\text{--}2)$$

We can observe the voltage across the resistor as the frequency of the voltage source is varied while making certain the output voltage is kept constant. An observation

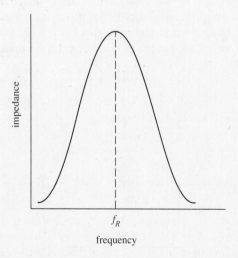

Figure 53-3. Impedance versus frequency in a parallel resonant circuit.

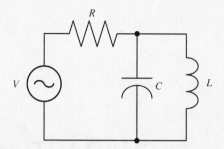

Figure 53-4. A parallel LC circuit used to determine f_R.

of minimum V_R indicates I_T is a minimum. The frequency at this point is f_R. By again varying the frequency above and below f_R while maintaining a constant voltage output, we can measure the value of V_R at different frequencies. We can calculate I_R from these values of V_R using formula (53-2). We can plot a graph of frequency versus I_R from these data.

We can also find the voltage across the parallel LC circuit (often called a *tank* circuit). Using this voltage V_t together with I_T, we can calculate Z_t, the impedance of the tank circuit:

$$Z_t = \frac{V_t}{I_T} \qquad (53\text{--}3)$$

The voltage V_t across the tank circuit is directly proportional to the Q of the circuit. If we change the circuit by placing a resistor in parallel with the tank circuit, we are loading the circuit, and the voltage across the tank circuit will decrease. This loading effect increases as the size of the parallel resistor decreases.

SUMMARY

1. In a parallel LC circuit, if the Q of the circuit is greater than 10, the resonant frequency of the circuit may be calculated from the formula

$$f_R = \frac{1}{2\pi\sqrt{LC}}$$

At this frequency $X_L = X_C$. (The parallel LC circuit is often called a tank circuit.)
2. In a high-Q circuit the impedance of the inductive branch equals (approximately) the impedance of the capacitive branch at resonance.
3. Because the voltage across each of the parallel branches in a high-Q circuit is the same and the impedances of both branches are equal at f_R the currents in both branches are equal. That is, $I_L = I_C$.
4. The currents, however, are approximately 180° out of phase, since one is capacitive and the other is inductive. They therefore tend to cancel each other, and the line, or total, current, which is the phasor sum of I_C and I_L, is therefore very low. In a very high-Q circuit, the line current is close to zero at f_R.

5. The very low line current means that the impedance of the parallel LC circuit is very high, since

$$Z = \frac{V}{I_T}$$

where V is the applied voltage and I_T is the total, or line, current.

6. For frequencies higher than f_R, the current in the capacitive branch is higher than the current in the inductive branch, and the circuit acts like a capacitive circuit.

7. For frequencies lower than f_R, current in the inductive branch is higher than current in the capacitive branch, and the circuit acts like an inductance.

8. The frequency-response curve of f versus Z of a high-Q parallel resonant circuit resembles the frequency-response curve of a series resonant circuit.

SELF-TEST

Check your understanding by answering the following questions:

1. The Q of a parallel resonant circuit in Figure 53–1 is 100. At resonance the current in the inductive branch is 50 mA, and the voltage across the tank circuit is 100 V. The current in the capacitive branch is _____ mA.

2. In the parallel resonant circuit of Figure 53–4, the line current at resonance is 0.05 mA. The voltage across the tank circuit is 2.5 V. The impedance of the tank circuit is _____ Ω.

3. For the circuit in question 2, at frequencies higher than f_R, the impedance of the parallel resonant circuit is _____ (higher/lower) than the impedance at f_R.

4. In the circuit of Figure 53–4, the coil resistance R_L is 12 Ω, $R = 10$ kΩ, $L = 30$ mH, and $C = 0.01$ μF. The resonant frequency of the circuit $f_R = $ _____ Hz.

5. The Q of the coil in the circuit of question 4 is _____ .

6. The voltage across R, in Figure 53–4, is 1.5 V at resonance. The line current in the circuit is _____ mA.

MATERIALS REQUIRED

Power Supply:
- ■ AF sine-wave generator

Instruments:
- ■ Oscilloscope
- ■ EVM or DMM

Resistors ($\frac{1}{2}$-W, 5%):
- ■ 2   33-Ω
- ■ 1   10,000-Ω

Capacitor:
- ■ 1   0.01-μF

Inductor:
- ■ 30-mH choke

PROCEDURE

A. Resonant Frequency and Impedance of a Parallel *LC* Resonant Circuit

A1. With the sine-wave generator and oscilloscope turned **off**, connect the circuit of Figure 53–5 (p. 380).

A2. Turn **on** the generator and oscilloscope. Adjust the scope to measure the voltage output of the generator. Increase the generator output voltage V to 4 V p-p. Maintain this voltage throughout the experiment. Set the generator frequency to 10 kHz. Adjust the scope to display two or three cycles of the sine-wave.

A3. Vary the generator frequency above and below 10 kHz and observe the voltage across the resistor V_R. At the minimum V_R the frequency will equal the resonant frequency f_R. Verify that $V = 4$ V p-p; adjust if necessary. Record f_R and V_R in Table 53–1 (p. 383).

A4. Note in Table 53–1 a series of frequencies above and below the resonant frequency f_R. Set the generator as close as possible to each frequency. At each frequency, measure the rms voltage across the resistor V_R and the parallel LC circuit (tank circuit) V_{LC}, checking periodically to make sure $V = 4$ V p-p. Record the frequency f, V_R, and V_{LC} in Table 53–1. After completing all measurements, turn off the generator and the oscilloscope and disconnect the circuit.

A5. Using the measured value of V_R for each frequency and the rated value of R, calculate the line current I at each frequency. Record your answers in Table 53–1.

A6. Using the calculated values of I from step A5 and the rms value of V (4 V p-p), calculate the impedance of the tank circuit at each frequency. Record your answers in Table 53–1.

B. Reactive Characteristics of a Parallel *LC* Circuit

B1. With the generator and oscilloscope **off**, connect the circuit of Figure 53–6 (p. 380). It may be assumed that the resonant frequency, f_R, of this circuit is the same as in Part A. Transfer the frequencies from Table 53–1 to Table 53–2 (p. 383).

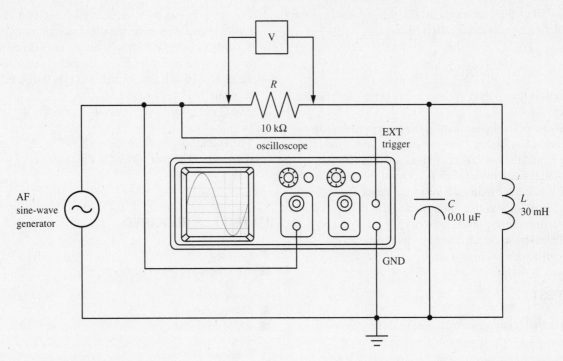

Figure 53–5. Circuit for procedure step A1.

B2. Turn **on** the generator and the oscilloscope. Set the generator voltage V to 4 V p-p and maintain this voltage throughout the experiment. Check V from time to time and adjust as necessary.

B3. For each frequency in Table 53–2, measure the voltage across the resistor in the capacitive branch AB, V_{R1}, and the voltage across the resistor in the inductive

branch CD, V_{R2}. Record the values in Table 53–2. After completing all measurements, turn **off** the generator and oscilloscope and disconnect the circuit.

B4. Using the measured values of V_{R1} and V_{R2} and the rated values of R_1 and R_2, calculate the currents in the capacitive branch I_C and the inductive branch I_L for each frequency. Record your answers in Table 53–2.

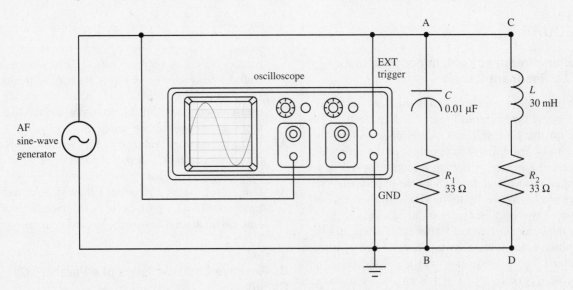

Figure 53–6. Circuit for procedure step B1.

ANSWERS TO SELF-TEST

1. 50
2. 50,000
3. lower
4. 9189
5. 144
6. 0.15

Characteristics of Parallel Resonant Circuits

Name _____ Date _____

TABLE 53–1. Frequency Response of a Parallel Resonant Circuit

Frequency Deviation	Frequency f, Hz	Voltage across Resistor V_R, V	Voltage across Tank Circuit V_{LC}, V	Line Current (Calculated) I, A	Tank Circuit Impedance (Calculated) Z, Ω
$f_R - 6000$					
$f_R - 5000$					
$f_R - 4000$					
$f_R - 3000$					
$f_R - 2000$					
$f_R - 1000$					
$f_R - 500$					
f_R					
$f_R + 500$					
$f_R + 1000$					
$f_R + 2000$					
$f_R + 3000$					
$f_R + 4000$					
$f_R + 5000$					
$f_R + 6000$					

TABLE 53–2. Reactance Characteristics of a Parallel *LC* Circuit

Frequency f, Hz	Voltage across Resistor R_1 V_{R1}, V	Voltage across Resistor R_2 V_{R2}, V	Current in Capacitive Branch (Calculated) I_C, A	Current in Inductive Branch (Calculated) I_L, A

Characteristics of Parallel Resonant Circuits

QUESTIONS

1. Explain, in your own words, the effects of changes in frequency on the impedance of a parallel *RLC* circuit.

_____

_____

_____

_____

_____

2. Explain, in your own words, the effects of changes in frequency on the total current of a parallel *RLC* circuit.

_____

_____

_____

_____

_____

3. What determines the value of *Q* in a parallel *RLC* circuit? What factor or condition determines if it is a high *Q*?

_____

_____

_____

_____

_____

4. On a separate sheet of $8\frac{1}{2} \times 11$ graph paper, plot a graph of frequency versus tank circuit impedance using your data in Table 53–1. Frequency should be the horizontal (*x*) axis and impedance should be the vertical (*y*) axis. Label all axes. Indicate the resonant frequency with a dotted vertical line.

5. On a separate sheet of $8\frac{1}{2} \times 11$ graph paper (not the sheet used for Question 4), plot a graph of frequency versus capacitive branch current I_C using your data in Table 53–2. On the same set of axes, plot frequency versus inductive branch current I_L from the data in the table. The horizontal (*x*) axis should be frequency, and the vertical (*y*) axis should be current. Label the axes and each curve. Indicate the resonant frequency with a dotted line.

6. Is the parallel *RLC* circuit of Part B resistive, capacitive, or inductive? Explain your answer.

_____

_____

_____

_____

_____

EXPERIMENT

54

LOW-PASS AND HIGH-PASS FILTERS

OBJECTIVES

1. To determine experimentally the frequency response of a low-pass filter
2. To determine experimentally the frequency response of a high-pass filter

BASIC INFORMATION

Frequency Filters

Electronic signals are often made up of more than one frequency. For example, the ordinary AM radio signal consists of a radio-frequency component containing the voice, music, and sound frequencies we hear when we turn on a radio. The carrier is a unique frequency assigned to the radio station by the Federal Communications Commission (FCC). It is combined with the audio component by a process called *modulation*. The modulated signal is then transmitted over the air by the radio station. The radio receiver contains circuits that can select a particular carrier frequency and reject all others. Once the modulated signal is in the circuitry of the radio, the carrier is separated from the audio component and eliminated, and only the audio component is allowed

to pass through to the speakers or earphones. The process of selecting and eliminating or rejecting particular frequencies or ranges of frequencies is called *filtering*. Filters are employed in such devices as FM and television receivers, security systems, computers, and motor controls.

There are different types of filters. Some are highly selective and permit a single frequency or a very narrow band of frequencies through, reducing or eliminating all others. These are called *narrowband filters*. Others, called *wideband filters,* pass a wide band of frequencies while rejecting all others. Filters are also classified as low-pass or high-pass. We examine the characteristics of the latter two types in this experiment.

High-Pass Filter

Filter circuits consist of capacitors, inductors, and resistors, in combination. An *LC* series resonant circuit is one example of a frequency filter, for these highly selective circuits favor one frequency, the resonant frequency, over all others.

A capacitor theoretically offers an infinite reactance to a zero-frequency signal—that is, to direct current. Therefore, if a capacitor is placed in series with a load resistance, it will pass the ac component of a complex signal but block the dc component of that signal. Figure 54–1(a) shows a 6-V p-p ac

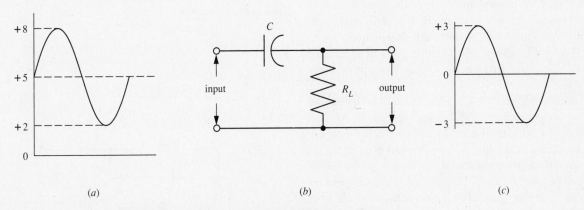

Figure 54–1. The dc blocking effect of a capacitor. (a) A 6-V p-p signal imposed on a +5-V dc axis. The resulting combined signal varies between +8 and +2 V. (b) The combined dc and ac signal is connected as shown to the input terminals. (c) The capacitor C blocks the dc component and permits only the ac component to pass through to the output. The signal across R_L is a pure 3-V ac voltage.

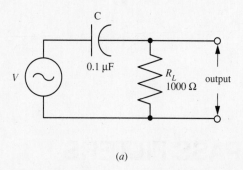

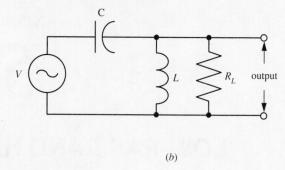

(a)

(b)

Figure 54–2. High-pass filters. (*a*) Capacitor *C* blocks the lower-frequency and passes the higher-frequency components of *V* through to R_L. (*b*) Adding inductor *L* modifies the filtering action of the circuit.

signal combined with 5 V dc. The resulting combined signal varies between +8 and +2 V. Figure 54–1(*b*) shows how a coupling capacitor *C* is connected in the circuit to block the dc but permit the ac component through. Figure 54–1(*c*) shows the pure ac sine wave across R_L. Note that the +5 V dc has been filtered by capacitor *C*.

The reactance of a capacitor varies inversely with frequency. This characteristic is used to pass some frequencies but reject others. In the circuit of Figure 54–2(*a*), the level of ac voltage V_R coupled by capacitor *C* to resistor *R* is

$$V_R = V \cos\theta = V \times \frac{R}{Z}$$

and

$$\theta = \tan^{-1}\left(\frac{X_C}{R}\right) \qquad \textbf{(54–1)}$$

The angle θ depends on the relative values of *C* and *R* and on the frequency *f* of the applied signal source. Assume there are three frequencies in the complex signal *V*, namely, 159 Hz, 1590 Hz, and 15,900 Hz. The reactance X_C of *C* at each of these frequencies is, respectively, 10,000, 1000, and 100 Ω. The voltage V_R delivered to *R* at 159 Hz is less than 10 percent of *V*. At 1590 Hz (the frequency at which $X_C = R$) more than 70 percent of *V* is delivered to *R*, and at 15,900 Hz more than 99 percent of *V* is delivered to *R*. Thus it is evident *C* will permit the higher frequencies to reach *R* with a minimum of reduction but will reduce, or attenuate, the lower frequencies. This is one type of high-pass filter.

The circuit of Figure 54–2(*a*) can be modified by adding an inductor *L* in parallel with the load resistor R_L to form another high-pass filter [Figure 54–2(*b*)]. The inductor has a low reactance at the low frequencies and effectively reduces the output voltage across R_L at frequencies for which X_L is less than $R_L/10$ (approximately). At higher frequencies the value of X_L increases. For frequencies at which X_L is greater than $10R_L$, the parallel impedance of X_L and R_L approaches R_L, stabilizing the load at the value R_L. The inclusion of inductor *L* modifies the response characteristics of the circuit, which is still a high-pass filter.

Low-Pass Filter

The reactance of an inductor varies directly with frequency. In Figure 54–3 this characteristic of *L* is used to pass the low frequencies through to R_L. If the input contains high and low frequencies, the high frequencies will be rejected, or blocked, from passing through to R_L. As in the case of the *RC* circuit in Figure 54–2, the *RL* circuit in Figure 54–3 is a voltage divider. The amplitude of the voltage delivered by *V* to R_L will depend on the inductance of *L*, the resistance of R_L, and the frequency of *V*. As frequency increases, X_L increases, and the voltage across R_L decreases. As frequency decreases, X_L decreases and the voltage across R_L increases. This is an example of a low-pass filter.

An example of another low-pass filter is shown in Figure 54–4(*a*). The addition of capacitor *C* in this circuit increases the filtering of higher frequencies. As the frequency *f* of the input signal increases, not only does X_L increase, reducing the output voltage V_{R_L}, but X_C decreases, thus increasing the amount of high-frequency signal bypassed by the capacitor. This, in effect, reduces the impedance of the output circuit, thus further reducing V_{R_L}.

If *L* is replaced by a resistor R_1, as in Figure 54–4(*b*), the result will still be a low-pass filter because of the bypass action of capacitor *C*.

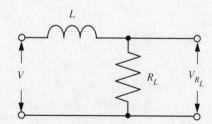

Figure 54–3. Low-pass filter. Inductor *L* blocks the higher frequencies and passes the lower frequencies through to R_L.

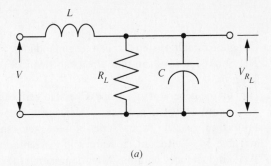

(a)

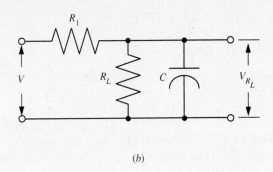

(b)

Figure 54–4. Other examples of low-pass filters.

Cutoff Frequency

Although a filter will pass or reject specific frequencies, it is not often necessary to eliminate or reduce the signal of the unwanted frequencies to zero. A value of frequency, called the *cutoff frequency*, is used to specify the point at which a frequency has been reduced, or attenuated, to a particular value. The cutoff frequency is defined here as the frequency whose output voltage is 70.7 percent of the maximum output signal. This, in effect, means the cutoff frequency has been attenuated by 29.3 percent of maximum.

SUMMARY

1. Complex electronic signals contain many frequency components. Filters are used to separate these frequency components, attenuating those that are unwanted and passing on those that are required.
2. Filters consist of combinations of capacitors, inductors, and resistors.
3. The circuits of Figure 54–2(a) and (b) show examples of high-pass filters. These circuits are basically voltage dividers. With increasing frequency, X_C decreases, and the output voltage across R_L increases.
4. The circuits of Figures 54–3 and 54–4 are low-pass filters. Here the voltage-divider action is affected by the increase in X_L with an increase in frequency. Accordingly, the output voltage across R_L is greater for low frequencies than it is for high frequencies. High frequencies are attenuated, and low frequencies are passed on to the output.
5. The terms low and high frequencies are relative. The frequency-response characteristic of a circuit depends on the circuit parameters—that is, on the values of L, C, and R, and on the circuit configuration. In all cases the value of output voltage for each frequency component can be calculated using ac circuit formulas.
6. The cutoff frequency is defined as the frequency whose output voltage is 70.7 percent of the maximum output voltage.

SELF-TEST

Check your understanding by answering the following questions:

1. Circuits that attenuate some frequencies and pass others are called _____ .
2. DC voltages can be separated from ac signals by means of a _____ .
3. In the high-pass circuit of Figure 54–2(a), the frequency at which the output voltage across R_L is 70.7 percent of maximum is the frequency at which $X_C =$ _____ Ω.
4. The cutoff frequency in the filter of Figure 54–2(a) is _____ Hz.
5. In the low-pass filter of Figure 54–3, the output voltage across R_L, V_{R_L}, is $V \times$ _____ $= V \times$ _____ .
6. In the circuit of Figure 54–3, if $R = 500 \ \Omega$ and $L = 2$ H, all frequencies higher than _____ Hz will be attenuated more than 29.3 percent; that is, the cutoff frequency is _____ Hz.

MATERIALS REQUIRED

Power Supply:
■ AF sine-wave generator

Instruments:
■ Oscilloscope
■ EVM or DMM

Resistor:
■ 1   10,000-Ω, $\frac{1}{2}$-W, 5%

Capacitor:
■ 1   0.1-μF

Inductor:
■ Large iron-core choke, 8 or 9 H (see Experiment 36 for type and specifications)

PROCEDURE

A. High-Pass Filter

A1. With the generator and oscilloscope **off**, connect the circuit of Figure 54–5(*a*).

A2. Turn **on** the generator and oscilloscope. Set the oscilloscope to measure rms voltage. Adjust the generator output for 7 V rms and 60 Hz. Maintain the 7-V rms output throughout the experiment; check from time to time and adjust the voltage as necessary.

A3. Examine Table 54–1 (p. 391). Measure the voltage across the resistor V_R for the series of frequencies shown from 60 Hz to 3000 Hz. Record each value in Table 54–1 in the "Without Inductance" column. After completing the measurements, turn **off** the generator and oscilloscope.

A4. Using the circuit of step A1, connect an inductor (8- or 9-H choke) across the resistor as shown in Figure 54–5(*b*).

A5. Turn **on** the generator and scope. Check to verify that V = 7 V rms; adjust if necessary. Measure the voltage across the resistor V_R for the series of frequencies shown in Table 54–1. Record the values in the table in the "With Inductance" column. After completing the measurements, turn off the generator and oscilloscope. Disconnect the inductor and capacitor.

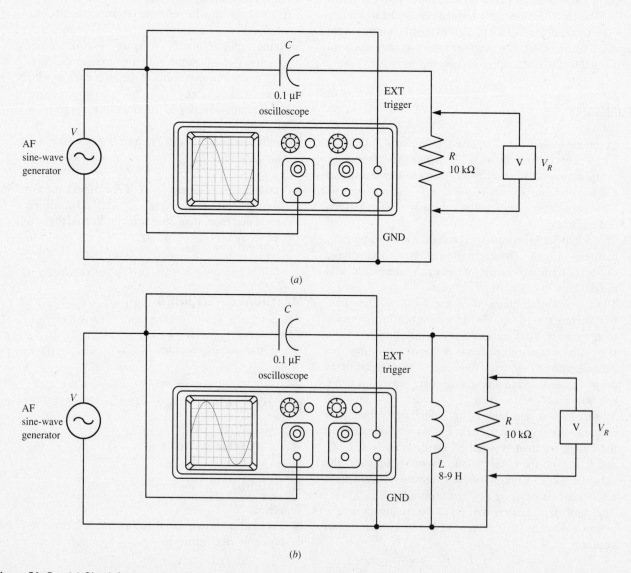

Figure 54–5. (*a*) Circuit for procedure step A1. (*b*) Circuit for procedure step A4.

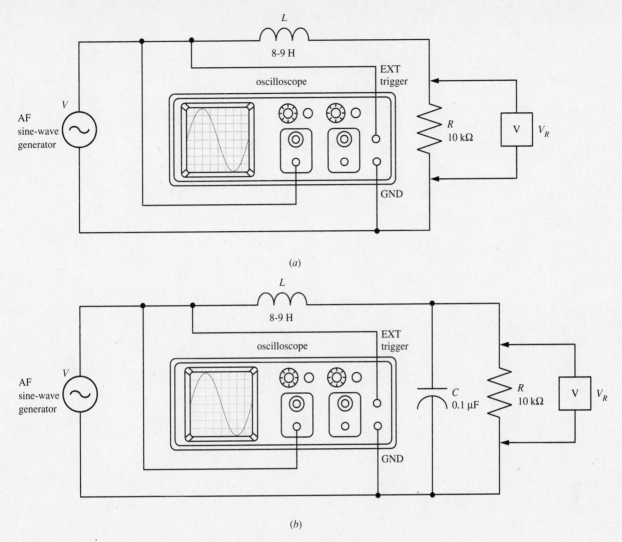

Figure 54–6. (a) Circuit for procedure step B1. (b) Circuit for procedure step B4.

B. Low-Pass Filter

B1. With the generator and oscilloscope **off**, connect the circuit of Figure 54–6(a).

B2. Turn on the generator and oscilloscope. Set the scope to measure rms voltage. Adjust the generator for an output of 20 Hz, 7 V rms. Maintain this voltage throughout the experiment; check it from time to time and adjust if necessary.

B3. Examine Table 54–2 (p. 391). Measure the voltage across the resistor V_R for the series of frequencies from 20 Hz to 2000 Hz. Record the values in Table 54–2 in the "Without Capacitance" column. After completing

the measurements, turn **off** the generator and oscilloscope.

B4. Using the circuit from step B1, connect a 0.1-μF capacitor across the resistor as in Figure 54–6(b).

B5. Turn **on** the generator and scope. Check to verify that V = 7 V rms; adjust if necessary. Measure the voltage across the resistor V_R for the series of frequencies shown in Table 54–2. Record the values in the table in the "With Capacitance" column. After completing the measurements, turn off the generator and oscilloscope and disconnect the circuit.

ANSWERS TO SELF-TEST

1. filters
2. blocking capacitor
3. 1000
4. 1592
5. R_L/Z; cos θ
6. 39.8; 39.8

TABLE 54–1. High-Pass Filter

Frequency, Hz	Applied Voltage V, V	Voltage across Resistor V_R, V	
		Without Inductance	With Inductance
60	7		
80	7		
100	7		
120	7		
140	7		
160	7		
180	7		
200	7		
300	7		
500	7		
1000	7		
2000	7		
3000	7		

TABLE 54–2. Low-Pass Filter

Frequency, Hz	Applied Voltage V, V	Voltage across Resistor V_R, V	
		Without Capacitance	With Capacitance
20	7		
30	7		
40	7		
50	7		
60	7		
80	7		
100	7		
150	7		
200	7		
300	7		
500	7		
1000	7		
2000	7		

QUESTIONS

1. Explain, in your own words, the difference between low-pass and high-pass filters. Your explanation should include a discussion of the components used in each type of filter and their effect on the filtering action.

_____

_____

_____

_____

_____

_____

_____

_____

_____

_____

2. On a separate sheet of $8\frac{1}{2} \times 11$ graph paper, plot a graph of frequency versus voltage across the resistor V_R for the circuit without inductance, using your data in Table 54–1. On the same set of axes, use data from the table to plot a graph of frequency versus voltage across the resistor for the circuit with inductance. Frequency should be the horizontal (x) axis, and voltage should be the vertical (y) axis. Use a different color or style of line for each curve. Indicate the cut-off frequency with a dotted vertical line.

3. Refer to your data in Table 54–1 and your graph for Question 2. Calculate the cutoff frequency for the circuit without inductance. Compare the calculated value with the indicated experimental value. Explain any discrepancy.

_____

_____

4. On a separate sheet of $8\frac{1}{2} \times 11$ graph paper (not the sheet used for Question 2), plot a graph of frequency versus voltage across the resistor V_R for the circuit without capacitance, using your data in Table 54–2. On the same set of axes, use data from the table to plot a graph of frequency versus voltage across the resistor for the circuit with capacitance. Frequency should be the horizontal (x) axis and voltage should be the vertical (y) axis. Use a different color or style of line for each curve. Indicate the cut-off frequency with a dotted vertical line.

5. Refer to your data in Table 54–2 and your graph for Question 4. Calculate the frequency at which the inductive reactance equals the resistance in the circuit without capacitance. Is this the same frequency determined from your data? Explain any discrepancy.

_____

_____

6. Discuss the effect on the frequency response of adding the inductance to the filter circuit of Figure 54–5. Refer to specific experimental data.

_____

_____

_____

_____

_____

_____

_____

_____

_____

_____

_____

_____

_____

_____

7. Discuss the effect on the frequency response of adding the capacitance to the filter circuit of Figure 54–6. Refer to specific experimental data.

_____

_____

_____

_____

_____

_____

_____

_____

_____

_____

_____

_____

_____

55

PHASE-SHIFTING CIRCUITS

OBJECTIVES

1. To determine experimentally the amount the phase angle can be varied in a circuit containing a fixed C in series with a variable, R
2. To determine experimentally the range of phase-shifting in a bridge circuit

BASIC INFORMATION

Series *RC* Phase-Shifting Circuit

The phase angle of a circuit is a function of the L, C, and R of the circuit. By varying the values of these components, the phase angle can be varied, or shifted. Phase-shifting circuits used for electronic control are an interesting application of ac circuit theory. In its simplest form, a phase-shifting circuit may consist of a capacitor in series with a variable resistor, as in Figure 55–1. The input voltage, V_{in}, is applied across the series circuit points FH, and the output voltage V_C is taken from the capacitor across points GH. As the value of R is varied, the phase of V_C varies with respect to V_{in}.

A phasor diagram of the circuit will clarify the phase relationships and range of phase variation as R is varied from maximum to minimum resistance. Figure 55–2(*a*) (p. 396) shows the phase relationships in the circuit for R greater than X_C. The applied voltage V_{in} is the reference phasor (FH).

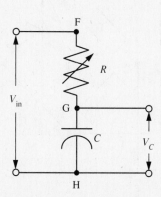

Figure 55–1. An *RC* phase-shifting circuit.

Since this is a capacitive circuit, current I leads the applied voltage by some angle θ. In the resistor R, voltage and current are in phase. The voltage across R, V_R, is drawn on the phasor I. The voltage V_C across C lags I by 90°. If a line HG is drawn from H perpendicular to V_R, HG_1 will represent the phasor V_C, and FG_1 will be the phasor V_R. From the right triangle FG_1H we see that

$$V_{in} = \sqrt{V_R^2 + V_C^2} \qquad (55\text{--}1)$$

This is consistent with our knowledge of the voltage relationships in a series *RC* circuit.

Angle $\varnothing_1$ is the angle by which the voltage V_C lags the applied voltage. From triangle FG_1H it is evident that

$$\tan \varnothing_1 = \frac{V_R}{V_C} = \frac{R}{X_C} \qquad (55\text{--}2)$$

Since R is greater than X_C, V_R is greater than V_C. Therefore, the value of V_R/V_C is greater than 1. If $\tan \varnothing_1$ is greater than 1, the value of $\varnothing_1$ must be greater than 45°. As R increases, $\varnothing_1$ increases. If we set a limit on the maximum value of R at $R = 10X_C$, then

$$\tan \varnothing_1 = \frac{V_R}{V_C} = \frac{10}{1}$$

and

$$\varnothing_1 = \tan^{-1}(10) = 84.3°$$

The maximum value of $\varnothing_1$ is 90° when R becomes infinitely large.

The phasor diagram of Figure 55–2(*b*) shows relationships resulting from $R = X_C$. Since $V_R = V_C$,

$$\tan \varnothing_2 = \frac{V_R}{V_C} = 1$$

$$\varnothing_2 = \tan^{-1}(1) = 45°$$

Figure 55–2(*c*) shows that $\varnothing_3$ becomes less than 45° when R becomes less than X_C. The limiting value of $\varnothing_3$ is 0°, which occurs when $R = 0$. What this means, simply, is that when there is no resistance in the circuit, the applied voltage and the voltage across C are the same and are in phase.

The maximum range of phase shift with the *RC* circuit of Figure 55–1 is from 0° up to but not including 90°.

The value of the voltage across C decreases as the phase angle between V_C and V increases.

Another characteristic can be seen from Figure 55–2. Since the angle FGH is always a right angle, point G must lie along

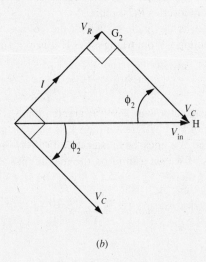

(a)

(b)

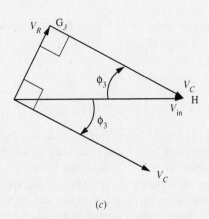

(c)

Figure 55–2. Phasor diagrams for *RC* phase-shifting circuits.

a semicircle whose diameter is FH. As R varies continuously from infinity to the value $R = X_C$, and finally to the value $R = 0$, G moves in a counterclockwise direction along the semicircle from point H to G_2 and finally to point F.

Phase-Shift Bridge Circuit

Angle ø can be shifted by almost 180° by means of a bridge-type phase-shift circuit, as in Figure 55–3. The applied voltage V_{in} appears across points FH, with point P as the center tap of the power source. That is,

$$V_{FP} = V_{PH} = \frac{V_{in}}{2}$$

The output voltage, V_o, is taken from the points PG. As R is varied from maximum resistance to 0 Ω, the phase angle ø of V_o varies from some small value up to 180°. The angle ø is a leading angle relative to the voltage FH.

The phasor diagram of Figure 55–4 (p. 397) shows the phase-shifting characteristics of the circuit in Figure 55–3. The reference phasor, FH, is the applied voltage V_{in}. The center tap is represented by P. Since current leads voltage in a capacitive circuit, the current phasor I is shown leading V_{in} by the angle θ_1. The current in R and the voltage across R are in phase and are therefore shown lying along the same line FG. The current I through the capacitor leads the voltage across the capacitor by 90°. This condition is shown as the voltage phasor V_C (GH). As R is varied, phasor PG, the voltage V_o, swings in a semicircle arc with P as the center and the phase angle between V_o and V_{in}. Point P is the center of the semicircle with the radius PF = PG = PH. The effect of R on this circuit can be shown from the relationships of the phasors in Figure 55–4. Since PG = PF, $\theta_1 = \theta_2$, and ø = $\theta_1 + \theta_2 = 2\theta_1$, or

$$\theta_1 = \frac{\text{ø}}{2} \qquad \qquad \textbf{(55–3)}$$

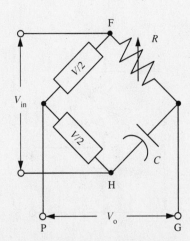

Figure 55–3. Phase-shifting *RC* bridge cirrcuit.

From right triangle FGH,

$$\tan \theta_1 = \frac{V_C}{V_R} = \frac{X_C}{R} \qquad (55\text{-}4)$$

Substituting for θ in formula (55-4), we obtain

$$\tan \frac{\phi}{2} = \frac{X_C}{R}$$

When $R = 0$,

$$\frac{X_C}{R} = \infty$$

$$\tan \frac{\phi}{2} = \infty$$

$$\frac{\phi}{2} = \tan^{-1}(\infty) = 90°$$

$$\phi = 180°$$

When $X_C = R$,

$$\tan \frac{\phi}{2} = \frac{X_C}{R} = 1$$

$$\frac{\phi}{2} = \tan^{-1}(1)$$

$$\frac{\phi}{2} = 45°$$

$$\phi = 90°$$

When R becomes very large, say, $R = 100 X_C$,

$$\tan \frac{\phi}{2} = \frac{X_C}{R} = \frac{1}{100}$$

$$\frac{\phi}{2} = \tan^{-1}(0.010)$$

$$\frac{\phi}{2} = 0.573°$$

$$\phi = 1.146°$$

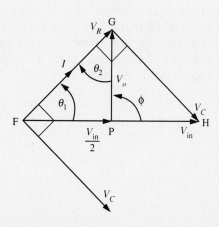

Figure 55–4. Phasor diagram showing the phase-shifting effects of the *RC* bridge circuit of Figure 55–3.

From Figure 55–4 the output voltage V_o is seen to be leading the applied voltage V_{in}. As R varies from a very large value $(100 X_C)$ to $0 \ \Omega$, the angle ϕ varies from 1.146° to 180°.

SUMMARY

1. In the simple *RC* circuit of Figure 55–1, as R is varied from $0 \ \Omega$ to its maximum value, the phase angle between the output voltage V_C and the reference phasor V_{in} varies from 0° up to some maximum value determined by the size of R and the value of X_C. The value in degrees of this phase angle ϕ is given by the formula

$$\phi = \tan^{-1}\left(\frac{R}{X_C}\right)$$

2. The angle ϕ is the angle by which the output voltage V_C lags applied voltage V_{in}.

3. In the simple *RC* phase-shifting circuit of Figure 55–1, $\phi = 0°$ when $R = 0 \ \Omega$; when $R = 10 X_C$, $\phi = 84.3°$. The angle ϕ approaches, but never equals, 90°, because R cannot be made infinitely large. Thus we can say that the range of phase variation in the simple *RC* circuit containing a fixed value C and a variable R is approximately 90°.

4. The value of the output voltage V_C in the simple *RC* phase-shifting circuit of Figure 55–1 decreases as R increases in resistance.

5. In the phase-shift bridge circuit of Figure 55–3, the range of variation of the phase-shift angle is increased up to but not including 180°.

6. In the phase-shift bridge circuit, the output voltage V_o is taken from the junction of R and C (point G) and from the center point P of a low-impedance circuit across the applied voltage V_{in}.

7. The phase-shift angle ϕ in Figure 55–3 is calculated from the formulas

$$\tan \frac{\phi}{2} = \frac{X_C}{R}$$

and

$$\phi = 2 \tan^{-1}\left(\frac{X_C}{R}\right)$$

8. When $R = 0 \ \Omega$, $\phi = 180°$. When $R = 100 X_C$, $\phi = 1.146°$. The larger the value of R with relation to X_C, the smaller is the phase-shift angle ϕ.

9. The amplitude of the output voltage V_o in the phase-shift bridge circuit remains constant over the range of variation of R. The output voltage is

$$V_o = \frac{V_{in}}{2}$$

10. Voltage V_o can be considered either leading or lagging with respect to the applied voltage V_{in}. In Figure 55–3, if V_o is taken from P to G, relative to V_{in}, V_o leads V_{in}. If V_o is taken from G to P, relative to V_{in}, V_o lags V_{in}.

Check your understanding by answering the following questions:

Answer questions 1 through 4 relative to Figure 55–1.

1. When $R = X_C$, the phase angle between V_C and V_R is _____ °.
2. For the circuit in question 1, V_C taken from G to H _____ (leads/lags) V_{in} by _____ °.
3. The maximum range of phase variation between V_C and V_{in} approaches but never equals _____ °.
4. (True/False) The value of the output voltage V_C is constant. _____

Answer questions 5 through 8 relative to Figure 55–3.

5. When $R = X_C$, the phase angle between V_o taken from G to P relative to V_{in} is _____ °.
6. For the circuit in question 5, V_o taken from P to G _____ (leads/lags) V_{in} by _____ °.
7. The maximum phase-shift between V_o taken from G to P and V_{in} is approximately _____ °.

8. If $V_{in} = 10$ V, the output voltage V_o taken from G to P is _____ V.

MATERIALS REQUIRED

Power Supplies:
■ Isolation transformer
■ Variable-voltage autotransformer (Variac or equivalent)

Instrument:
■ Oscilloscope

Resistors:
■ 2   33-Ω, ½-W, 5%
■ 1   500-kΩ, 2-W potentiometer

Capacitor:
■ 1   0.1-μF

Miscellaneous:
■ SPST switch
■ Polarized line cord with on-off switch and fuse

PROCEDURE

A. Series *RC* Phase-Shift Circuit

A1. With the line cord unplugged, line switch **off**, and switch S_1 **open**, connect the circuit of Figure 55–5. Set the autotransformer to its lowest output voltage. Connect the ground lead of the oscilloscope to point H of the circuit and connect the other oscilloscope input lead to point F. Also connect the lead from the external trigger terminal to point F.

A2. Turn **on** the oscilloscope and adjust it for voltage measurements. Plug in the line cord; line switch **on**. **Close** switch S_1. Increase the output of the autotransformer to about 6 V p-p. Adjust the oscilloscope to display three cycles, centered vertically and horizontally on the screen. The center cycle should be oriented as shown in row 1 of Table 55–1 (p. 401). The amplitude and position of this waveform (in relation to the display grid) will be used as the reference waveform for the next steps in the procedure. Once you have set the amplitude and position of the reference waveform, do not readjust any of the oscilloscope controls. Record the peak-to-peak voltage of this waveform in row 1 of Table 55–1.

A3. Disconnect the input lead to point F and connect it to point G. Leave the ground lead connected to H as is. Adjust the potentiometer to 0 Ω ($R_3 = 0$). Observe carefully the amplitude of the waveform and its new position. A shift of the waveform to the left of the reference wave indicates a leading angle with respect to the reference wave; a shift to the right indicates a lagging angle with respect to the reference waveform. Sketch the new waveform as accurately as possible in row 2 of Table 55–1. Measure and record the amplitude of the new wave and the number of degrees of shift from the reference waveform.

A4. Adjust the potentiometer (R_3) until the wave shifts 45° to the right of the reference wave. Measure the amplitude and record it in row 3 of Table 55–1.

A5. Adjust R_3 until the wave shifts 90° to the right. Measure and record the amplitude of the wave in row 4 of Table 55–1. Retain the circuit for Part B.

B. Bridge Phase-Shifting Circuit

B1. Disconnect the ground lead of the oscilloscope from point H and connect it to point P. Disconnect the input lead to point G and connect it to point F. The oscilloscope is now connected across R_1. Adjust the oscilloscope as in step A2 to establish a reference waveform. The autotransformer need not be adjusted. Measure and record the peak-to-peak amplitude in row 1 of Table 55−2 (p. 401).

B2. Disconnect the oscilloscope lead to point F and connect it to point G. The oscilloscope is now connected across P and G (P is still connected to the ground lead of the oscilloscope) and will measure voltage V_{GP} (G to ground). Adjust the potentiometer so that $R_3 = 0$. Sketch the new waveform in row 2 of Table 55−2, showing the proper angle of shift and amplitude. Record the amplitude and angle in row 2 of Table 55−2 in the "V_{GP}" column.

B3. Adjust R_3 until the waveform shifts 90° to the right. Sketch the waveform, and measure and record its amplitude in row 3 of Table 55−2 in the "V_{GP}" column.

B4. Adjust R_3 until the waveform shifts 180° to the right. Sketch the waveform, and measure and record the amplitude in row 4 of Table 55−2 in the "V_{GP}" column.

B5. Reverse the oscilloscope leads so that the ground lead is connected to G and the other oscilloscope lead is connected to P. The oscilloscope will thus measure voltage V_{PG}. With R_3 in the same position as in step B4, sketch the waveform as it now appears on the same graph as step B4 (row 4). Use the same axes as in step B4. Use a line of a different color or style to distinguish the waveform from that of step B4. Use the same style in steps B6 and B7 that follow. Record the phase-shift angle in row 4 of Table 55−2 in the "V_{PG}" column.

B6. Adjust R_3 to produce a phase shift of 90° to the left. Sketch the waveform in row 3 on the same axes as the waveform from step B3. Measure and record the phase-shift angle in row 3 of Table 55−2 in the "V_{PG}" column.

B7. Adjust R_3 to 0. Sketch the waveform in row 2 of Table 55−2. Use the same axes as in step B2. Measure and record the phase-shift angle in row 2 of Table 55−2 in the "V_{PG}" column. After all measurements have been completed, **open** S_1; turn **off** the scope; turn line switch **off**; and unplug the line cord.

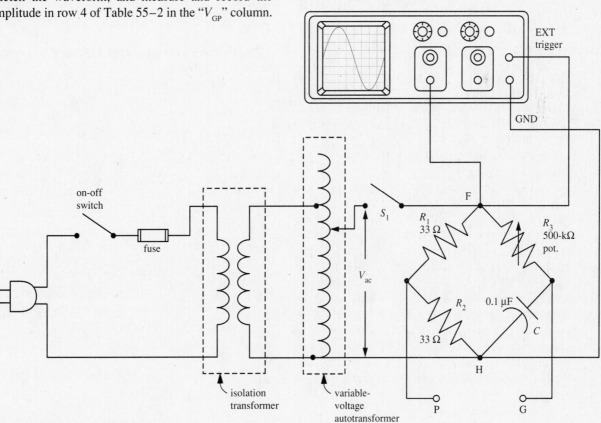

Figure 55−5. Circuit for procedure step A1.

ANSWERS TO SELF-TEST

1. 45
2. lags; 45
3. 90
4. false
5. 90
6. leads; 90
7. 180
8. 5

Experiment 55   Name _____ Date _____

TABLE 55–1. Series *RC* Phase-Shift Circuit

Row	Waveform	V p-p	Phase-Shift, degrees
1			0
2			
3			45
4			90

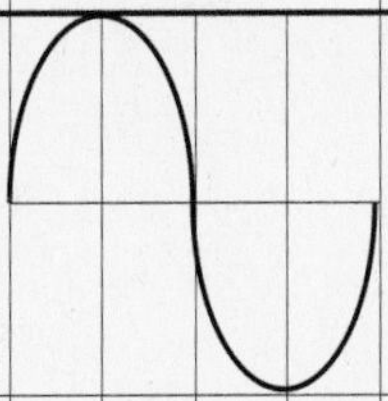

TABLE 55–2. Phase-Shift Bridge Circuit

Row	Waveform	V p-p	Phase-Shift, V_{GP}, deg.	Phase-Shift, V_{PG}, deg.
1	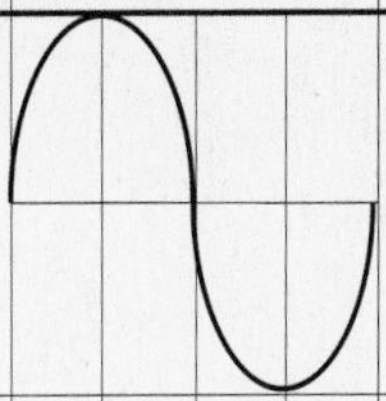		0	
2				
3			90	
4			180	

QUESTIONS

1. Explain, in your own words, the effect of varying resistance on the phase angle of a
   series *RC* circuit with a fixed capacitance value.

   _____

   _____

   _____

   _____

   _____

2. Explain, in your own words, two basic differences between the output of a simple
   series *RC* phase-shift circuit and an *RC* bridge phase-shift circuit.

   _____

   _____

   _____

   _____

   _____

3. Calculate the phase-shift angle for the minimum resistance ($R = 0\ \Omega$) and maximum resistance ($R = 500\ k\Omega$) used in Part A. Use Formula (55–2). Compare your answer with your data and waveforms in Table 55–1. State whether the phase-shift angles are leading or lagging.

4. Calculate the phase-shift angle for the minimum and maximum resistances used in Part B. Use Formula (55–4). Compare your answer with your data and waveforms in Table 55–2. State whether the phase-shift angles are leading or lagging.

5. What is the phase relationship between the two waveforms in row 2 of Table 55–2? Similarly, what is the relationship between the pairs of waveforms in rows 3 and 4 of Table 55–2? Are the relationships as expected? Explain.

_____

_____

_____

_____

_____

_____

_____

_____

_____

EXPERIMENT
56

NONLINEAR RESISTORS—THERMISTORS

OBJECTIVES

1. To observe the self-heating effect of current on the resistance of a thermistor
2. To determine experimentally the variation in resistance with time of current in a thermistor

BASIC INFORMATION

In all previous experiments the components that made up the circuits were assumed to have constant values. In particular, the values of resistors were assumed to remain unchanged despite changes in voltage and current. For the most part this assumption was acceptable for the purpose of the experiment.

However, there is a class of components known as *thermistors* whose resistance value changes with changes in operating temperature. As its name implies, a thermistor is a thermally sensitive resistor. It is part of a large family of materials called *semiconductors*. A semiconductor is a substance whose resistivity lies somewhere between that of a conductor and that of an insulator.

The thermistor is a two-terminal component that may be used in either ac or dc circuits. It is manufactured in a number of shapes, such as beads, rods, disks, washers, and flakes. Thermistors are often part of sensor devices used in temperature control, flow measurement, voltage regulation, and such.

Temperature Characteristic

A fundamental characteristic of the thermistor is its change of resistance with changes in temperature. A number of formulas have been developed to approximate the manner in which the resistance varies with temperature. One such approximate formula is

$$R = R_o \times e^k$$

and

$$k = B \left(\frac{1}{T} - \frac{1}{T_0} \right) \qquad \textbf{(56–1)}$$

where

R = resistance at any temperature T, in degrees kelvin (K)

R_0 = resistance at reference temperature T_0 (K)

B = a constant whose value depends on the thermistor material (determined from measurements at $0°$ and at $+50°C$)

e = 2.7183 (base of natural logarithms)

Formula (56–1) shows that the resistance variation is nonlinear. Thermistors are frequently referred to as *nonlinear resistors*.

Thermistors exhibit negative temperature coefficients (NTC) and positive temperature coefficients (PTC). The resistance of NTC thermistors *decreases* as their temperature *rises* and *increases* as their temperature *falls*. The resistance of PTC thermistors *increases* as their temperature *rises* and *decreases* as their temperature *falls*. Most materials used in circuits, such as copper conductors and the standard carbon composition and wirewound resistors, also exhibit the characteristic of a rising resistance when heated, although in most cases the effect is minimal. Therefore, their resistance can be considered relatively constant.

This discussion is concerned mainly with NTC thermistors. Figure 56–1 (p. 404) shows the variation in resistivity for three NTC thermistors, compared with the resistivity of platinum under similar temperature changes. Note the tremendous variation of specific resistance in thermistor 1, a change of 10,000,000 to 1 over a temperature range of 500°C, whereas that of platinum changes by a factor of less than 10 to 1.

Figure 56–2 (p. 404) shows the variation in resistance of another NTC thermistor. Here a range of 200°C causes a variation in resistance from 1000 Ω to approximately 2 Ω.

The resistance-temperature characteristic of thermistors has been utilized in electrical and electronic applications. In particular, thermistors are used as protective devices for electric-bulb filaments during the warmup period. They are used for temperature measurement, control, and compensation.

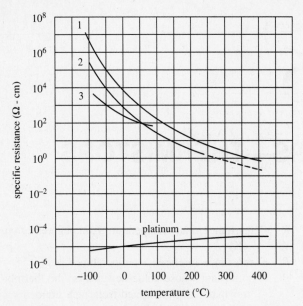

Figure 56-1. Temperature-specific resistance characteristics of three thermistors (labeled 1, 2, 3) compared with platinum.

Static Volt-Ampere Characteristic

There are secondary characteristics that arise from the relationship between temperature and resistance that thermistors exhibit. Among these is the static relationship between voltage and current. This secondary characteristic results from the self-heating that occurs when current flows through a thermistor. The static characteristic may be determined using the test circuit shown in Figure 56-3. In this circuit a rheostat R is used as a limiting resistor to vary the amount of current through the thermistor. As the resistance of R is decreased, the current I increases. As the resistance of R is increased, the current I decreases. A voltmeter is used to monitor the voltage across the thermistor for every value of current. Each time R is varied and a change in current is effected, enough time must be allowed for the voltage across the thermistor to reach a steady value. The resultant graph therefore reflects steady-state conditions of voltage versus current and is called the *static volt-ampere characteristic*.

The thermistor whose resistance-temperature characteristic is shown in Figure 56-2 has the static volt-ampere characteristic shown in Figure 56-4. A temperature of 25°C was the starting temperature for the tests from which the data for the graph were secured. The numbers on the graph—namely, 53, 73, 102, and so forth—indicate the thermistor temperature, measured after current had stabilized at the level shown on the graph. Thus, at 0.05 A, the voltage across the thermistor was 11 V, and the temperature was 53°C.

The graph for this thermistor shows that up to a temperature of 53°C, there is a linear relationship between voltage and current. That is, Ohm's law holds during this interval when only low current flow is involved. This is true because there is insufficient heat dissipated at this low level of current

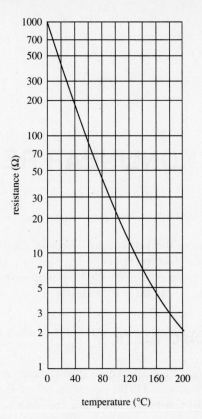

Figure 56-2. Temperature-resistance characteristic of a representative thermistor.

to affect the cold resistance of the thermistor. As current increases, however, the heat dissipated increases, thermistor temperature increases, and its resistance drops nonlinearly. The voltage across the thermistor now increases nonlinearly to a peak value V_M, designated the "self-heat" voltage. Beyond this value, the voltage across the thermistor drops nonlinearly with increasing current, as shown in Figure 56-4.

Applications based on the voltage-current characteristics of thermistors fall into four general categories, responding to changes in electrical parameters, in ambient temperatures, in dissipation (fluid and gaseous environments), and in radiation absorption. Devices falling into these categories include temperature alarm devices, pyrometers, flow meters, gas detectors, and microwave power regulators.

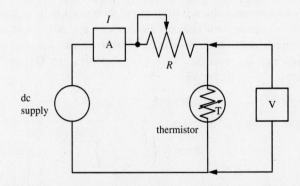

Figure 56-3. Circuit for determining the volt-ampere characteristics of a thermistor.

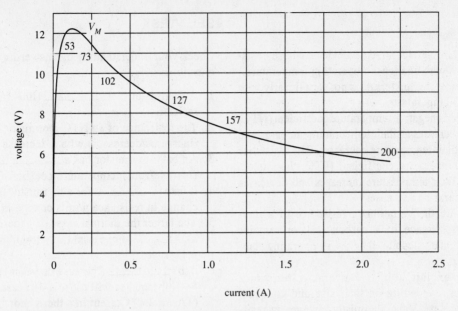

Figure 56–4. Static volt-ampere characteristics of a representative thermistor.

Dynamic Characteristic

A thermistor requires a certain time interval to react to a change in the external circuit. A change in current through a thermistor from one value to another does not result in an instantaneous change in temperature; hence, there is no instantaneous change in resistance. These changes do occur, but they require a definite time. The thermal mass of the thermistor element will determine the time interval. A small element will change temperature more rapidly than an element with larger mass.

Figure 56–5 shows a test circuit for determining the dynamic characteristic of a thermistor. The applied voltage V is an ac power supply whose output is adjustable, and R_L is the load resistance. A voltmeter set on ac volts is connected across the load R_L to monitor the voltage across R_L. When switch S is open, there is no voltage applied to the circuit. At the instant S is closed, the cold resistance of the thermistor and the value of the load resistor R_L constitute the voltage divider that determines the voltage across R_L. As current starts to flow, the temperature of the NTC thermistor increases, its

resistance decreases, and more of the source voltage appears across the load. Within a certain time interval, the circuit is stabilized. That is, the thermistor resistance is stabilized and the voltage across the load becomes constant.

Figure 56–6 is a graph showing the dynamic characteristic of a thermistor for two different values of supply voltage and load resistance. Note that with 48 V applied and a load resistance of 17 Ω, steady state is reached in approximately 75 s. With 115 V applied and a load of 50 Ω, steady state is reached in about 15 s.

The dynamic characteristic of a thermistor is utilized in audio devices, switching devices, voltage-surge protectors, and low-frequency negative-resistance devices.

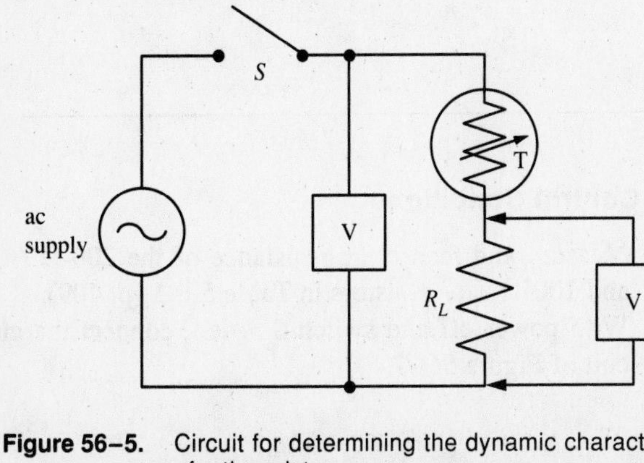

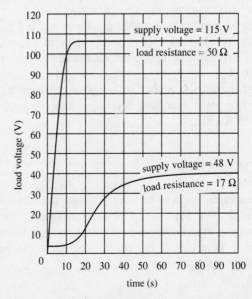

Figure 56–5. Circuit for determining the dynamic characteristic of a thermistor.

Figure 56–6. Dynamic characteristics of a representative thermistor.

SUMMARY

1. Thermistors are semiconductor devices whose resistance changes with changes in operating temperature.
2. The resistance of thermistors does not vary linearly with variations of temperature.
3. The resistance of negative temperature coefficient (NTC) thermistors decreases as their temperature increases and increases as their temperature decreases. The resistance of positive temperature coefficient (PTC) thermistors increases as their temperature increases and decreases as their temperature decreases.
4. Compared with the variation in resistance of a thermistor, the resistance of a carbon resistor remains relatively constant despite changes in operating temperature.
5. Thermistors exhibit unique temperature-resistance characteristics, depending on their size and construction. Thus, one particular thermistor changes in resistance from 1000 Ω to 2 Ω over a temperature change of 200°C. Another changes from 10,000,000 Ω to 1 Ω over a temperature variation from −100°C to +400°C.
6. Because of the self-heating effect of current in a thermistor, this device changes resistance with changes in current.
7. A secondary characteristic of a thermistor is the change in current it exhibits in an electric circuit. A graph of current versus voltage (Figure 56–4) of a thermistor, called a static volt-ampere characteristic, shows that current does *not* vary linearly with applied voltage.
8. A thermistor does not undergo instantaneous changes in resistance with changes in temperature. A certain time interval, determined by the thermal mass of the thermistor, is required to accomplish the resistance change. A thermistor with a smaller mass will change more rapidly than one with a larger mass.
9. The dynamic characteristic of a thermistor is a graph of the time it takes a thermistor to stabilize its resistance as a function of the applied voltage and current (Figure 56–6).
10. Thermistors are used in temperature alarm devices, pyrometers, switching devices, voltage-surge devices, and the like.

SELF-TEST

Check your understanding by answering the following questions:

1. A thermistor is a _____ (linear/nonlinear) resistor.
2. The resistance of an NTC thermistor _____ (increases/decreases) with a decrease in temperature.
3. An NTC thermistor has a _____ (positive/negative/zero) temperature coefficient.
4. It takes _____ for a thermistor to undergo a change in resistance with a change in temperature.
5. The larger the thermal mass of a thermistor, the _____ (more/less) time it will take to change its resistance.
6. A(n) _____ (increase/decrease) in current in an NTC thermistor will cause a decrease in its resistance.
7. (True/False) Current in a thermistor varies linearly with voltage across it. _____

MATERIALS REQUIRED

Power Supply:
■ Variable 0–15 V dc, regulated

Instruments:
■ DMM
■ VOM (analog or digital)

Resistors:
■ 1 100-Ω, 5-W
■ 1 200-Ω, 5-W

Miscellaneous:
■ Thermistor, 100 Ω cold resistance (25°C). Resistance ratio $R_{25°}/R_{50°} \approx 3$ similar to Panasonic ERT-D2FGL1O1S or equivalent

■ SPST switch
■ SPDT switch
■ Optional electronic timer with alarm

PROCEDURE

This experiment can be performed by two students working as a team. One student can be the timekeeper; the other student can read the voltmeter. If an electronic timer with an alarm is available, the entire experiment can be performed by one student.

A. Control Conditions

A1. Measure and record the resistance of the 200-Ω (R_1) and 100-Ω (R_2) resistors in Table 56–1 (p. 409).
A2. With power **off** and switch S_1 **open**, connect the circuit of Figure 56–7.

A3. Turn power **on**; **close** S_1. Switch S_2 should be in position 1. Adjust the power supply voltage to 15 V. Maintain this voltage throughout part A. Check the voltage from time to time and adjust if necessary.

A4. Once the power supply has been set at 15 V, measure the voltage across R_1 and record it in Table 56–1. Immediately throw S_2 to position 2 and measure the voltage across R_2. Record the value in Table 56–1. (Neglect changes in the polarity of the measurements.)

A5. Maintain power **on** for 5 min. After 5 min, measure the voltage across R_2 (S_2 should still be in position 2) and record the value in Table 56–1. Throw S_2 to position 1, measure the voltage across R_1, and record the value in Table 56–1. After measuring the voltages, **open** S_1 and turn **off** power. Disconnect R_1 from the circuit.

B. Dynamic Characteristic of a Thermistor

The timing element in this part is critical. Before performing the steps that follow, review them completely, including Table 56–2 (p. 409). Although the voltage measurements are important, changes in their polarity can be neglected and need not be recorded. Use a DMM for measuring voltages. An electronic timer with an alarm would be useful for accurate timing of the voltage measurement. Switch S_2 makes it easy to measure the voltage across the thermistor and R_2, but it requires reading the voltmeter quickly and switching immediately from one switch position to another to read the second voltage.

B1. Measure and record the cold resistance of the thermistor in Table 56–2. Also record the room temperature.

B2. Connect the thermistor in the circuit of Part A as in Figure 56–8. Power should be **off** and S_1 **open**. Switch S_2 should be in position 1.

B3. Turn power **on**; **close** S_1. Adjust the power supply voltage to 15 V. Maintain this voltage for the following three steps (through step B6).

B4. As soon as the supply is set at 15 V, measure the voltage across the thermistor V_T, and record it in Table 56–2. Immediately set S_2 to position 2 and measure the voltage across resistor R_2, V_2, and record it in Table 56–2. These will be the voltage measurements at $t = 0$.

B5. Repeat step B4 15 s later (that is, at $t = 15$ s), then 15 s later at $t = 30$ s, followed by measurements at $t = 45$ s, $t = 1$ min, $t = 2$ min, $t = 3$ min, $t = 4$ min, and finally at $t = 5$ min.

B6. After making the voltage measurements at $t = 5$ min, turn off the DMM, set S_2 to position 1, and change the function switch of the DMM to the ohmmeter scales. Open S_1 and quickly turn on the DMM to read the hot resistance of the thermistor. Record the value in Table 56–2. After making the measurement, change the function switch of DMM back to the voltage scales.

B7. Adjust the voltage of the power supply to 9 V and maintain this voltage for the balance of Part B. Check periodically and adjust if necessary.

B8. After the thermistor has had a chance to return to room temperature (at least 5 min after power is turned off), repeat steps B4 through B6 for the time sequence $t = 0$, 15 s, 30 s, 45 s, 1 min, 2 min, 3 min, 4 min, and 5 min. After $t = 5$ min, measure the hot resistance of the thermistor using the procedure of step B6. Record all voltage and resistance measurements in Table 56–2.

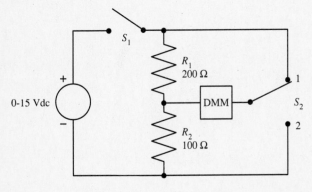

Figure 56–7. Circuit for control measurement; procedure step A2.

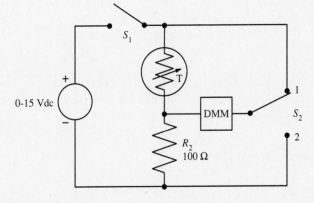

Figure 56–8. Circuit for procedure step B2.

ANSWERS TO SELF-TEST

1. nonlinear
2. increases
3. negative
4. time
5. more
6. increase
7. false

TABLE 56–1. Control-Circuit Measurements

	Resistance, Ω			Voltage, V	
	Rated Value	Measured Value		t = 0	t = 5 min
R_1	200		V_1		
R_2	100		V_2		

TABLE 56–2. Dynamic Characteristics of a Thermistor

	Thermistor Resistance R_T, Ω
Cold (room temp. = ___ °C)	
After 5 min at 15 V	
After 5 min at 9 V	

Power Supply, V		Time t								
		0	15 s	30 s	45 s	1 min	2 min	3 min	4 min	5 min
15	V_T									
	V_2									
9	V_T									
	V_2									

QUESTIONS

1. Explain, in your own words, the effect that current has on a thermistor over a period of time.

_____

_____

_____

_____

_____

2. Explain, in your own words, the difference between negative and positive temperature coefficients of a thermistor.

_____

_____

_____

_____

_____

3. Refer to your data in Table 56–1. Is the temperature coefficient of the resistor in this experiment negative, positive, or zero? Support your answer with specific references to your data.

_____

_____

_____

_____

4. Refer to your data in Table 56–2. Is the temperature coefficient of the thermistor in this experiment positive, negative, or zero? Support your answer with reference to specific data in the table.

_____

_____

_____

_____

_____

_____

5. What specification of the thermistor is an indicator of the maximum current that the thermistor should carry?

_____

_____

_____

6. Refer to your data in Table 56–2. Calculate the maximum current through the thermistor and the minimum current through the thermistor for the 15-V circuit.

57

NONLINEAR RESISTORS—VARISTORS (VDRs)

OBJECTIVES

1. To determine experimentally the volt-ampere characteristic of a varistor
2. To determine experimentally the relationship between voltage and resistance of a varistor

BASIC INFORMATION

In the previous experiment you discovered that there is a class of resistors called thermistors, whose resistance is dependent on temperature. There is another group of nonlinear resistors whose resistance is voltage-dependent. These voltage-sensitive resistors are called *varistors*, or VDRs. The current in a varistor varies as a power of the applied voltage, and for a particular varistor it may increase by many orders of magnitude when the applied voltage is only doubled.

The older popular varistors were made of silicon carbide mixed with a ceramic binder, fired at a high temperature to produce a solid fused blank, and then covered with a metallic coating. Electrical contact is made to this coating. One drawback to the use of silicon carbide varistors is that they are limited to low-power circuits.

Modern varistors are made from metal oxides and are usually referred to as MOVs. The most common oxide used is zinc oxide; these varistors are called ZNR varistors. Metal-oxide varistors are fabricated in much the same manner as silicon carbide varistors. Varying the thickness of the MOV makes different operating voltages possible.

One important advantage of the MOV over the older silicon carbide varistor is its higher resistance. Since the MOV is typically connected across the line or across a load, the high resistance means that a much lower current is drawn by the MOV in its normal, or standby, condition.

The MOV provides an economical and reliable means of protecting against voltage surges, or "spikes." These spikes are transient conditions that occur rapidly and without warning. The rapid response time (as little as 50 ns) of the MOV makes them excellent circuit and device protectors.

Volt-Ampere Characteristics

Because the MOV is now the most widely used varistor, only its characteristics are discussed. The MOV has a bilateral and symmetrical volt-ampere characteristic. This means that current direction through the MOV does not affect its behavior. Figure 57–1 shows the symmetrical nature of its volt-ampere curve. The MOV is not polarized and can be used in either ac or dc circuits. It protects equally well against positive as well as negative spikes.

Varistors possess negative temperature coefficients (as did the NTC thermistors in Experiment 56). This means the resistance of the MOV decreases as its temperature increases. During its conduction stage the voltage across the MOV remains relatively constant despite tremendous increases in current through the MOV. Although the MOV does not behave according to Ohm's law, at conduction the static resistance of the MOV can be defined as V/I. The dynamic, or changing resistance, of the MOV is defined as the incremental change of the instantaneous voltage across the MOV divided by the incremental change of the instantaneous current through the MOV. This can be written as

$$\text{Dynamic } R = \frac{dv}{di}$$

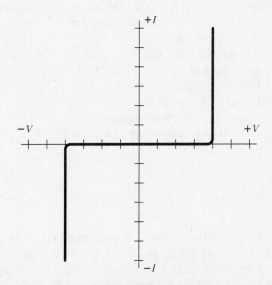

Figure 57–1. Symmetrical volt-ampere characteristic of an MOV.

Electrical Specifications and Ratings

Specifications for MOV devices are often given in terms of maximum ratings. The following ratings are typical of those given by MOV manufacturers.

Rated (applied) Voltage.   This is the maximum continuous voltage that can be applied across the varistor. For ac circuits this is usually given as an rms value.

Rated Peak Pulse Current—One Time.   This is the maximum peak current that can be applied for one 8×20-μs pulse with rated line voltage also applied. The 8×20-μs pulse unit is a standard that specifies that the pulse wave will reach its peak in 8 μs and fall to one-half its peak value in 20 μs.

Clamping Voltage.   Peak voltage across the MOV with a specified waveform (usually 8×20-μs pulse) and a peak pulse current applied.

Energy.   The maximum energy in joules capable of being handled by the MOV under transient conditions.

Power.   Maximum power dissipation in watts.

Operating ambient temperature.   A range of minimum and maximum temperatures under which the MOV can operate under specified ratings.

SUMMARY

1. Varistors are voltage-dependent resistors (VDRs). They are nonlinear devices whose resistance is dependent on the voltage across them.
2. Modern varistors are fabricated from metal oxides, usually zinc oxides. They are referred to as MOVs or, more specifically, as ZNR varistors.
3. Varistors are rated for their continuous power dissipation at a specified temperature and for the maximum voltage they can tolerate.
4. Other electrical specifications include the following:
   (a) Rated peak pulse current—1 time
   (b) Clamping voltage
   (c) Energy
   (d) Operating ambient temperature
5. Varistors have negative temperature coefficients.

6. During conduction the voltage across a varistor remains relatively constant despite tremendous changes in current through the varistor.
7. Varistors are nonpolarized devices that are unaffected by pressure or vibration. They are bilateral and have a symmetrical volt-ampere characteristic curve.
8. Varistors are used as protective devices in lightning arrestors and in ac and dc circuits to protect components against voltage surges.

SELF-TEST

Check your understanding by answering the following questions:

1. The resistance of a varistor _____ (increases/decreases) as the voltage across it increases.
2. (True/False) Most modern varistors used are fabricated from silicon carbide. _____
3. Varistors operate independent of current direction because they are _____ (polarized, nonpolarized) devices.
4. If the temperature of a varistor increases, its resistance _____ .
5. (True/False) A standard pulse waveshape is often specified as 8×20-μs. _____
6. The clamping voltage specified for an MOV refers to the _____ voltage across the MOV with a specified waveform and a peak pulse current applied.

MATERIALS REQUIRED

Power Supply:
■ Variable 0–15 V dc, regulated

Instruments:
■ DMM
■ VOM

Resistors:
■ 1   100-Ω, 5-W

Miscellaneous:
■ 6-V metal-oxide varistor (MOV) (GE V12ZA1 or equivalent)
■ SPST switch

PROCEDURE

1. With power **off** and switch S_1 open, connect the circuit of Figure 57-2(a). Set the power supply for a 0-V output.
2. Turn **on** power; **close** S_1. Slowly increase the power supply voltage until the voltmeter measures 2 V. Watch the ammeter carefully as the voltage is increased; be prepared quickly to increase the current range.
3. With 2 V across the MOV, measure the current and record the value in Table 57-1 (p. 415).

4. Increase the voltage across the MOV in steps of 2 V, 4 V, 6 V, 8 V, 10 V, 12 V. At each step measure the current and record the value in Table 57–1. After completing the measurement, **open** S_1 and turn **off** the power supply.

5. Reverse the leads from the power supply to the circuit so that the polarity to the MOV is reversed, as in Figure 57–2(*b*). Reverse the ammeter connection also, if necessary.

6. Repeat steps 3 and 4, but record the current measurements with the negative voltages across the varistor. Also show the current measurements as negative values.

7. For each voltage (negative and positive values) calculate the static resistance of the MOV using Ohm's law. Record your answers in Table 57–1.

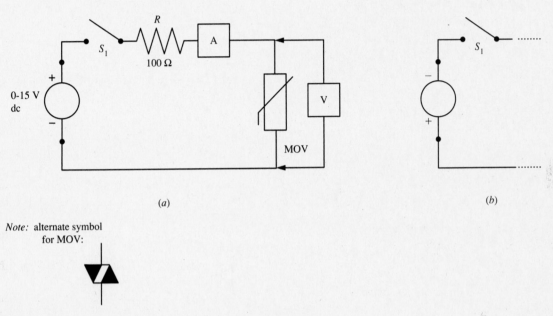

(*a*)

(*b*)

Note: alternate symbol for MOV:

Figure 57–2. (*a*) Circuit for procedure step 1. (*b*) Circuit for procedure step 5.

ANSWERS TO SELF-TEST

1. decreases
2. false
3. nonpolarized
4. decreases
5. true
6. peak

TABLE 57 – 1   Volt-Ampere Characteristic of a Varistor (MOV)

Voltage across MOV V_V, V	Current I_V, A	Calculated Static Resistance of MOV R_V, Ω
+2		
+4		
+ 6		
+8		
+10		
+12		
−2		
−4		
−6		
−8		
−10		
−12		

QUESTIONS

1. Explain, in your own words, the relationship between voltage and resistance of a varistor.

_____

_____

_____

_____

_____

_____

_____

_____

2. On a separate sheet of $8\frac{1}{2} \times 11$ graph paper, plot a graph of voltage across the
   varistor versus current through the varistor using your data in Table 57–1. Draw
   the axes similar to those shown in Figure 57–1. Label all axes.

Nonlinear Resistors—Varistors (VDRs)   **415**

3. Refer to your data in Table 57–1 and your graph of Question 2. Discuss the voltage-current characteristics of the MOV used in this experiment.

_____

_____

_____

_____

_____

_____

_____

_____

_____

4. Does the varistor in this experiment have a positive, negative, or zero temperature coefficient? Refer to specific data to support your answer.

_____

_____

_____

_____

_____

_____

_____

_____

WIRING METHODS

SCHEMATIC DIAGRAM

A schematic diagram is an electronic blueprint. It tells the technician where and how the electrical parts are connected in a specific circuit, such as a television or computer circuit. There are a few simple guidelines for reading and interpreting schematic diagrams:

1. Parts are shown by their electrical symbols (see Figure 1–1).
2. The leads that extend from a part such as a resistor or capacitor are shown by straight lines.
3. (a) If two or more parts make an electrical connection at a common point (terminal), this point is indicated by a heavy dot on the schematic. Sometimes the junction point is identified by a letter or letters such as TP3 (for test point 3).
   (b) If two lines in a schematic diagram cross one another and no dot appears at the point of crossing, they are not electrically connected at that point.

BREADBOARDING

A schematic diagram is a set of graphic instructions to the technician for assembling and wiring the parts of a circuit. Schematics are used in the experimental laboratory as well as in the production and servicing operations. The physical product created on the production line is in permanent form. The experimental circuit in the laboratory is usually assembled in temporary form so that circuit changes may easily be made until the product design is final.

Experimental circuits are assembled on a *breadboard*. This is a device that ensures simple, rapid circuit connection and disassembly. Moreover, the breadboarding technique allows component parts to be used over and over again, permitting a wide variety of circuit arrangements for study and experimentation. The experiments in this manual will be facilitated by and are intended for breadboarding devices.

Handwiring

Components in electrical and electronic circuits may be connected by handwiring. In this process, component leads are secured to terminal posts. The pigtails (leads) of parts

that must be electrically connected are secured at a common terminal post. If the parts are too far apart, their terminal leads may not be long enough to reach a common post. In that case, the pigtails of these remote parts (which must be joined electrically) are secured to separate terminal posts. A wire is then used to connect these terminal posts. The handwiring method of interconnecting components is used both for permanent wiring in a chassis and on a temporary breadboard.

Printed Circuits

Modern, mass-produced electrical and electronic devices use boards on which the copper conductors are bonded to a non-conducting base material. This base is usually a plastic, ceramic, glass, or phenolic board on which the components are mounted. Interconnection is made by means of conductive copper surfaces, which are etched on the board.

Making a Printed Circuit Board

The term *printed circuit* is a misnomer, since the circuit is actually not printed on a board. In its original form the board consists of a sheet of thin conducting material, usually copper, bonded to one or both sides of a nonconducting base. The process of making the circuit involves removing parts of the copper sheet that will not be part of the circuit. The excess material is usually removed by machining and chemical processes.

A printed circuit board can be made of simple circuits in the shop or laboratory. You should refer to a good instruction book before attempting to make your own board, but the general process is as follows:

1. A layout of the physical circuit is made on paper. This layout cannot have lines crossing unless these lines are actually connected. It is usually not possible simply to use a schematic diagram as a layout, although the schematic must be followed to make the circuit function as designed.
2. Once the layout has been made, it must be transferred to the surface of the board. This can be done by carefully drawing the layout on the board with a pen, using *resist* ink. Resist materials are those that will resist the etching action of the chemicals used to remove the excess copper

from the board. There are also paste-on tapes of various sizes and shapes that can be applied directly on the board to duplicate the circuit layout.

Some boards are available with a photosensitive coating over the copper. Using a negative of the layout placed over the board, the arrangement is exposed to light. The board is then developed using photographic chemicals. The result will be a "photo" of the layout directly on the copper surface.

3. After the layout has been successfully transferred to the surface of the copper, the copper around the layout must be removed. This is done by immersing the board in an acid or alkaline etching solution. The solution eats away the copper around the resist material and leaves the copper under the resist intact. The base material—plastic, glass, or the like—is not affected by the etching solution. Once the copper around the layout is removed down to the base material, the board is removed from the etching solution.

4. The board is now washed thoroughly to remove all traces of the etching solution. Next the board is dried. The board will now have the circuit layout clearly outlined against the base material. However, in most cases the resist ink, paste-on elements, or photographic image still remain over the copper.

5. The layout medium is removed by using solvents or chemicals that will not damage the underlying copper. The process of removing the resist leaves the copper dirty or oxidized and not suitable for connecting components. The surface of the copper is therefore rubbed carefully with very fine steel wool.

6. In most cases, components are connected to the printed circuit board by inserting their leads through holes in the board that pass through the copper. Holes are therefore drilled through the boards for these leads. The copper around these holes is usually coated with a very thin layer of solder (a process called tinning) to make it easier for the solder to join both the lead and the copper on the board.

7. After the board has been cleaned, drilled, and tinned, it is ready to receive the components, resistor, capacitors, semiconductors, and the like, according to the schematic diagram. The components are usually mounted close to the surface of the board on the noncopper side. Their leads are bent and passed through the holes drilled for them. The leads are then soldered to the copper on the copper side of the board.

Solderless Breadboards

Although handwiring and printed circuits are used for permanently connected circuits, the solderless breadboard is used only for temporary connections, such as in an experiment laboratory or a design and engineering prototyping activity.

Solderless breadboards are produced commercially by many companies. They are essentially similar in design and use. The typical board consists of rows of hundreds of holes, called *tie points*. These tie points contain metal inserts designed to grip and make a solid electrical connection with the component leads. The tie points are connected together by means of buses running under the boards. The spacing of the tie point holes is such that it will accommodate inserting the pins of a DIP (dual inline package) integrated circuit or switch.

Solderless breadboards are often combined or mounted with power supplies and switches to form complete experimental units.

B

FAMILIARIZATION WITH HAND TOOLS USED IN ELECTRONICS

After an experimental electronic circuit has been breadboarded, its operating characteristics are tested by the technician. Design changes may be indicated and are effected as required. When the technician and engineer are satisfied that the circuit is performing as it should, it is ready for prototype assembly.

The necessary printed circuit boards are designed and produced together with the housing in which the board will be mounted. In this phase of product preparation and assembly, the technician may use forming, drilling, and cutting power tools. This phase will also require mechanical hand tools such as the scribe, punch, hammer, screwdriver, wrench, hacksaw, and file.

HAND TOOLS USED IN ELECTRONICS

Electrical assembly follows the preparation of the housing and mounting of the boards and other components, such as controls, switches, jacks, binding posts, and the like. This is the stage that requires electrical interconnection of components. It is concerned with the preparation and soldering of conductive wiring between parts. The common hand tools used in electrical assembly include diagonal pliers, long-nose pliers, soldering aid, wire stripper, soldering iron, soldering pencil, soldering gun, knife, and heat sink.

Diagonal pliers, commonly called dykes, cutters, or diagonals, are used for cutting soft wire and component leads. They should not be used for cutting hard metals such as iron or steel. Some diagonals have a small, notched cutting surface for stripping the insulation from a wire. This stripping hole will accommodate #22 wire, which is a common hook-up wire used in electronics.

Long-nose pliers are used to hold wire so that the stripped end may be twisted around a terminal post or pushed through a hole in a printed circuit board or terminal. Long-nose pliers sometimes have cutting edges so that the pliers can serve for both gripping and cutting wires.

Needle-nose pliers are a variation of the long-nose. Their long, thin jaws can get at difficult-to-reach spots. Bent-nose pliers are similarly designed to reach into narrow places inaccessible to other tools.

The soldering aid is a useful tool that simplifies soldering jobs. A standard aid has a sharp metal pointer at one end and a slotted V-type grip at the other. One function of the pointer end is to help clear solder out of terminal eyes on solder lugs. The gripping end is useful in unwrapping wire and component leads when these are being unsoldered from terminal posts.

The wire stripper removes insulation from hook-up wires. There are different types of strippers, ranging from the simple type found on diagonal pliers to automatic multisized strippers that can handle wires of different diameters. The automatic hand-held stripper is popular with electronics technicians. In addition to mechanical, there are also thermal and chemical strippers.

The soldering iron is still a standard electrical hand tool, although many technicians favor the soldering gun. A heating element inside the iron utilizes power from the power line. The heat is channeled to the tip. The tip is applied to and heats the area to be soldered.

Soldering irons are rated by the amount of power they dissipate and thus, indirectly, by the amount of heat they can develop. The low-wattage soldering iron (25 or 35 W) is used for soldering or unsoldering components from a printed circuit board, or in delicate soldering applications requiring low heat levels. Interchangeable tips in a wide variety of shapes and sizes add flexibility to these soldering irons.

The soldering gun has gained wide popularity because of its fast heating characteristic. A trigger switch on the gun handle applies power to the heating and soldering tip and heats it in 30 s (approximately). Unlike the soldering iron, which is slow heating and must be continuously hot, the soldering gun is heated only at the moment it is to be used. Between soldering operations it is left off. Popular sizes are 100- and 125-W guns. The guns are used only for heavier wires and components.

A pocket knife is useful in many small tasks requiring cutting and scraping. It can be used for scraping and cleaning the terminal ends of wires and components, in preparation for soldering.

The soldering heat sink is a small metal clip used to prevent overheating during soldering or unsoldering of heat-sensitive electronic parts. The heat sink is clipped onto the lead between the body of the part and the terminal point at which heat from a soldering iron is applied. It absorbs heat and reduces the amount of heat conducted to the component.

Desoldering tools simplify the job of heating and removing solder from terminal boards, printed circuits, and other soldered connections when it is necessary to remove components. Terminal and printed circuit holes must be free of solder before the leads of a new component can be inserted. Desoldering tools used by the technician take two forms. The first is a spring-loaded vacuum suction tool; the second, a suction device in the form of a hollow rubber ball attached to a stainless steel or plastic tube. In operation the open end of the tube is placed on a heated solder point or solder hole. With the spring-loaded device, the spring is released and the molten solder is sucked into the tube, clearing the joint or hole of solder. With the ball-type tool, the ball is squeezed and as it is released the solder is again sucked into the tube. The ball-type device is often attached to a special soldering iron so that the heating and vacuum processes are combined.

CAUTION: A word of caution about the use of hand tools. They should never be used in a live circuit—that is, in a circuit to which electric power is applied. Most hand tools are made of metal, and metal is a conductor of electricity. Failure to observe this safety precaution may result in electrical shock, damage to the tool, or destruction of the circuitry.

APPENDIX

C

SOLDERING TECHNIQUES

Soldering is generally required to assure permanent electrical connections. Before being soldered, however, the connection must also be physically strong. In other words, solder is not meant to hold a connection together. Its purpose is mainly that of providing a good, low-resistance path for current to travel from one part of a circuit to another.

Wires or wires and terminals are wrapped or twisted together, then solder is melted into the heated joint. When the heat is removed, the solder and wire cool, making the soldered joint look like a solid piece of metal. It is not possible, after proper soldering, to separate the wires at a joint except by breaking them or by unsoldering them.

Solder is an alloy of lead and tin. It has a low melting point and comes in wire form for electronics use. Electronics solder is made up of about 60 percent tin and 40 percent lead, though the composition may vary for certain applications. Rosin-core solder is used for soldering electronic components. The rosin is a flux that flows onto the surface to be soldered, preventing high-resistance oxides from forming in the connection, thus assuring a more perfect low-resistance path. Acid and soldering paste should not be used in electronics, since they can corrode the metals used in wiring.

Proper soldering requires the following:

- Clean metallic surfaces
- Sufficient heat applied to the joint to melt solder when solder is applied to the heated wire surface.

If the wires or terminals to be soldered have not been preheated sufficiently, the molten solder will not adhere to their surface. The joint may look well soldered, but the chances are that it is not. Cold-solder joints cause very high resistance electrical contact. Defects arising from poor soldering are difficult to discover when troubleshooting.

Soldering is not always used in modern manufacturing processes. Solderless wire-wrap connections use wire and terminals especially designed to provide strong mechanical and low-resistance connections without the addition of solder.

TINNING

To ensure maximum transfer of heat from the iron to the surfaces at the joint, the tip of the iron must be tinned. A new soldering tip or a tip that has been used for a long period of time must be cleaned by scraping it with a knife or emery cloth, steel wool, a wire brush, or fine sandpaper. If the tip is badly pitted, it may be necessary to file it. This technique applies to *copper* tips.

Many modern tips are gold-plated or iron-bearing. A gold-plated tip should be cleaned by wiping it against a wet sponge. Iron-bearing tips can be cleaned with a wire brush. These tips should never be filed or cleaned with sandpaper or emery cloth.

After the tip is cleaned, it is heated. Rosin-core solder is permitted to melt completely around the tip. The small pool of solder is allowed to stay on the tip for a couple of minutes; then the excess is wiped off with a cloth or sponge. Solder is again applied, coating the tip. It is immediately wiped off. The tip is now adequately tinned. If the iron is overheated and permitted to discolor before solder is applied, it will be difficult to tin.

It is necessary not only to tin the iron; the surfaces to be soldered should also be cleaned and tinned. Tinned surfaces assure good electrical connections and proper bonding when soldered. Wire may be tinned by placing it on the tip of the iron and heating it sufficiently so that the wire will melt solder. The strands of stranded wire should be twisted together tightly before it is tinned. Terminals should also be tinned.

The tip of a soldering iron tends to become dark and dirty when in use. To keep the iron at maximum-heat-transfer efficiency, the tip should be cleaned periodically by wiping it on a damp sponge. A wire brush will also clean the soldering tip. Soldering irons used for extended periods, such as when used in production and servicing facilities, are often connected to thermostatically controlled holders that cycle power to the irons, thus preventing them from overheating.

HEAT SINKS

Heat sinks are used to avoid heat damage to heat-sensitive components during soldering. Solid-state devices such as transistors and integrated circuits are extremely sensitive to heat and can be permanently damaged if they are overheated. Any small metal clip, an alligator clip for example, may be used as a heat sink. It is clipped onto the lead between the component and the point at which the soldering iron is applied. The clip acts as a heat load and reduces heat transfer to the component.

After the iron is removed, the heat sink should be kept in place until the joint has cooled.

Wherever they can be used, heat sinks are a good practice. However, they are not practical in many physical applications, where solid-state devices have been soldered into circuit boards in such a way that they do not allow space for heat sinks. In such cases, sockets are soldered in place and the devices, such as transistors and integrated circuits, are plugged into the sockets.

APPENDIX

D

TRANSFORMER CHARACTERISTICS

IDEAL TRANSFORMER

A transformer is a device for coupling ac power from a source to a load. The source of power is connected to the primary winding; the load, to the secondary winding. In the process, certain transformations take place that are related to the construction and materials of the transformer and to the number of turns of wire in the primary and secondary windings.

Transformers find many uses in electronics. There are power transformers, audio transformers, radio-frequency transformers, isolation transformers and many other types.

A conventional transformer consists of two or more windings on a core, magnetically coupled. An ac voltage applied across the input or primary winding causes current in the primary. This sets up an expanding and collapsing magnetic field which cuts the turns of the secondary winding, inducing an ac voltage in the secondary. When a load is connected across the secondary, current flows in the load.

The core around which the primary and secondary are wound may be iron, as in the case of low-frequency power and audio transformers. An air core may be employed for coupling higher-frequency circuits. The core material and the geometry of the windings determine the characteristics of coupling.

An alternating current in a winding of an iron-core transformer magnetizes the core first in one direction and then in the other. It is this moving magnetic field in the core that cuts the windings of the other coil, inducing a voltage in it.

Transformer T in Figure D–1 represents an ideal transformer. In an ideal iron-core transformer, one in which there are no power losses; 100 percent of the source (primary) power would be delivered to the load. Formula (D–1) shows this power relationship.

$$V_P \times I_P = V_S \times I_S \qquad \textbf{(D–1)}$$

Here V_P and I_P are the primary voltage and current, respectively, while V_S and I_S are the secondary voltage and current. In a lossless transformer, the ratio between the primary voltage and the voltage induced in the secondary is the same as the ratio a between the number of turns N_P of the primary winding and number of turns N_S of the secondary winding.

$$\frac{V_P}{V_S} = \frac{N_P}{N_S} = a \qquad \textbf{(D–2)}$$

If $a = 1$, there are as many turns in the primary as there are in the secondary, and the voltages appearing across the primary and secondary are equal. This type of 1:1 transformer is called an *isolation transformer*.

If a is greater than 1, a lower voltage appears across the secondary than across the primary. This is called a voltage stepdown transformer.

If a is less than 1, a higher voltage appears across the secondary than across the primary. This is a voltage stepup transformer.

Formula (D–1) for an ideal transformer may also be written as

$$\frac{V_P}{V_S} = \frac{I_S}{I_P} \qquad \textbf{(D–3)}$$

From formulas (D–2) and (D–3) we have

$$\frac{V_P}{V_S} = \frac{I_S}{I_P} = a \qquad \textbf{(D–4)}$$

The last formula states that current and voltage in the windings of a transformer are inversely related. Therefore, a voltage stepup transformer is also a current stepdown transformer, whereas a voltage stepdown transformer is a current stepup transformer.

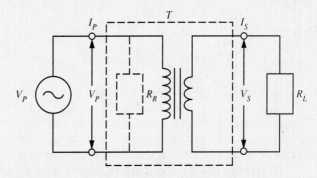

Figure D–1. Schematic representation of an ideal transformer.

POWER LOSSES IN A TRANSFORMER

The ideal transformer does not exist because there are power losses that do not permit 100 percent transfer of power from the source to the load. One such loss is related to the resistance of the windings and is the I^2R or heating loss. Thus, there are I^2R losses associated with the resistance of the primary winding (primary loss) and also with the resistance of the secondary winding (secondary loss). When there is no load on the secondary winding, there is no current and hence no power loss in the secondary. However, there is a magnetizing current in the primary and hence some I^2R loss in the primary.

Eddy currents are present in the core of an iron-core transformer. A circulating eddy current is induced in the iron core by the changing magnetic field. Eddy currents heat the transformer and act like an I^2R loss. Eddy currents thus rob the source of power and represent another power loss.

Another loss associated with a transformer core is *hysteresis loss*. The hysteresis effect results from the fact that magnetic lines of force lag behind the magnetizing force that causes them. Hysteresis can be understood by considering the fact that when a magnetizing force is removed from an iron-core magnetic circuit, a portion of the flux remains within the iron. This residual magnetism can be removed by applying to the iron a magnetizing force opposite in direction to that of the initial force. The energy required to demagnetize the iron acts as a core loss, which is associated with the reversal of magnetizing current in the winding.

One other loss must be mentioned. This is associated with the magnetic leakage that exists in a transformer. Not all the lines of magnetic force will link the turns of the secondary winding. The lines of force thus lost to the magnetizing circuit constitute magnetic leakage.

EFFECT OF LOAD CURRENT ON PRIMARY CURRENT

Primary current in the transformer of Figure D–1 depends on the load current in the secondary, as is evident from the formula

$$\frac{I_S}{I_P} = a$$

or

$$I_P = \frac{I_S}{a} \qquad \textbf{(D–5)}$$

Therefore, as load current I_S increases due to a decrease in load resistance R_L, primary current must increase. The increase in primary current is explained by the assumption that the change in load impedance (as is evidenced by an increase in load current) appears also as a reflected impedance in parallel with the primary. Since power in the primary

and secondary was assumed to be equal, and since power can be dissipated only in a resistance, the reflected impedance must be the resistance R_R.

In addition to the primary current I_P resulting from the reflected load impedance, the primary supplies the current for the iron-core losses and the magnetizing current. The phasor sum of these two currents (the magnetizing current is out of phase with the voltage) is called the *exciting current*. The exciting current is about 3 to 5 percent of the rated output of the transformer. This explains why I_P will normally measure more than that predicted by formula (D–5).

RESISTANCE-TESTING TRANSFORMER WINDINGS

Resistance (ohmmeter) tests of the individual windings of the small transformers used in electronics are used to determine the continuity of the windings. An ohmmeter test also establishes the resistance of each winding. The technician then compares the measured resistance with the rated value to determine if a suspected transformer is in fact defective.

The following considerations may be helpful in analyzing the results of resistance measurements.

Winding Measures Infinite Resistance. This winding is "open." The break may be at the beginning or end of the winding, where the connection is made to the terminal leads. This type of break can be easily repaired by resoldering the leads to the winding. If the discontinuity is elsewhere, the transformer must be replaced.

Winding Resistance "Very" High. A winding whose resistance measures "very" high compared with its rated value may be open, or there may be a cold-solder joint at the terminal connections. If the condition cannot be corrected, the transformer must be replaced.

Winding Resistance "Very" Low Compared with Rated Value. Turns of the winding must be "shorted" somewhere on the transformer, or the winding may be shorted to the frame. However, a small difference in resistance between the rated and measured values may be insignificant. For example, if a primary rated at 120 Ω measures 100 Ω, the difference may be attributable to the inaccuracy of the meter. In cases of doubt, other tests (which will not be discussed here) are required.

Resistance between Windings. The windings of transformers, other than autotransformers, are insulated from each other. There should be infinite resistance between insulated windings, as long as the transformer is not connected in a circuit. If the insulation between two windings breaks down, there will be a measurable resistance between these windings, signifying a defective transformer.

Factors that Determine Resistance of a Winding. The resistance of a winding depends on the diameter of the wire and the number of turns. Thus, resistance varies inversely as the square of the diameter and directly as the number of turns. Large diameters are required for high-current windings, smaller diameters for windings that carry less current. The primary of a voltage stepdown transformer has more windings than the secondary. Moreover, a voltage step-down is also a current stepup transformer. Hence, the secondary must carry more current than the primary. Therefore, the resistance of the primary will be higher than that of the secondary. How much higher depends on the transformer ratio and on the diameter of the two wires.